DOUBT AND CERTAINTY

ALSO BY TONY ROTHMAN

Instant Physics

A Physicist on Madison Avenue

Science à la Mode

Censored Tales (U.K.)

Frontiers of Modern Physics

The World Is Round

ALSO BY E.C.G. SUDARSHAN

Wolfgang Pauli and the Spin Statistics Theorem (with I. M. Duck)

Classical Dynamics (with N. Mukunda)

Fundamentals of Quantum Optics (with John Klauder)

Introduction to Elementary Particle Physics (with Robert Marshak)

Doubt and Certainty

THE CELEBRATED ACADEMY

Debates on SCIENCE, MYSTICISM, REALITY,

In General on the KNOWABLE and UNKNOWABLE,

with PARTICULAR FORAYS into Such Esoteric Matters as

the MIND FLUID, the BEHAVIOR of the STOCK MARKET,

and the Disposition of a QUANTUM MECHANICAL SPHINX,

To Name a Few

Tony Rothman and George Sudarshan

With illustrations by Shannon K. Comins

§ Helix Books

PERSEUS BOOKS   Reading, Massachusetts

Library of Congress Catalog Card Number: 98-87634

ISBN   0-7382-0006-9

Perseus Books is a member of the Perseus Books Group

Jacket design by Rose Traynor
Text design by Jean Hammond
Set in 10-point Minion by Carlisle Communications, Ltd.

123456789 - DOH -0201009998
First printing, September 1998

Find Helix Books on the World Wide Web at
http://www.aw.com/gb/

CONTENTS

FOURTH DEBATES:

Why Do Things Happen? *Causality, Synchronicity and All That* 68

FIFTH DEBATES:

Does Time Go Forward? *Past and Future*   94

SIXTH DEBATES:

Why Is There Left and Right? *Chirality, Complexity and Emergent Properties*   126

SEVENTH DEBATES:

Is the Universe Weird? *Quantum Mechanics*   153

EIGHTH DEBATES:

Is There an Answer? *Theories of Everything*   191

KNOWLEDGE AND WISDOM

But scientists, who ought to know
Assure us that it must be so.
Oh, let us never, never doubt
What nobody is sure about.

— *Hilaire Belloc*

If you have recently walked through the nearest Barnes and Noble superstore or equivalent, you may have noticed that its bookshelves hold various descriptions of the universe: one aisle displays scientific discussions; poetic and legendary accounts are found to the east of the café; Western religious treatises occupy a far wall, gradually being squeezed out by Eastern philosophy and the New Age. But the fact that the varying descriptions are scattered over the four corners of the vast store induces a sense of fragmentation, not unlike the puzzlement you experience in a supermarket when you discover low-cal sweetener is in "diet foods" and not next to sugar.

The superstore is of course a microcosm, a reflection of society at large. Although in recent years the science section has been greatly expanded, at the same time we have witnessed a general dissatisfaction with traditional science's inability to provide a spiritual foundation for life. A decade ago, in his *Closing of the American Mind,* Allan Bloom wrote: "All that is human, all that is of concern to us, lies outside of natural science."

For those of us schooled in the scientific outlook, it is a harsh judgment, but Bloom's discontent—or similar ones—seems rife with the public. To our nonscientist friends, the rules of science evidently appear arbitrary, merely the general consensus of a fraternity of dead white European males. In that case the rules aren't to be taken too seriously. And so a large number of people have come to regard science as just another belief system, neither more nor less sound than astrology, alchemy or ESP. The problem has been compounded over the years by repeatedly exaggerated claims on the part of scientists for what science can deliver in the way of ultimate understanding or happiness. As a result, the public has become largely agnostic about science.

All this has led at millennium's close to widespread attempts to make science more spiritually meaningful. This is most evident in the "New Age" section of the superstore, where writers combine scientific terminology with, typically, "Eastern" mystical thought. But it is also evident in the science aisle, where one can find—under "physics"—books on synchronicity, not to mention treatises on the compatibility of the Bible with general relativity. Scientists tend to turn pale—sometimes crimson—at such sightings. Yet scientists themselves have not been immune from the general trend: not far from synchronicity you will spy books by reputable physicists on God and the laws of nature, on immortality, on the consciousness of the universe, and by biologists on the physics of angels.

There can be little doubt that the millennium amalgam (some would say millennium muddle) has produced a great deal of confusion in the public's mind about what constitutes science. Not long ago, one of us, a cosmologist, found himself at a popular Austin bar with a colleague, a string theorist. We fell into a conversation with a woman who seemed quite interested in what we did (what exactly cosmologists and string theorists do will be the subject of some debate in *Doubt and Certainty*). After we had given her an idea of our research, she asked, to our great surprise, "What about those quantum field theorists?"

"Who do you mean?"

"Deepak Chopra."

That the public equates quantum field theory with Deepak Chopra indicates that the time has come to sort things out.

But what exactly needs to be sorted out? Well, the present climate does raise questions. How much of the millennium synthesis is meaningful and how much is pure nonsense? Which of the New Age trends, if any, are compatible with physics? More generally, what does science describe? Should its domain be broadened? Is reductionism the only approach to science? Can we learn anything from parallels between physics and Eastern philosophy?

It would be grandiose to say that *Doubt and Certainty* answers all these questions. Actually, it answers none of them. Rather, *Doubt and Certainty* represents our examination of more specific scientific questions regarding, say, the nature of quantum mechanics or cosmology, which in turn often lead to broader issues. We say "our" examination but want to stress that we have tried to present more than our own point of view. To be sure, the examination is not carried out in exactly the traditional expository format of a popular-science book. Rather, the ruminations have been organized into a series of debates that take place during your recent visit to the renowned Academy and which include protagonists ranging from semilegendary figures like Pythagoras to contemporaries such as Deepak Chopra, Steven Weinberg and Stephen Hawking, not to mention the authors. When the debates don't erupt into out-and-out warfare, they center mostly around physics (for better or worse the authors' primary field of expertise), but often make rash excursions into other areas of science, art, philosophy and theology.

As the twofold title implies, the main tension running throughout the debates is that between doubt and certainty. The "certainty" factions of the Academy will emphasize what is known, what you can rely on or, to be tautological, what is certain. The authors, physicists that they are, have some sympathy for certainty and believe that scientific progress is not entirely illusory. Some things are true and will remain true.

But wisdom is more important than knowledge. Too often, writers and scientists present speculation as fact and fail to stand back and examine their own assumptions. Hence the first half of the title: "Doubt." This side of the debate presents a complementary view to the other. We stress "complementary." If the protagonists of certainty emphasize what is known, what is rock solid, then their antagonists in their philosophical doubt emphasize what is less than certain, what is unclear and, most of all, that to explain a phenomenon is not necessarily to understand it. Which view is correct? Both are. All experience is part of the One.

We should point out that what most people believe to be uncertain in science is fairly certain, and what they believe is certain is not. When we speak of uncertainty we are not speaking of Heisenberg's uncertainty principle, which is not only precise, but certainly certain. Neither do we expect anyone to disprove Einstein's theory of relativity or quantum mechanics. No experiment has ever been found to violate either theory, and it is unlikely one will. That these theories describe the way nature operates is as close to a factual statement as one will find in science. As we said, some things are true and will remain true.

What, then, can we possibly mean by uncertainty or doubt? Unfortunately, the term *uncertainty* is ambiguous. It can connote the unknown. Much is unknown in science at the present time—why do roses have the shape they do? As the debates progress to ever more modern topics, they will necessarily touch on much of what is uncertain in the sense that we simply don't know the answer. For the most part we give those topics short shrift, since hardly anything can be said other than "we don't know."

But uncertainty can also connote philosophical doubt that applies even to established theories. It is these doubts that mostly concern the academicians of the Academy throughout their arduous deliberations. Do theories describe what scientists claim they describe? Is determinism a meaningful concept? Does mathematics reflect the world as well as scientists say it does? What is doubtful about established scientific theories lies not so much in the results, but in their basic assumptions about reality. When the unknown vanishes, doubt remains. That is good. Only by questioning assumptions do we arrive at true understanding.

Understanding? Questions, especially the infernal ones, are often easier to ask than to answer. The reader should not expect pat resolutions to the debates that follow. More often than not we will be able to do little more than set out the arguments and abandon you to your own resolutions about the specific scientific questions under discussion, as well as the broader issues.

Throughout *Doubt and Certainty,* the reader will detect the flavor of Chinese and Indian philosophy. To an extent this is due to the choice of subject matter; several of the debates deal directly with topics relating to New Age beliefs, such as synchronicity and "quantum consciousness." We do have some sympathy with attempts to connect philosophy in general with physics. In part this is because science has relied on philosophical principles. In part, as some academicians will argue later, it is because physics is only an incomplete description of reality and must be supplemented by other modes of perception. "Truth is One, though the wise call it by different names."

The reference to Eastern works in particular is understandable. Far more than in the West, Eastern "religious" philosophy approaches "natural" philosophy. The Upanishads are filled with statements like the one just quoted, which resonate strongly with the physicist. Throughout the Indian scriptures a physicist is struck by the concern with ultimate Reality and the perceiving of it. And one only has to read the opening words of the *Tao te Ching*, "The Tao that can be expressed is not the true Tao," to feel that the Taoists, like the physicists, understood that the essential mystery of the universe always lies beyond. Contrast this with the view of the Socratics, who felt one could perceive reality by logic alone.

There is also a personal reason for our Orientation. The senior author (Sudarshan) is by birth and upbringing Indian and has made long study of the Indian philosophy. The junior partner (Rothman), who has been responsible for the writing, has attempted to convey a little of the flavor of the metaphors that are launched (usually from East to West) during conversations between the two of us.

Nevertheless, *Doubt and Certainty* is meant to represent a debate and, as physicists, we are well aware that much exception can be taken to statements like "The quantum field is just another label for the field of pure consciousness or pure potentiality." Most physicists would dismiss them outright. However, the encounter in the Austin bar leads us to believe, as a colleague wrote in the pages of *Physics Today,* the time for mere polemics has passed. We hope that *Doubt and Certainty* represents a genuine exchange of ideas.

Dialogues, of course, have a long and distinguished tradition in science and philosophy, dating back to Galileo, the Bhagavad Gita and the Upanishads. However, they are not the most economical way of transmitting information and the modern time-limited reader often finds them too discursive. Moreover, it is difficult to involve the reader in a debate on modern topics without a significant amount of background. We have attempted to deal with these obstacles by introducing novelistic tendencies—a narrative element to compress and deliver material. Each debate typically begins with an expository section to provide background, then works its way into the debate proper and somewhere along the line (usually toward the end) includes a "Practical Exercise," which might be tried out in the Real World. The result is this hybrid, which sometimes strikes us as a cross between Plato's *Republic* and the *1001 Nights.*

Placing real persons in a fictitious debate poses a few scholarly and stylistic problems. On the side of scholarship, there is perhaps a slight danger that the reader will really believe Deepak Chopra made such and such a remark to Pythagoras. On the stylistic side, any sense of "dialogue" is surely disturbed by excerpting written passages not originally directed at an interlocutor. We have attempted to err on the side of scholarship by quoting the participants accurately and fully, and explaining the context of the remarks even when it lessens an exchange's believability. In the case of legendary and historical figures, where language is absent altogether or might prove difficult for the modern reader, we have often invented or paraphrased. In a dialogue between someone living and dead, such as that between the reader and Dirac in the Second Debates, one can safely assume it is invented. Nevertheless, whenever any ambiguity might arise, we have made a note in the Commentaries, which are found at the end of the book.

Although we have tried hard not to sacrifice scientific accuracy, we emphasize that *Doubt and Certainty* is not aimed at specialists. We have avoided virtually all mathematics. We have not avoided tackling some difficult topics. In our attempt to keep the discussion at the simplest level possible, we have sacrificed details that some "educated nonscientists" might demand. Occasionally, we have added an asterisk (*) to mark a short explanation at the bottom of the page. For those who crave higher knowledge, however, references and more detailed discussions of certain topics (marked by a dagger (†) in the main text) can be found in the Commentaries.

Along the route to understanding, *Doubt and Certainty* plows some extremely well-plowed territory. We till the soil again knowingly and hope only that the points of view expressed are sufficiently provocative to warrant a reexamination of such basic questions. At the opposite extreme we do not claim that *Doubt and Certainty* will satisfy those searching for a unified field theory of existence. If *Doubt and Certainty* proves enjoyable and enhances a worldview it will have served its purpose.

Finally, we would like to thank Shannon Comins for her excellent illustrations and for putting up with frequent changes. We are also indebted to Alyce Wilson for rendering the line drawings; to Dilip Kondepudi for supplying much of the reference material for the Sixth Debates; to the Center for Particle Theory at the University of Texas for their hospitality and support while the book was being written; and to the visitors in RLM room 9.326 for some great arguments.

In the same way, O Arjuna, when duality comes to an end, they and I live in one place, with one name.
    —*Jnaneshwar's Gita, Chapter 14*

Tony Rothman
trothman@titan.iwu.edu

George Sudarshan
sudarshan@utaphy.ph.utexas.edu

THE QUESTION

In the fourth century A.D., the Taoist philosopher Pao Ching-Yen wrote, "The Confucians say that Heaven created the people and planted lords over them. But why should illustrious Heaven be brought into the matter and why should it have given such precise instructions?"

Pao Ching-Yen was asking why illustrious Heaven should muddy itself in the affairs of lowly humans, but in a general sense his question is the central one of these debates: Does Heaven give precise instructions? To what extent can humans understand the workings of the universe? Is the universe describable?

The Chinese certainly entertained doubts on the subject. Indeed, the word for nature in Chinese (*tzu-jan*) literally means "spontaneous," and when news of Newton's laws reached China in the seventeenth century, so the story goes, Chinese scholars laughed at the very idea that nature should be subject to law. [†]

But four hundred years have lapsed since Chinese scholars mocked Newton's laws and the scientific world has traveled down a much different path. Today it would be difficult to find a natural scientist who did not hold a deep belief in *some* rules and regulations. Physicists, extremists in such matters, subscribe almost without exception to a rather rigid caste system: that there exists a hierarchy of explanation, that emotions can be boiled down to chemistry, that chemistry in turn is a manifestation of physics and that, ultimately, all phenomena can be understood in terms of a few basic laws and principles. Today's leading advocate of reductionism—the belief in an unbroken, vertical chain of explanation—is probably the eminent physicist Steven Weinberg, who writes in his *Dreams of a Final Theory:* "We search for universal truths about nature and when we find them, we show that they can be deduced from deeper truths." As might be inferred from the title of his book, Weinberg's best guess is that there does exist a final theory—a single, consistent theory that explains all the fundamental interactions of nature—and that humans are capable of finding it.

Yet, the physicist's view is far from universally admired. In his *The Same and Not the Same,* the eminent chemist Roald Hoffmann declares, "I tell you, that even in two 'hard natural science' fields as close to each other as chemistry and physics, even *there* there are concepts in chemistry which are not reducible to physics. Or if they are reduced they lose much that is interesting in them."

Indian tradition provides still another perspective: it acknowledges the existence of laws, but maintains that there is a yet higher knowledge. From the Mundaka Upanishad:

> The illumined sages say Knowledge is twofold, higher and lower.
> The study of the Vedas, linguistics, rituals, astronomy and all other arts
> Can be called lower knowledge. The higher
> Is that which leads to Self-realization.

But who is right? The Chinese? The physicists? The chemists? The Indian philosophers? Are the various viewpoints complementary or mutually exclusive? Can we produce a final theory, and if we do, does it necessarily follow that we have explained everything? By the same token, does having a working theory mean we have understood it? What do we mean by understand? what do we mean by final? what do we mean by explain?

Suddenly, while still stranded in the prologue, we have been transported back to a state of know-nothingness. This is the plan. For it is true that the first step to wisdom is renunciation. Let us agree therefore to surrender pride and be unafraid to reexamine perennial questions; we shall then be like the lotus leaf floating clean and dry on the water.

The Academy gates open.

IS THE UNIVERSE DESCRIBABLE?
Laws, Principles, Theories and Models

Thou, inconceivable, incomprehensible, ineffable,
Invisible, inscrutable, impalpable, unfathomable;
Thou, without beginning, without time;
Thou, unclouded wisdom, unerring vision, true existence . . .
  —*Gregor Narekatsi, A.D. 1000*

We shall not cease from explorations
And the end of all our exploring
Will be to arrive where we started
And know the place for the first time.
  —*T. S. Eliot*

Grounds for Debate

According to tradition, sometime around 388 B.C., Dionysus, the tyrant of Syracuse, became so enraged at the corrupting influence of one Plato that he had the philosopher placed aboard a Spartan ship leaving port with orders to murder him or sell him into slavery at Aegina. Luck was with Plato, for at Aegina an admirer managed to save his life with a ransom and spirit him on to Athens, Plato's birthplace. Once in Athens, Plato's high-minded rescuer, Anniceris, refused to accept repayment of the ransom, and instead Plato used the money to buy a garden in the sacred grove dedicated to the legendary hero Academus. Shortly thereafter Plato founded his famous Academy, which for several centuries illuminated the Grecian landscape.

But each day ends with the setting of the sun. By the time B.C. had turned to A.D., the Academy had entered a period of eclipse, and nothing is known of it except through rumor and innuendo until it reemerged into the light five hundred years later, only to be dissolved by the Emperor Justinian in 529.

Or so the story goes.

Throughout history one catches glimpses of the Academy in different locations and circumstances, sometimes in the woods of India under the leadership of Prince Siddhartha, who became the Buddha; sometimes in Fez, Morocco, as Karueein, the oldest surviving university. The lure of the Academy is a powerful one, and scholars, scientists,

1

philosophers, artists and earthier sorts eagerly flock through its gates from all corners of the globe for the chance to debate fundamental issues. You may have yourself wandered onto the grounds (it is permissible to ignore the inscription about geometry above the portals), perhaps as one of the thousands of tourists who visit the legendary domain annually, and encountered the scientific faction engaged in lively discussion.

Whatever the academicians were debating, the question, Is the universe describable? never arose. The participants instead confined themselves to highly technical points and, like most visitors, you quickly lost interest or were pummeled into stupefaction. Through the thicket of extraterrestrial vocabulary, you did notice that the words "theory," "law" and "principle" figured prominently, as they did in the prologue above. Nevertheless, in listening in on the academicians it struck you, as it does many, that scientists have a surprising amount in common with the ancient church representatives at the Council of Nicea: capable only of endless disagreements and schisms.

"Zapomnyat" and "Zapomniec"; A Preliminary Contemplation on the True Nature of "Theory"

It is said that on witnessing Academy disagreements, the Theban Sphinx became so despondent that nothing seemed to be known with certainty that she hurled herself off a cliff and perished. In other versions of the story, she became so incensed at having been subjected to technical disputes that she found the nearest physicist at the Academy café and ate him.

That would be a sensible reaction. The topics debated at the Academy are far removed from daily life. They are . . . academic. One might reasonably put aside such luxuries and devote oneself to caring for the sick and the homeless. However, in encountering an Academy physicist today, people do not tell him to join the Peace Corps; they demand his views on UFOs and alien abductions. These interrogations take place with alarming frequency, alarming enough that you have almost certainly carried out such an ambush yourself. The ensuing arguments, while not profound, are predictable.

In fact, your first Academy dispute undoubtedly began when, baffled by the squabble before you, you innocently remarked, "I have a theory that the government is hiding aliens in Area 51," and were met with incomprehension.

The jaw of the cornered academician dropped, not only at the idea of UFOs in Area 51, but at the ill-considered use of the word "theory." On the streets, people speak loosely; when you proclaimed, "I have a theory that the government is hiding aliens in Area 51," you almost certainly meant no more than you had an idea or, being conspiracy minded, you held a deep suspicion. Once aliens were discovered in Area 51 most people would shrug, "There are aliens in Area 51" and no longer refer to it as a theory. Trial lawyers, on television, declare, "That is a very nice theory, but what does it have to do with the facts?" The everyday usage of "theory" is for an idea whose outcome is as yet undetermined, a conjecture, or for an idea contrary to evidence.

But scientists use the word in exactly the opposite sense. On Academy grounds stands a sacred temple consecrated to theories. Inscribed on the altarpiece is a description of "theory": it refers only to a collection of hypotheses and predictions that is

Figure 1.1

amenable to experimental test, preferably one that has been successfully tested. It has everything to do with the facts. Any physicist will tell you that Einstein's theory of relativity does not refer to some arbitrary conjectures Einstein dreamed up while smoking his pipe, which you are at liberty to reject as it pleases you. It refers to a complete mathematical construct that says some things happen in nature and other things don't. The most famous prediction of Einstein's theory is that no material object may attain the speed of light. This is not an assumption of the theory but an outcome of the theory; it has been tested perhaps millions of times and has never been found to be violated. Neither has any other prediction of relativity.

For these reasons your interlocutor, once having recovered, merely shrugged off your claim that "the government is harboring aliens in Area 51"; it does not amount to a substantial theory. If you could prove it, you would have succeeded in establishing only a single fact (an interesting fact, one that might result in presidential action, but one with little predictive power; and one of the hallmarks of a theory is its power to predict the outcome of experiments or further observations). If you could not prove it, you would undoubtedly find an excuse to explain your failure, such as declaring that the government is holding the aliens in another dimension. Of course, there is no way to test such an assertion, and so to a traditional scientist it does not rate as a theory.

Scientists are highly discriminatory in this area. The creationist contention that the Earth was created in 4004 B.C.—or thirty seconds ago—with all fossil records and memories intact is also not susceptible to any experimental test. If all dating methods indicate that fossils are billions of years old, then as far as science is concerned, they are billions of years old. Science deals only with appearances.

These points are basic ones, but lack of their appreciation results in a deep misunderstanding between scientists and nonscientists. Your academic interlocutor may have suggested you commit to memory the

Aphorism:
"Zapomnyat' " means to remember in Russian,
but "zapomniec" means to forget in Polish.

The Transient Nature of All Things; the Preliminary Contemplation Continued

In many fairy tales the hero reaches a fork in the road at which stands a stone bearing the inscription "If you go to the right you will die by one means, if you go to the left you will die by another means, and if you go straight ahead you will die all the same." Such fairy tales are usually not read to small children at bedtime. However, a similar situation arose in the above argument. While being berated by the physicist for your views on theories, several objections to his line of thought came to mind. Depending on which of these objections you raised, the debate probably went along one of two paths—and you died all the same.

You might have protested that scientists have merely defined "theory" on their own terms. Other types of theories exist that may not yield to experimental test, such as mystical experiences, ESP or occult phenomena. One is at liberty, of course, to define "theory" any way one likes. However, if one is going to try to prove ESP by experiment, it is unfair to term it a science, then resort to nontestable explanations when the experiment fails ("The subject was experiencing bad vibrations today and was unreceptive"). By the same token, if one is going to term creationism a "science," it behooves the practitioners to accept the scientific definition of theory. Otherwise, they are merely talking about something else. Mystical experiences, depending on what one means, offer some room for debate, which the authors are agitating to begin below.

A more interesting question along the same path is whether science itself has abandoned the traditional experimental basis for theories. Many scientists hold the opinion that the newest theories (in particular the highly publicized "string theories") lack any possibility of experimental verification and have effectively become metaphysics or religion. We find a genuine issue here, which the academicians will take up in detail in the Eighth Debates.

Along the same path lies the question whether history (or even economics) can be considered a science. When a scientist requires that a theory be experimentally testable, he or she implicitly demands that a result be reproducible, which requires many experiments. History, by definition, deals with events that have occurred only once. Politicians frequently argue that "massive nuclear deterrence prevented a third world war." The statement certainly has a peculiar ring to it. The event (the absence of WWIII) is unique. The experiment cannot be reproduced. It is therefore unclear whether one can assign definite causes to the event, in particular to an event that never took place.

Conceptual difficulties arising from unique events are rampant in cosmology, the branch of physics that deals with the origin of the entire universe. If "universe" truly means "everything," this is an event that can have taken place only once. Yet cosmologists frequently make such statements as, "the odds of the universe being created in a completely uniform manner are infinitely remote." If the universe was created only once, can any meaning be attached to such a statement?

The two directors of the Academy Division of Sports and Recreations schedule the question, If history is not a science, is cosmology a science? for the Ninth Debates.

At the fatal fork in the path, you may have resolved to experience death in a more traditional fashion and remarked, "All scientific theories are overturned sooner or later." Raising this objection during a scientific debate is obligatory. During an "Aliens in Area 51" debate, the moment generally comes when one or more physicist-academicians vetoes the idea of UFOs on the grounds that, according to Einstein, nothing can travel faster than the speed of light. The time required for a UFO to reach Earth from the nearest star would be prohibitively long (over 100,000 years from Alpha Centauri at the velocity of the space shuttle) for UFOs to zip back and forth, transporting abductees to home base for analysis.

During the debate, such a remark may well have elicited your reply that Newtonian physics—which describes the motion of objects—was thought to be correct for three hundred years until it was supplanted by relativity and by quantum mechanics. Why should anyone believe relativity to be correct? At this juncture Academy visitors cite *Star Trek* and warp drives, at which the physicist pulls his hair, if he has any.

Recovering his composure, he will, with a 97 percent probability, answer that the prohibition of faster-than-light travel is a law of nature, and it cannot be violated. Here, another essential term has been introduced and the discussion of theory becomes intertwined with an argument over laws and principles. But suddenly the carillon in the Academy tower begins to play. Like most visitors, you glance at your watch, struck by how many hours have already trickled away since you passed through the Academy's portals. You also realize you are coming down with a headache and request directions to the nearest analgesic vending machine. They are located on every column, and after taking two aspirin, you return to the debate, fortified.

The Metaphysics of Laws and Principles Is Briefly Described

The opening deliberations have reminded all concerned that the terms "laws" and "principles" are impossible to avoid in any scientific discussion. This was not always true. The concept that the universe should operate lawfully is a comparatively recent development, having originated at the time of Kepler, Galileo and Newton. As we have mentioned, in China the concept does not seem to have taken root at all before modern times, the Chinese words for law, *li* and *fa,* having been applied exclusively to human affairs.

The world of contemporary Western scientists could not be farther from the world of the ancient Chinese. The scientists' world is rigidly lawful. A handful of laws and principles provides the very foundation for a scientist's outlook and, according to textbooks, on the same laws and principles scientists build their theories. The crucial assumption of special relativity, for example, is that the speed of light is constant to all observers,* 300,000 kilometers per second. On the basis of this postulate, Einstein showed that the mass of any object increases with speed until it actually becomes infinite at the speed of light; and the closer the speed gets to light-speed, the more energy it takes to accelerate the object. For this reason, no material object can reach light-speed, much less travel faster; to push it above the speed limit would require an infinite amount of energy.†

Each time particle physicists perform an experiment in their accelerators, they verify the correctness of Einstein's predictions. If Newtonian physics were true and Einsteinian physics false, the Superconducting Supercollider, the enormous accelerator of 85 kilometers' circumference that was never built in Texas, could have been built inside a basement and for several thousand dollars instead of ten billion.

If you are in an intransigent mood, you might insist, "Einstein will yet be proven wrong, just as Newton was." Warp drives do not accelerate starships faster than the speed of light; they somehow bypass it altogether, perhaps by creating wormholes through spacetime. It is true that one cannot prove a law of nature in the same way that one proves a mathematical theorem. Traditionally, one can do only one of two things: perform experiments until you are convinced that there are no exceptions to the rule (or until all the opponents of the theory have died off). This is the experimentalist approach. The other is to show that one theory is logically connected to another theory that you already believe. This is the reductionist approach.

It is true that all theories have holes in them. Newtonian mechanics is applicable only to objects traveling at velocities small compared to the speed of light. Special relativity does not take into account the effects of gravitation. Nevertheless, within their applicable domains, it does appear that these theories describe how nature operates, if only in the restricted meaning of predicting the outcome of experiments to arbitrary accuracy. Relativity, as we have said, has been subject to millions of experimental tests and each one has confirmed the theory. The same can be said for quantum mechanics. This is about as certain as things get in science.

The physicist's debating quiver is full of laws. As Apollo's chariot passed high in the sky above the First Debates, at least one physicist-academician vetoed the idea of UFOs on the grounds of the law of conservation of energy, which says that energy can be neither created nor destroyed. More precisely, she remarked that the amount of energy a UFO would need to travel from the nearest star to Earth was more than our civilization has used since the dawn of time, making the entire enterprise doubtful. Apart from the laws there are also famous principles: Mach's principle, the Heisenberg uncertainty principle, the principle of equivalence, et cetera.

---

* Precisely, constant to all observers moving at constant velocity, or inertial observers.

Indeed, when you bolstered your argument that nothing is certain by referring to *Star Trek* and the new developments sure to come in physics, for instance the transporter, one physicist objected: "Transporters are impossible because they violate the Heisenberg uncertainty principle." Heisenberg's uncertainty principle states that it is impossible to measure precisely both an object's position and velocity at the same time and hence it would be impossible to reassemble the atoms of Captain Picard or Data with the required precision to ensure they were reconstituted in some manner resembling their previous selves.

However, at this point you may have noticed that the uncertainty principle sounds suspiciously like a law of nature and have rightly questioned your opponents on the precise difference between a law and a principle. Traditionally, a natural law is a statement about the way the universe operates that can be experimentally tested, for instance conservation of energy. A principle is a more general statement, one that does not necessarily yield to experimental test, but that one hopes is nevertheless true and that guides the seeker to a precise statement about laws.

But, as you observed, scientists carelessly use the terms "principle" and "law" interchangeably. The Heisenberg uncertainty principle is misnamed; it is really a law of nature, which can be subjected to experimental test. The principle of equivalence, which states that all objects fall at the same rate in a gravitational field, is also a law of nature. To test it one merely measures, as accurately as possible, the acceleration of falling objects.

Yet, there are principles of another category. Reductionism is one of these, and some visitors claim to have overheard the term during the previous conversation. The belief that all phenomena can be reduced to a few laws and principles is traditionally said to have begun with the presocratic Thales, whom Plato wished to exclude from the Academy, and who believed that water was the primary substance.* Since then reductionism has had undoubted success in guiding physicists in their search for new theories, but it is the theories that are experimentally tested, not reductionism.

At this, New Age tourists surround the authors, taking exception to the remark about reductionism's success, and the authors reply that they suspect reductionism will be a featured contention throughout, in particular in the Sixth Debates, which deal in part with complexity and emergence.

There are other guiding principles in science, such as the principle of mechanism, which will figure prominently in the Fourth Debates, on causality, but in the twentieth century, another principle has gradually dominated the search for fundamental theories. This is sometimes referred to as the principle of beauty, which states that the most mathematically beautiful theories are the correct ones. In the famous words of academician Keats,

"Beauty is truth, truth beauty,"—that is all
    Ye know on earth, and all ye need to know.

---

* Thales was also the first businessman to engage in futures trading. One winter he optioned at extremely low rates the use of all the olive presses for the next harvest season. When the harvest arrived there was suddenly a rush of requests for the presses. Thales rented them out at high rates and cleaned up, thus demonstrating that academicians could become rich if they desired (a demonstration whose effect on subsequent academic history has been negligible).

For particle physicists, the principle of beauty has come to mean the principle of symmetry, which states that correct theories shall display certain mathematical symmetries not unlike those found in regular geometric designs. (Symmetry is scheduled for the Third Debates.)

The principles of reductionism and beauty are clearly of a different nature than conservation of energy, and so your confusion, that scientists use the terms "principle" and "laws" interchangeably, is fully justified. If you encountered Steven Weinberg at the Academy, he would almost certainly argue that the success of reductionist theories validates the principle.

A Practical Exercise: Are Principles More Sacred Than Laws?

It seems indisputable that scientists hold principles to be more fundamental than laws. A decisive experiment would cause most physicists to abandon relativity, or perhaps conservation of energy. But even if this were to happen, many physicists would continue to believe that nature is inherently symmetrical. As theorist Michio Kaku puts it, " . . . nature, at the fundamental level, does not just prefer symmetry in a physical theory, nature demands it." Nevertheless, such principles are verified only indirectly, through the success of the theories based on them and, as an experimentalist will argue in the Third Debates, there must be a certain amount of faith involved in adhering to them despite the scientist's frequent abhorrence of metaphysical statements, an abhorrence evidenced by Steven Weinberg's remark, "I know of no one who has participated actively in the advance of physics in the postwar period whose research has been significantly helped by the work of philosophers."

There is buried in such statements the conviction that we have honed the methods of science over the millennium and are now closer to the Truth, not only in terms of facts, but in terms of principles. For example, tourists at the Academy are often surprised at the disdain physicists bear for the Academy's early director, Plato, and his most famous pupil, Aristotle, whom physicists accuse of setting back the course of science one thousand years, if not two. The general reason for the contempt is explained by F. M. Cornford, who has dropped by for a few minutes: "Knowledge, Plato thought, was to be found not by starting from 'facts' observed by the senses, framing generalizations, and then returning to the facts for confirmation, but by turning away and escaping as fast as possible from all sensible appearances." In other words, the universe can be understood by pure logic, without recourse to experimentation, a notion exactly contrary to the whole concept of modern science.

The specific reason for the contempt is Plato's contention that the planets move in circular orbits around the Earth, a contention that Aristotle enshrined and that went unchallenged by philosophers for over fifteen hundred years. Modern physics was born only when Copernicus in the sixteenth century proposed that the Earth was not the center of the solar system, when Kepler discovered that the orbits of the planets were ellipses and when Newton explained that elliptical orbits resulted from his law of gravitation.

Late in the afternoon of your day at the Academy, on your search for the coffee machine, you encounter a modern physicist and Aristotle crossing paths. You are surprised

Figure 1.2
A modern physicist
demonstrates Kepler's
Second Law to
Aristotle. Because
planets sweep out
equal areas in equal
time, the areas of the
triangles are the same.

that the physicist begins an argument before introducing himself or saying hello, but he launches right in and loses no time in challenging Aristotle, the great proponent of circular motion: "By what right do you decree that the orbits of the planets should be circles?"

Aristotle responds by declaring that "circles are the most perfect shape and circular motion is the most perfect motion."

The physicist will have nothing of this and properly objects, "But the planets don't move in circles; they move in ellipses."

Aristotle, taken by surprise at this intelligence, scratches his chin. "But are they really moving precisely in ellipses?"

"Well," admits the modern physicist, "there are slight deviations from ellipses due to the gravitational pull of the other planets. But to a high approximation, the orbits are ellipses, not circles. Furthermore, each planet moves faster or slower depending on whether it is nearer or farther from the sun. Nevertheless, the area swept out by a line joining the planet to the sun will always be the same for equal time periods."

Aristotle has a difficult time understanding this, so the modern physicist draws a picture. Nevertheless, the Greek has difficulty swallowing Kepler's second law and asks, "But why does the planet behave this way? This seems far from perfect."

The modern physicist replies, "There is a quantity in physics called angular momentum, which in this case is the product of a planet's mass, velocity and distance from the sun. Physicists have discovered that angular momentum is conserved; in an isolated system it cannot change. The conservation of angular momentum results in the equal-area law, which we call Kepler's second law in honor of the astronomer who discovered it."

This news really bothers academician Aristotle, who after fierce cogitation exclaims, "Why should angular momentum be conserved?"

"Ah," replies the modern physicist. "This is a consequence of an underlying symmetry of nature: that space is the same in all directions."

"What do you mean by that?" asks the Greek.

"That space has the same properties north, south, east or west. To put it another way, that in an ideal universe you cannot tell which direction you are facing. One can show that a lack of preferred direction results in the conservation of angular momentum."

Aristotle scowls. "Philistine! This is no different from declaring that circles are the most perfect shapes."

After a pause, the modern physicists replies with some temperance. "Well, we do have experimental evidence that orbits are ellipses, not circles."

"Yet, you have already claimed that the orbits are not quite ellipses. Does this mean that space is not quite perfect?"

"Well yes, but that is merely because the other planets have gotten in the way."

"But you have never observed the solar system empty of planets. What grounds do you have for assuming that in a vacuum space has no preferred direction?"

"This is the simplest, most natural assumption."

"But nature abhors a vacuum. You can never observe the solar system without planets."

And so on, until you depart for coffee.

Aristotle was mistaken about the planetary orbits. However, he might have said, "The most perfect motion is rest," but given that the planets were moving, he picked the most perfect—the most symmetric—orbit he could think of: a circle. The declaration that space is the same in all directions—*isotropic,* to use the technical term—is no different in this regard; it is not based on any observations that can actually confirm it. Thus the modern approach may be more successful empirically, but is no different in character from the ancient approach. To claim that momentum is conserved *because* space is uniform is a teleological statement—a statement that says nature behaves as it does to achieve certain goals—no different from Aristotle's statement about vacuums. Modern academicians who claim that physics is devoid of philosophical considerations are not being entirely forthright.

The festive day has passed into twilight. From all that has been said thus far, a reasonable impression would be that there is much agreement—among scientists anyway—about what is allowed by nature and what isn't, but less agreement about what it all means and how we got there. But at the moment you are probably more concerned with how you got into the Academy and how you might get out in time to make the evening's concert. You glance at your watch, pondering the consequences of being late, but to no avail. At the debating palazzo the two authors are just getting up a head of steam and call for silence, and you begin to perceive that you have, like Theseus, wandered into a labyrinth not of your own devising. After sufficient ringing on glasses, the noise level subsides and the evening session begins.

The authors, in the most agitated manner, begin to expound on

The Confusion of Models with Reality

In the controversy over aliens in Area 51 one word was made conspicuous by its absence. The word is "model" and it has great bearing on the question, Is the universe describable? A theory is a statement about how the universe operates which, ideally, rests on as few assumptions as possible and applies to the widest class of phenomena possible. For example, Newtonian mechanics rests on the famous three laws of motion, with a couple of other unspoken assumptions, and for three hundred years the theory was thought to explain everything. Nevertheless, when applied to a specific problem, Newton's equations are frequently too difficult to solve. One therefore makes simplifications to produce a tractable system. Such models bear the same resemblance to the real system as a Tinkertoy engine bears to a real locomotive, or a Brancusi bird bears to a real one. Much is omitted. Only the outlines remain.

From this point of view, a theory is general, a model is specific. A theory has few assumptions; a model has as many additional assumptions as necessary to produce an answer.

Neurobiologist Jean-Pierre Changeux, who has been visiting the Academy for these debates, goes further. In his *Conversations on Mind, Matter and Mathematics* with mathematician Alain Connes, Changeux maintains: "Proceeding hypothetico-deductively, as physicists do, biologists build thought objects, or models, that they test against external physical reality. These models are simplified representations of an object or a process; they are coherent, noncontradictory, minimal and subject to validation by experiment."

But now you observe that the distinction between theory and model has apparently been blurred. Would not Changeux's first sentence make equal sense if "theory" were substituted for "model"? Does this not show that scientists often use the terms "theory" and "model" interchangeably, as they do "law" and "principle"? Inarguably true. This is the first reason Changeux's remarks are of interest. The second reason is that they raise the question, Do any scientific models satisfy all of Changeux's requirements? A few examples for the starlit evening:

At the opening of the twentieth century, Max Planck produced his famous blackbody theory of radiation. With this theory he explained, quite successfully, the spectrum of radiation* that would be emitted by a perfect emitter and absorber of radiation. (For historical reasons, such perfect absorbers are known as black bodies. Many real objects behave like black bodies: black velvet, the Earth, human beings, most solid objects.) In order to account for the behavior of the black body, Planck modeled it as a system of oscillators, in other words, as a system of invisible springs that absorb and emit radiation.

The extremely peculiar thing about Planck's theory is that nowhere does he say what is oscillating. Are the oscillators real? Are they imaginary? Furthermore, a blackbody absorbs any radiation incident on it, and then reemits it in the characteristic blackbody spectrum. (The Earth receives sunlight, which is peaked in the visible part

---

* A spectrum is a graph that gives the intensity of radiation emitted at each frequency (or the intensity of each color, when dealing with visible light).

of the spectrum and then reemits blackbody radiation, which is peaked in the infrared part of the spectrum.) But nowhere did the original theory explain how light at one frequency is converted into light at another frequency; more technically, how the blackbody equilibrates. Indeed, if the blackbody consists of ideal oscillators, such behavior would be impossible.

In terms of academician Changeux's remarks, this is highly contradictory (impossibility being the highest state of contradiction). The model is coherent in the sense that it makes definite predictions, but is incoherent in the sense that the fundamental objects in the model (the oscillators) are left totally undescribed.

You point out that we have switched from using the term "theory" for Planck's work to "model." The switch was unconscious, actually, but seems reasonable. Only when quantum field theory was invented, three decades after Planck introduced his quantum springs, did physicists come to understand that they represented oscillations of the electromagnetic field itself. It was at this stage that a "full" theory was developed; however, to this day there remains no mechanism to explain how a blackbody equilibrates.

A similar, if more extreme, situation involves the "shell model" of the nucleus, invented by Maria Goeppert-Mayer and J. H. D. Jensen. In atomic physics, the atom is often pictured as a nucleus, consisting of protons and neutrons, surrounded by shells, or orbits, of electrons. The number of electrons occupying each shell, particularly the outermost, determines the chemistry of a given element.

The nucleus of the atom has proven much more difficult to describe in terms of fundamental laws than the atom itself, though a number of attempts have been made to model it. The question that the shell model attempted to address was, Why are certain nuclei exceptionally stable? For instance, the helium nucleus, with atomic mass 4, is exceptionally stable, as is the oxygen nucleus with mass 16.

The shell model attempted to describe the nucleus in analogy to the atom itself. In other words, in the atom a central nucleus provides the force that holds electrons in their orbits. When the outer shell of an atom is filled—when no more electrons are allowed in that orbit—the atom is inert; it does not participate in chemical reactions. Could the same apply to the nucleus?

The shell model postulated that some central force holds the neutrons and protons of the nucleus in orbit. Although the model successfully accounted for the "magic numbers" of helium, oxygen and others, it had a serious defect: there was no basis for assuming that a central force within the nucleus existed. There is no nucleus within the atomic nucleus. For this reason the shell model was termed a model—there was simply no explanation for why it worked. (You may notice Changeux's requirement that the model be self-consistent does not even seem applicable here; one might say it's not even wrong.) Interest in the shell model gradually waned and for many decades the situation remained unchanged. Only recently have physicists claimed to have a better understanding of the shell model in terms of interactions among the neutrons and protons themselves.

By now the debating palazzo is filled with participants and curious onlookers. The reductionists, who are present in force, seize the pause in the exposition to argue that in the shell model we have a typical example of the way science progresses: explanations become more and more fundamental. Models become incorporated into more basic theories. True enough, but it is not obvious that this chain of explanation continues unbroken down to the level of quarks or strings. It may surprise you to know that there does not exist a consistent model of even the hydrogen atom—the simplest atom of the periodic table. And it is far from clear that a consistent model can be produced. To understand this, it is helpful to follow the development of the theory historically.

Hydrogen gas, when excited by an electric current, emits a certain sequence of bright lines, which are easily viewed through a prism or an inexpensive diffraction grating. To explain the frequency of the hydrogen lines was one of the outstanding puzzles of late-nineteenth-century physics and more than any other problem led to the development of quantum theory.

Most introductory texts claim the problem was solved by Niels Bohr, who invented a model of the atom based on quantum principles, a model that exactly predicted the frequency of the hydrogen spectral lines. However, this is not entirely true. Even in 1914, a year after Bohr published his paper, it started to become clear that the frequency of the hydrogen lines did not exactly fit Bohr's formula.

A further problem had been evident for a quarter of a century before Bohr published his model: several of hydrogen's spectral lines were not single lines but closely spaced pairs, or doublets. Bohr's model could not account for this; neither could the new version of quantum mechanics developed by Heisenberg and Schrödinger in 1925 and 1926, respectively. Their quantum mechanics did not take into account Einstein's theory of relativity, nor did it take into account that the electron can be thought of as spinning "up" or "down"; such spin gives rise to the doublet structure. P. A. M. Dirac wedded quantum mechanics and special relativity, included the properties of electrons, and produced a startlingly new theory, called relativistic quantum mechanics, which explained the doublets.

However, even here there were small disagreements with observation, first observed by Willis Lamb, in which some spectral lines were slightly redder than predicted by Dirac's theory. To explain the "Lamb shift" required a new theory, quantum electrodynamics, or QED, which is associated with the names Richard Feynman, Julian Schwinger, Sin-Itir Tomonaga and Freeman Dyson. QED posits that the electromagnetic field is continuously producing electron–anti-electron pairs, which quickly annihilate one another. These "virtual pairs," as they are called, distort the system's energy enough to account for the Lamb shift. Since QED took care of the Lamb shift in the late 1940s, the only corrections have been due to more accurate calculations.

The reductionist faction, again taking the floor, reasonably argues that physics has come ever closer to explaining the hydrogen atom. But QED has its own problems. First, all the calculations in QED are approximations. What's more, these approximations suffer from a well-known disease of infinities. The terms in the equations of QED often become infinite, exactly like what happens when you divide a number by zero. Although physicists can get rid of the infinities by a process termed "renormalization," by general consensus renormalization is a sleaze. No one really understands why the procedure works. Academician Dirac once declared: "It seems to be quite impossible to put this theory on a mathematically sound basis . . . the remarkable agreement between its results and experiment should be looked on as a fluke."

It is certainly true that with renormalization, agreement between QED and observation is striking—to eleven decimal places. But the real question is whether it is possible to construct an underlying theory. What is not usually acknowledged is that QED is not actually a theory of the entire hydrogen atom, the system it was designed to explain. A hydrogen atom consists of an electron and a proton. Yet in QED the proton is merely treated as the stationary source of the electric field and not as a particle that actively participates in the interactions—for example in the giving off of light. To construct an underlying theory beyond QED, one needs to include the proton as an active member of the interaction.

At hearing this a voice is raised: the quark model solves the problem by regarding the proton as a combination of three more fundamental particles known as quarks; it then treats any reaction involving protons as reactions between quarks. We reply that no one has successfully calculated how three quarks interact to make a proton, which forms the nucleus of hydrogen, and therefore one cannot appeal to quark theory to explain the hydrogen spectrum.

Even if one managed to explain how the proton emerged from quarks, a difficulty would remain. A hydrogen atom consists of an electron and a proton tied together by the electromagnetic field. Physicists refer to such a system as a bound state. To date no one has produced a bound state from any predictions by QED. In terms of QED, it is unclear whether a hydrogen atom even exists.

That might seem to be a severe shortcoming of the theory, yet there is a further ambiguity. In QED electrons never exist alone, but always in the company of photons—particles of light. Therefore if a hydrogen atom is to be composed solely of an electron and a proton, not only is it unclear whether a hydrogen atom exists, it is also unclear what is meant by such an object.

We see that quantum field theory is a bit like walking through mud—if you walk fast enough you can get out; if you walk slowly you sink in. What lessons are to be drawn from this situation? In an obvious sense physicists have produced in QED a theory that works but which they don't understand. In this, quantum field theory is very much like acupuncture. Most people would agree that acupuncture works, or at least sometimes works, but no one has an accepted explanation for it.

Particle physicists routinely claim that QED is the most successful theory of all time because its predictions are verified to eleven decimal places.* Such numerical agreement with experiment is impressive. On the other hand, the infinities that plague the equations are in precisely the quantities, like the energy of the system, which one normally demands be finite. Therefore, QED seems inconsistent with the basic postulates of quantum mechanics on which it is founded. Feynman, with his New York accent, might appear at the Academy gates and say, "What's your problem, boys? You can calculate and get good results." Dirac, in his taciturn manner, would reply . . . well, we have already quoted Dirac.

---

* Due to observations of the binary pulsar, certain predictions of general relativity are now verified with even better precision than those of QED.

One remains uneasy that QED has described the underlying reality of the universe. It is an excellent computational scheme but the range of phenomena it predicts is limited. The range of phenomena it describes is also limited. Perhaps it should be termed a model and not a theory.

If there has been a fundamental confusion between the model and reality, the confusion is not limited to QED. The Directors of Sports and Recreations have scheduled for the Ninth Debates such questions as, Are current astronomical observations determining whether the universe will expand forever, or whether certain models will expand forever?

As the preliminary discourse on models winds down, there is a general chorus (with some dissent) to remember the famous

Aphorism:
The Tao that can be expressed is not the true Tao.

The participants break for a well-deserved intermission.

Other Modes of Perception

Having had enough of this rigmarole and desperate to catch the second half of the concert, you spend the intermission searching for an exit from the Academy. However, the place seems much larger than it did when you arrived and you quickly become lost. There is nothing for it but to return to the debating grounds, where you remind the authors that their entire exposition has emphasized the scientific description of the universe, one based on theories, laws, models and principles.

To be sure, there are other modes of description: art, literature, music, psychology. Most relevant for these debates are popular attempts to connect quantum physics and consciousness and to interpret mystical experiences in terms of quantum mechanics. The most famous of these efforts is Fritjof Capra's 1975 best-seller *The Tao of Physics*. In that book, Capra drew analogies between physics and Eastern religions, principally Buddhism, but in the end admitted they were merely analogies and even went so far as to say that similar parallels could be drawn with Western mystical traditions.

Capra's work has spawned many imitators, and *The Tao of . . .* has become an almost obligatory title for any New Age science book. (Alternate titles for these debates were *The Tao of the Tao* or *The End of the Tao*.) Behind the New Age interest in mysticism is the conviction that Western science has produced a split between mind and heart and that this has done more harm than good. Scientific thinking is rational, but Ultimate Reality—the Brahman as it is called in Indian tradition—is accessible only to direct experience, which transcends intellectual thought. On the other hand, the New Agers see in modern physics—in quantum mechanics especially—a link between the spiritual and the rational.

Steven Weinberg will have none of this and says so forthrightly: "Those who seek extrascientific messages in what they think they understand about modern physics are digging dry wells." The authors also find in New Age works many misunderstandings about both modern physics and Eastern religions, with the result that their conclusions seriously mislead. Whether the academicians are capable of sorting things out in these debates remains to be seen, but we can begin by asking, Are mystical experiences compatible with physics? Are mystical traditions theories in the scientific sense of the word?

Both the New Age and Indian mystic literatures answer yes. Eknath Easwaran, in his introduction to the Upanishads, says, "The Upanishads are not philosophy" and calls them "the supreme science." Rammurti S. Mishra, in his commentary on the Yoga Sutras, writes, "Psychic operation, like physical law, has definite order and constitution. If we know the constitution of the mind, the operation of the mental mechanism and psychic processes, we can use them for our development physically, mentally and spiritually." He goes on to write, "Study of psyche is more important than study of any other subject since all subjects depend on mind"; moreover, "psychology includes all sciences, arts and disciplines." And not only did Deepak Chopra entitle a recent book, *The Seven Spiritual Laws of Success*, to emphasize the nearness of his beliefs to science, but he does not blush when making statements such as, "The physical laws of the universe are actually this whole process of divinity in motion, or consciousness in motion."

Before joining the fray, you may want to pick up a few technical terms. "Upanishad" means something like "sitting down near" and suggests the teachings given by a master to his students in a forest academy. In form the Upanishads are long poems or hymns, sometimes written as dialogues. Unlike Western religious documents, the central concern of the Upanishads is the nature of Reality and the relationship of Mind to Reality. The date of composition of the Upanishads is thought to be before 1000 B.C. and they are considered the most important of the Hindu spiritual documents.

The Indian preoccupation with laws of the mind found practical expression in the Yoga Sutras of Pantanjali. Pantanjali was the semilegendary compiler of the sutras, and the sutras themselves are exercises, expressed as highly compressed aphorisms, whose purpose is set forth in a famous line at the very opening of the collection:

Yoga chitta vritti nirodhah.

Yoga, the union of the mind with the Brahman, or the Supreme Consciousness, comes about when all interference (*vritti*) is stilled from the mind (*chitta*).

The remainder of the sutras sets forth explaining how the union with the Supreme Consciousness of the universe is to be achieved through meditation. Because the mind is involved in this process it is not "objective" in the sense of traditional sciences, the outcome depends on the observer. Physics, since its inception, has attempted to remove the observer from consideration. Quantum mechanics, in a way that will be more fully explored in the Seventh Debates, apparently opens the door for the participation of observers in determining the outcome of measurements, in defining the very nature of Reality. Hence the mystical–quantum connection.

What do scientists have to say about all this? Steven Weinberg's view is certainly the predominant one. Although over the past twenty years physicists have grown accustomed to speaking openly about the interpretation of quantum mechanics in Academy corridors, very little is said about it in lecture halls and classrooms. Its status is a bit like Violeta's in *La Traviata*: not entirely welcome in respectable society. Once again Weinberg for the majority: "So irrelevant is the philosophy of quantum mechanics to its use, that one begins to suspect that all the deep questions about the meaning of measurement are really empty. . . . "

Weinberg goes on to admit to "some discomfort in working all my life in a theoretical framework that no one fully understands," and he does concede there is something very mysterious about quantum mechanics. This goes beyond the defective mathematics of QED. Quantum mechanics does not appear to contain any defective mathematics. Within its domain, its predictions are experimentally verified. And yet simple questions like, How did this particle get from here to there? have no answers in quantum theory.

Clearly if a description of the universe entails more than merely predicting the outcome of an experiment, that is, also conveying an understanding of why that outcome has taken place, quantum mechanics (or rather its practitioners) have failed. Despite the failure, physicists do not like to hear talk of another reality, unattainable through physics, and mention of New Age philosophy runs the risk of excommunication from the Academy's School of Natural Sciences.

Nevertheless, in the spirit of these debates we ask whether in any sense the mystical traditions may be considered science. Because such traditions deal with the mind, one must reconcile oneself to the idea that accounts will be somewhat subjective. You may assume the lotus position while one of the authors recounts a helpful

True Story

I was hanging around with the Maharishi at the Maharishi International University in Iowa and was told that some people had come from Switzerland and were practicing special yoga instructions with a view to achieving *siddhis* (certain attainments set forth in Pantanjali's textbook). "What is going on in Switzerland?" I asked them. "We cannot tell you. Talk to Maharishi." I talked to Maharishi and he said, "If you want to find out, why don't you go to Switzerland?"

We went together. There were about one hundred people practicing special "siddhi" meditation. They described their experiences. Maharishi then said to them, "If you have any questions about your experiences, ask the professor." Curiously enough I could provide an interpretation.

In terms of what?

In terms of yoga philosophy. I had studied it before but not practiced it. In the subsequent month I also practiced.

And what did you discover?

I found a clear distinction between two different classes of attainments. Those that pertain to knowledge, I seemed to able to reach. Those that in my mind violated physical laws, such as people levitating, people being able to see around obstacles, people becoming very heavy—with none of these had I any luck.

I was there only two or three weeks. It was quite clear to me that the prescriptions for the various mediations were quite successful but not to all the people all of the time.

I compare this favorably with how to create a complex computer program. I can do many things with a computer, but at some point something always goes wrong and

I need help. This does not mean the handbook is not good. Computer science and proving theorems about computability is not quite the same as actually doing the computation itself.

A Comparison of Meditative Experiences with Experimental Physics

At this juncture scientists/academicians vigorously object that such stories mean nothing whatsoever; it is all subjective experience. There is, however, a specific sense in which meditative experiences may be compared to science. If one examines Kashmir Shaivism, an aesthetic meditative system, one finds extremely specific instructions for achieving various states of consciousness. We are first told that there is a junction, or transition point, between each of the three states of consciousness and the next: the waking state, the dreaming state and the state of dreamless sleep. We are next told how to experience this junction:

> You should [focus on] the center of any two movements, any two breaths. After some time when that concentration is established, then whenever you go to bed to rest you will automatically enter the dreaming state through that junction.

Then specific predictions are made:

> . . . Here you do not lose awareness even though you feel intoxicated. . . . Here the aspirant does not experience moving about nor does he hear or see . . . he cannot move any part of his body. . . . At that moment the aspirant hears hideous sounds.

Meditative states are communicable and reproducible, and this procedure thus conforms to the standard definition of science. Theoreticians object that there is no fundamental theory explaining such states. It is here we make a remarkable observation: most physicists are experimental physicists. Most physicists who write books are theoretical physicists.

It is a striking fact that of the dozens of popular expositions on physics that can now be found on the shelves of local bookstores, perhaps one appears to be by an experimental physicist. An alien learning about Earth from television might conclude that everybody is beautiful and virtually everyone is a doctor, a lawyer or a cop. This is termed a selection effect. That the public hears the voice of theorists, not experimentalists, is also a selection effect.

Unlike theorists, who spend their days making explanations, experimentalists spend their days fixing equipment and, on the rare occasions when an experiment actually works, making observations. Biology is very much an experimental endeavor. So is psychology, for which a universally accepted theory does not exist. So is plasma physics, the science behind the effort to attain nuclear fusion. Most of the behavior of tokamaks, the prototype fusion reactors, is understood not by theory but by trial and error.

Meditative experiences are carried out by practitioners and in this sense are more akin to plasma physics than to quantum physics; they are experientially understood, not theoretically understood. If tokamak physics is too complicated to be successfully modeled on Newtonian physics, it seems fruitless to attempt to model the mind on the basis of quantum mechanics.

This is not to say that one cannot model meditative states. The sutra we already quoted tells us that yoga is the elimination of the random motions of the mind. One can view this in terms of a fluid model: a body moving in an ordinary fluid, like water, experiences friction that slows it down. To slow down, a body must lose energy. On a microscopic level this corresponds to the object absorbing sound waves, known as phonons, from the liquid and reemitting other phonons. But if there are no phonons in the liquid, then this process cannot take place and the object moves without loss of energy. Now, certain fluids cooled to near absolute zero become "superfluids," fluids without friction; they have no phonons in them. Because all friction in the system has vanished, the body is free to move without slowing.

In this model of the mind fluid, which can be made quite precise, the random motions of the mind correspond to the phonons, and the quelled state of the mind corresponds to the object moving in the superfluid without friction. Does the model require that the mind be a fluid? No! You cannot pour the mind into a glass and drink it. The essential point is that disturbances can be removed and the superfluid properties (the meditative state) can be verified in practice.

Is this a mere analogy? Does a nucleus-within-the-nucleus exist? No. But if you can predict the behavior of the nucleus on the basis of the shell model, it serves a purpose. Yet another model of the nucleus, the "liquid drop model," explains certain other properties of the nucleus by regarding it as a drop of liquid. Is the nucleus a drop of liquid? No. Does it have some properties that resemble a liquid drop? Yes. Are the liquid-drop model and the shell model entirely compatible? Probably not. Do we use both of them? Yes. Chemists sometimes represent molecules in terms of two-dimensional drawings, sometimes as three-dimensional computer graphics, sometimes as Tinkertoy models. Are any of these the molecule?

Analogies are used all the time in science. Do the objects of the analogies need to exist? Perhaps the better question is, Do we care whether they exist? What scientists do care about is whether they can predict. To the extent that you can reproduce precise states of mind, the practice of meditation can be viewed as scientific. To the extent one can model mind properties, and conceivably predict mind behavior, one can view meditative practices as scientific.

However, it strikes us as somewhat arbitrary to single out quantum mechanics as the mind model. In any case, one needs to bear in mind that one is not describing the same system as quantum mechanics, or quantum electrodynamics. These theories deal with the electromagnetic field. A model of the mind describes the mind. As Don Juan might have said, this is a separate reality.

In Which the First Debates Are Summarized

The day was undeniably long. Some time ago you became wearied by all that was transpiring and finally gave up the attempt to find your way back to the Real World tonight. So, you curled up next to a fountain on the lawn and, to the soft gurgling of water, fell asleep, fully intent on leaving at sunup. But as you passed from the waking to the dreaming state, you saw many of the topics that arose in these First Debates returning time and again. You could not help but mentally review the discussions.

Perhaps the central points were semantic. Visitors and academicians discovered they often use the word "theory" in completely different senses; scientists use it to denote a system of assumptions and predictions that can be experimentally tested. Nevertheless, the question arose whether certain contemporary theories deserve to be called theories. The term "law of nature" finds fairly consistent usage among scientists, as a general statement about nature that is never violated. However, it appeared that often the word "principle" is used interchangeably with "law," and at one point it became unclear whether modern principles are more "scientific" than Aristotelian principles.

Given that scientific theories are to be experimentally tested, the authors argued that certain mystical experiences can be viewed as science in an observational sense, and can be modeled. But the authors warned against assuming that models describing mystical practices represent the same thing as quantum physics.

A large portion of the debates was spent contemplating the relationship of theories to models. Whether scientific theories describe the real universe, whether they describe models, or whether the theories are the models was not resolved, but it did seem undeniable that scientific models are often mutually inconsistent. Scientists jump from one model to the other without gluing together the joints. The Bhagavad Gita makes the point in another context:

> As the Soul in this body passes through childhood, youth and old age, so (after departure from this body) it passes on to another body. The sage is not bewildered by this.

Perhaps the sage should be. A Westerner, Roald Hoffmann, put it somewhat differently, when asked how well chemical models describe molecules: "Maybe it's not that different from the way we approach romance in our lives, equipped with a piecewise reliable set of images from novels and movies."

Is the universe describable?

IS NATURE UNREASONABLY MATHEMATICAL?

Mathematics, Theories and Reality

A man who could give a convincing account of mathematical reality would have solved very many of the most difficult problems of metaphysics. If he could include physical reality in his account, he would have solved them all.

—*G. H. Hardy*

Mathematics is a kind of primate behavior as languages, musical systems and penal codes are.

—*Leslie A. White*

The Question

It is told that every year the king of Travancore invited scholars of the scriptures to gather for learned discussions and recitations of sacred verses. The rest of the time they spent in contemplation. All their needs were met, they were housed and fed. One very learned scholar, a mere youth, was apparently staying inside the temple, amusing himself with a pack of cards. Discovering the activity, the elders complained to the king, "This man is committing sacrilege. He is playing cards!" So the king summoned him and asked, "Is the accusation true?"

"Your Highness, the accusation is not true. What is true is that I had a pack of cards and that I was using it in contemplation. The aces remind me that all the universe is undivided, but it can appear different to different people. The twos point to the primary duality of the observer and the observed. The three refers to the three-limbed things. In any process, there are three limbs—the actor, the acting, and the action, the seer, the seeing, and the seen. The four refers to the four cardinal directions. The five refers to the five elements. . . ." At the end the accusers begged his pardon and the king rewarded the young scholar.

We see that the profane can become sacred and that an ordinary pack of cards can have meaning far beyond the mere numbers or faces inscribed on the deck. It is possible that you have had a similar experience if you have ever strolled along a tropical beach and chanced upon the shell of a chambered nautilus. For years, if not centuries, mathematicians have recognized that the spiral pattern of the chambers can be described by a famous sequence of numbers, called the Fibonacci sequence.[†] Pondering the nautilus shell, many have been struck by wonder that its pattern corresponds so closely to a mathematical sequence, a product of the human mind.

If you have never seen a nautilus shell, you may have had the same experience when you were awoken on the morning after the First Debates, not by the sun or the spray of water, but by shouting from nearby fields. As you rolled over and got to your feet in order to discover the source of the commotion, you found yourself staring into the face of a giant sunflower. Like the nautilus shell, the sunflower's spiral pattern of florets is also described by the Fibonacci sequence.

Strangely, the relationship of mathematics to the Real World is exactly the topic of the debate that erupted before dawn. The academicians have been at it for some time now, arguing with one another on their morning constitutional, even as the sun's first rays flood the Arcadian landscape. The academic taste for discourse both astonishes and unnerves you. Do they not have chores to attend to, destabilizations to engineer in emerging democracies?

No.

But the reason for the continued discussion is easy to understand. Yesterday's deliberations left unresolved whether the universe is describable. Nevertheless, as is clear from the tone of the general aubade, people will never cease trying to describe it. In truth, most people give up the attempt at about the same time they cease trying to find the meaning of life. For those who persist, the chief weapon in the attack depends on profession. For artists, it may be words or music or paint. For physical scientists, the chief weapon in the war for Reality is mathematics.

But to what extent can mathematics describe the world? Is it the same world as the world of music and literature? If not, what does it describe?

If yesterday's questions struck you as esoteric, then these must strike you as positively useless. In any case, mention of the Real World evidently reminded you to find a telephone in order to make excuses for last night. Unfortunately, phone links at the Academy to the outside operate only on major holidays, such as Aristotle's birthday, and your tourist bus departed yesterday.

There is little for it but to listen while we entertain you with the necessary background.

The Conventional Wisdom of Physicists

In a 1959 lecture entitled "The Unreasonable Effectiveness of Mathematics in the Natural Sciences," physicist Eugene Wigner remarked that "it is difficult to avoid the impression that a miracle confronts us here." Wigner meant the striking ability of mathematics, an abstract system, to describe the real world, and he considered this miracle to be "quite comparable . . . to the two miracles of the existence of the laws of nature and of the human mind's capacity to divine them." Wigner's lecture remains more famous than read, but his phrase "the unreasonable effectiveness of mathematics" has become a mantra to most physicists. Few doubt that mathematics' ability to describe the nat-

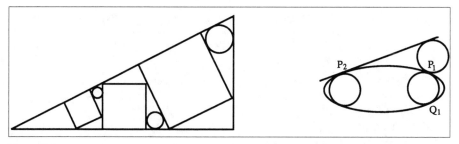

Figure 2.1

Two *sangaku* problems. In the first, the three circles are inscribed within their respective right triangles. The problem is to show that the radius of the middle circle is the geometric mean of the radii of the larger and smaller circles. In the second problem, two circles of equal radii are inscribed in an ellipse. P_1 and Q_1 are the points at which the right-hand circle touches the ellipse. A third circle of the same size also touches the ellipse at point P_1. One is to show that the tangent line drawn touches the ellipse at point P_2 if and only if $P_1Q_1 = \sqrt{2}\, b$, where b is the minor axis of the ellipse.

ural world is truly remarkable, and it should be considered closer to a matter of divine intervention than coincidence.

Closely allied with this point of view is what has become known as the principle of beauty, which made a brief appearance in the First Debates. The theories most likely to be the correct ones are mathematically the most beautiful. The name of Paul Dirac, a hero to many modern physicists, inevitably comes up when the subject of beauty arises. Pondering the future course of physics in 1963, Dirac wrote:

> There is one other line along which one can still proceed by theoretical means. It seems to be one of the fundamental features of nature that fundamental physical laws are described in terms of a mathematical theory of great beauty and power, needing quite a high standard of mathematics for one to understand it. You may wonder: Why is nature constructed along these lines? One can only answer that our present knowledge seems to show that nature is so constructed. We simply have to accept it. One could perhaps describe the situation by saying that God is a mathematician of a very high order, and He used very advanced mathematics in constructing the universe.

Dirac concludes: "Just by studying mathematics we can hope to make a guess at the kind of mathematics that will come into the physics of the future." Underlying the view of Wigner and Dirac is of course a strong reductionist philosophy: nature can be boiled down to physics, and physics to mathematics. The idea of mathematical beauty strikes many nonscientists as foreign, but it is not really so different from geometric beauty.

In medieval Japan the two coincided. From about 1639 to 1854, when Japan was by imperial edict totally isolated from the West, a sort of home-grown mathematics flourished. Professional mathematicians, samurai, farmers, women and children would solve the most difficult geometry problems, inscribe the solutions on wooden tablets, and hang the tablets in Shinto and Buddhist temples. Two typical problems are shown in the illustration above; in both, geometric design reveals a beautiful mathematical relationship.

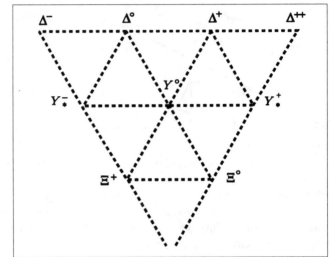

Figure 2.2
Baryon decouplet. If a
quantity called
hypercharge is plotted
versus another quan-
tity termed *isospin,*
certain heavy particles
fall into this famous,
highly symmetric dia-
gram. In 1962 the dia-
gram was incomplete.
Where would you put
the missing particle?

Just as in these *sangaku,* rich mathematical theories reveal hidden, elegant connections. This gives some idea of what Dirac meant by mathematical theories of "great
beauty and power." For physicists of the twentieth century, mathematical beauty has
come to mean mathematical symmetry, and fundamental physical theories are those
which display these symmetries. Already the academicians are preparing for tomorrow's debates on symmetry, and one vocal proponent of mathematical beauty cannot
resist giving a famous example to bolster the claim that it is at the bottom of everything.

The masses of the proton, electron and other known subatomic particles span an
extremely wide range; at present there is no explanation for the particle masses and
physicists have exerted much effort in finding one. Symmetry has played an important
role in this search. The most celebrated incident took place in 1962 when Murray Gell-
Mann, Susumu Okubo and Yuval Ne'eman predicted a particle known as the Omega-
minus (Ω^-). Gell-Mann drew a nice symmetric, geometric pattern (see Figure 2.2), not
unlike a *sangaku.* Each position in the pattern except for one empty spot contained a
known particle. Gell-Mann put his finger down on that spot and said, "There is a particle." He was correct. Experimentalists later found an actual elementary particle that
corresponded to the empty spot. "Few people can any longer doubt that the universe is
unreasonably symmetrical," says the academician.

Is Mathematics Real?

By this stage several questions might well have occurred to you. The stance that the
universe is unreasonably mathematical seems to presume that mathematics is prior,
that it exists "out there." Mathematicians unquestionably feel that way. Andrew Wiles,
who recently proved Fermat's Last Theorem, was not long ago overheard to remark on
Academy grounds that if another universe existed "the mathematics would be the
same." Virtually all modern mathematicians share his point of view. G. H. Hardy, in his

famous *Apology*, said, "I believe that mathematical reality lies outside us, that our function is to discover or *observe* it, and that the theorems which we prove, and which we describe grandiloquently as our 'creations,' are simply our notes of our observations. Alain Connes, conversing with Changeux, repeats the claim almost verbatim: "[The mathematical world] exists apart from us because, as all mathematicians agree, its structure is independent of individual perception."

In this mathematicians are surprisingly Platonic. Plato regarded numbers, along with his famous forms, as the closest thing to Reality. He declares, "Numbers have the power to lead us to reality" and also, "Geometry is the knowledge of the eternally existent."

"It cannot be denied," reply his fellow banqueters.

Plato, in turn, was deeply influenced by the Pythagoreans. The Pythagoreans, to the extent we have any reliable information about them, evidently believed that "number" was the substance of all things. Whether they meant material objects are actually composed of numbers, or whether everything can be described by numbers is difficult to say. But they did understand that simple musical intervals could be explained in terms of the first four integers,* and from this observation formulated the doctrine of the "music of the spheres": the heavenly bodies literally produced music whose pitch depended on their speeds, which were proportional to their distances from the Earth, which were in turn proportional to musical intervals. To explain why humans didn't hear such music was more difficult.

Of all the major mathematicians, only Henri Poincaré seems to have veered from the Pythagorean-Platonic tradition. "The axioms of geometry," he says to the gathering, "are neither *a priori* truths, nor experimental facts, but mere 'conventions.'" The mathematicians jeer Poincaré, who merely answers, "I am certain that the opportunity will arise within the week for me to change your opinions."

On the morning constitutional, Einstein has been having similar thoughts. As you may know, the Euclidean, or plane, geometry we learn in high school is only one of several consistent geometries (we discuss these further below). Evidently, to Einstein, like Poincaré, the existence of non-Euclidean geometries proves that mathematics is an invention of the human mind. "We come now to the question," he says, siding with Poincaré, "what is *a priori* certain or necessary in geometry or its foundations? Formerly we thought everything; nowadays we think—nothing. Already relativity shows that our concept of distance is logically arbitrary; there need be no things that correspond to it, even approximately."

"Why didn't somebody tell us?" retort the mathematicians, who stubbornly cling to the idea of mathematical reality and refuse to admit the heretics Einstein and Poincaré to the mathematical ghetto at the cafeteria. The heretics might argue, by way of compromise, that the existence of non-Euclidean geometries indicates the presence of many mathematical realities. But mathematician Connes insists that they merely show the consistency of a single "higher reality." That is, Euclidean geometry assumed

---

* An octave is produced by two strings whose lengths are in the ratio 2:1, a third in the ratio 4:3, and the perfect fifth in the ratio 3:2.

a certain postulate.* Showing that the postulate did not always hold resulted in the non-Euclidean geometries. Plane geometry is just one aspect of Connes's "higher reality," which connects all the possible geometries into a consistent whole and which also exists "out there" with its offspring.

Jean-Pierre Changeux proves stubborn: "For me [mathematical axioms] are expressions of cognitive facilities, which themselves are a function of certain facilities connected with the use of human language." So mathematics is merely some derivation of human language.

Mathematicians everywhere are scoffing. And for us convinced scientists the position is difficult to swallow, but Changeux is far from the most extreme in this regard. Leslie White, an anthropologist whom we have already quoted in one of the epigraphs to the Second Debates, goes on to claim that mathematical reality is no less cultural than "a code of etiquette, traffic regulations, the rules of baseball, the English language or the rules of grammar." This position strikes the authors as absurd. Although different cultures at different times have used different mathematical symbols and may have lacked certain mathematical concepts, we can think of no instance in which the existing mathematical concepts of one culture were invalid in another.

"This question about the reality of mathematics seems as unanswerable as the existence of God," we catch you muttering under your breath. Likely true. Both authors are somewhat agnostic on the issue but would claim, at least, that mathematics has a more universal character than any human language or set of traffic laws.

Nevertheless, you have observed an oversight during the first exchange of the morning: Dirac's assertion that God is a mathematician was not adequately addressed. Even if mathematics exists "out there," it does not necessarily follow that our world fits mathematics like a glove. Mustering all your courage, you suggest that the debaters turn their attention to the question, Did God create the world in the image of mathematics? which may be a more tractable issue.

We agree.

However . . .

Is God a Mathematician? The Age of the Claim Is Revealed

It is important to understand that Dirac's claim is not new. Scientists as a rule are ahistorical in their thinking, tending to assign credit to the most visible members of their generation. Hence the famous

Aphorism:
Either you do the calculation or you get the credit.**

Nevertheless, physicists have been strongly influenced by the mathematicians' faith in mathematical reality, and theoretical physicists, if not exactly Platonists, are certainly

---

* This is the one that states it is possible to draw only one line parallel to a given line through a point not on the line.
** Clearly this maxim has universal applicability.

Pythagoreans. As a young man of twenty-four, Johannes Kepler wanted to understand why the orbits of the known planets had the particular sizes they did. On July 19, 1595, divine inspiration struck. Under the spell of both Plato and Pythagoras, Kepler realized that the orbits of the planets should fit inside the five perfect, or Platonic, solids.* Of his insight, Kepler later wrote: "Geometry is unique and eternal, a reflection of the mind of God. That mankind shares in it is because man is an image of God." Kepler published his Platonic model of the solar system in 1596 as the *Mysterium Cosmographicum*. It was, of course, wrong.

However Kepler, a Pythagorean to the core, remained convinced that nature must be described by simple ratios of integers. Although he failed to fit the planetary orbits into the Platonic solids, he then began a long attempt to associate ratios of the sides of the solids with the Pythagorean musical intervals. This mad endeavor led him to discover his famous (and correct) third law of planetary motion: the square of a planet's orbital period is proportional to the cube of its average distance from the sun (p^2 is proportional to d^3). The title of the book in which Kepler published his third law, *The Harmony of the World,* was to be taken literally: he really did believe in the music of the spheres. Note the prominence of integers in Kepler's formula.

Kepler's laws led directly to the creation of Newtonian mechanics, one of the most successful physical theories ever devised. Is it strange, then, that natural scientists began to rely on the effectiveness of mathematics? At the same time, the simplicity of Kepler's third law, and others, exerted a powerful effect on scientists. In particular, the magic of integers has never lost its sway. The most famous example of the past few centuries involves that of Jacob Balmer, a Swiss schoolteacher and numerologist. His Habilitation** thesis was entitled "The Prophet Ezekiel's vision of the Temple broadly described and architectonically explained." In it, Balmer disclosed the secret numerical relationships in the Temple, just as today his spiritual descendants explain the secrets of the Great Pyramids. Balmer wrote only one other paper that anyone remembers. It was not much different, except that instead of explaining the proportions of the Temple, he attempted to explain the spectrum of hydrogen.

In the First Debates we mentioned that to explain the puzzling sequence of hydrogen lines was one of the outstanding puzzles of late-nineteenth-century physics. Balmer, after some struggle, produced a simple formula involving two integers that gave the sequence of spectral lines as closely as experiments could measure them.*** To explain the miraculous agreement of Balmer's formula with nature now became a great endeavor; it was exactly this endeavor that led Bohr to devise his model of the hydrogen atom and to introduce quantum mechanics into atomic physics. Bohr's model produced Balmer's formula, lock, stock and barrel.

---

* In other words, the planetary orbits should be bounded by the regular tetrahedron, the cube, the octahedron, the dodecahedron, and the icosahedron. This explained why there could be only five planets.

** In some European countries the doctorate is not the highest degree awarded; in these places the Habilitation is necessary to become a professor.

*** Balmer's formula says that the frequencies of the spectral lines of hydrogen are given by $f = R(1/n^2 - 1/m^2)$. R is some number that Balmer chose to fit the data. Then substituting $1, 2, 3 \ldots$ for the integers n and m gave all the spectral lines of hydrogen.

With the triumphs of quantum mechanics and relativity, the metaphor of God as mathematician perhaps reached an all-time high. Early in the twentieth century, another famous physicist, James Jeans, wrote, "Nature seems very conversant with the rules of pure Mathematics" and "The Great Architect of the universe now begins to appear as a pure mathematician." Pythagoras was alive and well.

In 1938 Arthur Eddington, one of the great astrophysicists of the twentieth century, began a famous 1938 lecture with the words: "I believe there are 15,747,724,136,275,002,577,605,653,961,181,555,468,044,717,914,527,116,709,366, 231,425,076,185,631,031,296 protons in the universe and the same number of electrons." This extraordinary statement might be said to be the culmination of Eddington's "Fundamental Theory," which he developed late in his career and in which he attempted to derive the constants of nature directly from mathematical laws, without recourse to observation.

It was pure nonsense. The values Eddington deduced by logic for the natural constants were incorrect, and it is fair to say that no one has a good idea of how many protons there are in the universe. (The universe may even contain an infinite number of protons, which is larger than Eddington's number.)

Eddington's Fundamental Theory, like those of Kepler and Balmer, contained distinct numerological overtones, and by proposing it, Eddington tarnished his career in the eyes of many physicists, who petitioned to have him barred from the Academy altogether. The reaction is instructive: on the one hand, the universe is mathematical; on the other hand, it should not be *too* mathematical.

We see that physicists have long constructed theories based on purely mathematical considerations. Many have been successful; others have failed miserably. The simple truth is, not all mathematical constructs result in physically meaningful theories. Judging from the 4,000 pages of the *Physical Review* that appear every month it seems absurd to believe that even a sensible fraction of the ideas within corresponds to physical reality.

"Calloo callay!" you cry, preparing to abandon the Second Debates as resolved. Another question answered.

Not so fast.

Although many mathematical theories might not produce anything physical, the question of greater interest is the reverse: Do all physical theories merely uncover preexisting mathematical concepts? Is it inevitable that natural science follows mathematics?

Before debating this question, we wish to take a short excursion. And before the excursion we suggest an exercise. Recall that the prediction of the Omega-minus was the most famous incident that established physicists' belief in symmetry. On the other hand, while we take the following excursion, ask the nearest academician whether the prediction was any different in *kind* from Eddington's. Or Kepler's or Balmer's.

Then duck.

A Short Meditation on the Alliance Between Jungians and Physicists

The short excursion we take before addressing the question, Do physical theories uncover preexisting mathematical concepts? is to point out that followers of the psychologist Carl Jung and physicists, despite their mutual hostility, share much the same views on mathematics and reality.

It is, for example, always a surprising thing that people can communicate. Communication sometimes takes place in a way that we would not consider formal communication. Our ancestors who scrawled scenes of hunting on the walls of caves are gone but their signs and symbols remain and are appreciated. Jungians contend that we appreciate these signs and symbols because all humans, and perhaps even animals, share the same brain structure, which provides the basis for interpreting these symbols. This is the celebrated collective unconscious. Evolutionary biologists might substitute "DNA" or "genome" for "collective unconscious."

But is there a difference between interpreting a sign from the cave dwellers and a sign from nature? It is not much of a leap to extend the idea of communication between people to communication with nature. At least this is what the Jungians attempt to do. They refer to the Indian concept of *prakriti,* which is "the matrix of all physical and psychic being," the basic energy from which both the mental and physical world take shape. Because *prakriti* encompasses both the mental and physical world, in the Jungian view there is little difference between interpreting signs from ancestors and interpreting signs from nature. In other words, we are able to understand nature in terms of its symbolic content. We understand nature because its signs and symbols are within us. Why do people put great value on precious stones? Because precious stones have the same radiant colors seen in highly meditative states, which are blissful. In the same way mathematics is not foreign to us but is a consequence of our thought processes. Nature is a representation of mathematics.

Thus, physicists' view that physics is a transcription of nature's mathematical signs does bear a certain kinship to the Jungian view.

"I would prefer to be exiled from the Academy than admit to such a statement, sir."

Yes. Astrologers also frequently rely on Jung. When physicists point out that the gravitational force of Venus acting on a newborn infant may be less than the gravitational force of the pediatrician standing at the foot of the bed, astrologers counter that the basis of astrology lies not in gravity but in synchronicity. Just as the collective unconscious represents the primordial psychic pattern of all humanity, synchronicity represents the pattern of all the events in the cosmos. Synchronicity, in the same way as the casting of the yarrow stalks in the *I Ching* represents the grand pattern of the cosmos, may be said to be a geometric way of looking at the universe.

It is possible that Jung is simply wrong and that synchronicity does not exist. (The Directors of Sports and Recreations have scheduled synchronicity for the day after tomorrow, in the Debates on Causality.) In any event, it is interesting that both Jungians and mathematicians, at heart, are Platonists (despite Plato's remark that "I have hardly ever known a mathematician who was capable of reasoning").

A Contemplation: Does Physics Rediscover Preexisting Mathematics? Are There Mathematical Concepts Not Used in Physics?

On your Academy tour you undoubtedly visited the sculpture garden whose collection includes Michelangelo's renowned statue of Pegasus and Bellerophon attacking the chimera. Today the debate passed through as well and you were reminded of the remark the Florentine supposedly made when asked how he sculpted a horse. He answered that the horse was inside; the sculptor merely carved away everything in the block of marble that did not resemble a horse. This is essentially the Platonic view of the relationship between the physical and mathematical worlds: As scientists create their theories, they are merely uncovering more and more existing mathematics.

Tacitly this position assumes that the fund of mathematical concepts is greater than the number of physical concepts. Consequently, in the sculpture garden, attention now turns to two questions: Are there more mathematical concepts than physical concepts? If so, does each discovery in physics correspond to a preexisting, Platonic, mathematical concept?

Wigner himself conceded that, surely, only a fraction of all mathematical concepts is used in physics. This seems to be true. Perhaps the most famous is the spherical cow. Mathematics can certainly describe things that don't exist and that cannot exist. We have already given a few examples of mathematical theories that correspond to nothing physical; the incomprehensible size of the *Physical Review* is an *a priori* demonstration that most theories are of this sort. If you browse the Web under "mathematics," you will find dozens of websites devoted to number theory. Number theory is the most abstract branch of mathematics; it deals with the properties of numbers themselves: prime numbers, integers, rational numbers, irrational numbers and so on. Fermat's Last Theorem is the most famous problem from number theory, and surely one of the most useless.[†] Most of number theory has never found application to anything outside pure mathematics, which is precisely the way mathematicians want it.

At this, one of the academicians climbs atop Pegasus, with raised arms and cries, "A moment!" She maintains with conviction that every mathematical idea sooner or later finds its way into physics, and that the history of physics is replete with examples of arcane concepts that later turned out to be useful: matrices, complex numbers, quaternions, knots . . . *ad infinitum.*

Perhaps not quite *infinitum.* As the crowd gathers 'round, several members of the Academy reply temperately, "We cannot refute the idea that within the next thousand years every concept in number theory will find its way into physics, but it does seem unlikely." Others standing nearby mention projective geometry, which has contributed little to the natural sciences. One problem here is that the only branches of mathematics most scientists have ever heard of are those that have contributed to their fields. Nevertheless, from this exchange we tentatively conclude that although many mathematical concepts have been extraordinarily useful in science, a large number have not.

Figure 2.3
In the plane geometry taught at high schools, triangles have 180 degrees and parallel lines never meet. On a sphere, triangles contain more than 180 degrees and parallel lines, such as two lines of longitude, may intersect. On a saddle, triangles contain less than 180 degrees and parallel lines diverge. Such geometries are often referred to as flat, closed, and open, respectively, although in the Ninth Debates this terminology will be questioned.

However, the academician's presence atop the statue forces attention to the related issue: Do the natural sciences merely rediscover existing mathematical concepts? Wigner believed that the concepts used in physics "were developed, in many if not most cases, independently by the physicist and recognized then as having been conceived before by the mathematician."

This of course is exactly Michelangelo's block of marble, or perhaps the Texas experience. Travel in Texas is quite convenient. You need to get to Tyler from Austin and check a roadmap. Lo and behold, you find a previously unknown highway between Tyler and Austin. Isn't it fantastic that wherever we want to go the roads are already there? Isn't it fantastic that the road engineers have already seen where we want to go? Physicists are drones and want to use the pathways mathematicians have already worked out.

One reason physicists seem to be like Texas travelers, rediscovering preexisting roadways, is that until the nineteenth century, many mathematicians were also physicists. You find what you need, or if you can't find it, you invent it. Nowadays, the professions have diverged, but scientists tend to obtain their maps from AAA. Usually the AAA marks out the express route or the scenic route, but if you should stumble onto a previously unmarked road, they say, "Oh, we knew about this but it is not very convenient and rarely traveled."

The example of non-Euclidean geometries is instructive in this context. For thousands of years mathematicians thought that Euclidean geometry—"plane geometry"—was the single true, eternal and Platonic geometry, existing independently of human perceptions. However, in the nineteenth century, Nikolai Lobachevsky, Janos Bolyai and Bernhard Riemann discovered geometries which were as internally consistent as that of Euclid, but which described the geometry of curved surfaces. For example, two lines of longitude on a globe begin parallel at the Equator but intersect at the North

Pole. In Euclidean geometry, two parallel lines never intersect. Also, the two lines of longitude each make 90-degree angles with the Equator, then intersect at some angle at the North Pole; this means that the triangle they define has interior angles whose sum is greater than 180 degrees, which is impossible in Euclidean geometry (see Figure 2.3).

A Texan traveling along a road has no way of knowing whether the surface he is on is Euclidean or otherwise except by making experiments. If he constructs a triangle with vertices Austin, Dallas, and Houston, surveys the angles, and finds that they total more than 180 degrees, he can only conclude that his ranch is curved. The AAA replies, "We knew that! There is no contradiction." There may be no contradiction, but clearly the outcome depended on a correspondence to physical reality (travel on a globe), not on mathematical consistency.

The matter is additionally befogged when Poincaré steps up to the statue and— However, the authors cut him off, knowing what he is about to say, and reply that he will be allowed to express his objections in the Ninth Debates, on cosmology.

We have argued that there are mathematical concepts not used in natural science. Now we raise a potentially more serious objection to the "unreasonably effective" view of mathematics: the goalposts keep shifting. Virtually nothing is known about Pythagoras and his character is mostly legendary, so everyone is astonished to encounter him at the Academy golf course, where he extols the "unreasonable effectiveness of integers" in describing the natural world.

You see, the Pythagoreans knew that the first four integers can describe the musical intervals; indeed they considered the power of the integer sacred and swore their most binding oath by it. But the Pythagoreans probably did not even know about negative numbers. Negative numbers (as we conceive of them) seem first to have been mentioned much later, in a Western work around A.D. 275, by the Greek mathematician Diophantus, who apparently solved an equation, got a negative number for an answer, and denounced the solution as absurd. So when you suggest that Pythagoras extend the "unreasonable effectiveness of integers" to the "unreasonable effectiveness of positive and negative integers," he accuses you of being out of your mind, and rejects your application to the brotherhood (recommending that you be sacrificed altogether).

In the midst of your argument with Pythagoras, Kepler passes by, asking permission to tee off. But overhearing the conversation he, with his usual enthusiasm, praises simple geometric relationships and refers to the "unreasonable effectiveness of geometric figures in explaining the natural world." A fellow traveler patiently explains to him that the Newtonian physics born of his planetary laws requires a more advanced sort of mathematics, namely differential equations. After the visitor spends some time teaching him what a differential equation is, Kepler scoffs at the idea that nature requires such complex mathematics to describe it.

At the mention of complex mathematics, nearby academicians launch into a discussion of complex numbers* and how important they became for physics by the nine-

---

* A complex number is a number of the form $x + iy$, where x and y are real numbers and $i = \sqrt{-1}$. The $\sqrt{-1}$ is referred to as "imaginary."

teenth century. Kepler scratches his head; although the concept was introduced by Cardano around 1545, in the early seventeenth century, the meaning and very existence of complex numbers was still hotly debated. Complex numbers result from taking the square root of a negative number and, as everyone knows, the square root of a negative number is a figment of the imagination.

While you are watching this conversation, Werner Heisenberg strides onto the green and argues that you need the mathematical construction of matrices to describe the quantum mechanical aspects of nature. Erwin Schrödinger overhears the remark and, finding Heisenberg's formulation of quantum mechanics repulsive, objects that you can do the same thing without matrices. Heisenberg replies that Schrödinger's formulation of quantum mechanics is "bullshit" and they nearly come to blows. But you are left wondering, are matrices effective or merely redundant?

What are we to make of this? Are we to declare with Leopold Kronecker, "God made the integers, all else is the work of man?" Where does the "unreasonable effectiveness of mathematics" begin—and where does it end? Ilya Prigogine has remarked that "the integers are due to non-equilibrium processes"; if the universe were perfectly uniform, nothing would be differentiated and the concept of "two things" would not exist. Only because matter eventually clumped into discrete objects could the concept of integer make any sense.

The remark is only slightly facetious. It is clear that, in some sense, the "unreasonable effectiveness of mathematics" is a put-up job. Just as all theories find the end of their utility, so do the mathematical concepts needed to express them. When integers fail, you invent real numbers. When algebra fails, you invent calculus . . . It is a bit like falling in love. At first you say, "This is it!" Then after some time you recognize that your love has feet of clay. Finally you say, "I'll look elsewhere." It is well to remember the

Aphorism:
In Russian the word brak means both "marriage" and "defective products."

The idea that natural science is the "excavation of preexisting mathematics" is a corollary to reductionism—that concepts are reduced to deeper and deeper explanations, mathematics being possibly the lowest stratum. However, this is certainly a one-way street, an after-the-fact explanation. Once we have a theory, we find the mathematics to fit it. It does not seem to work the other way around. As Jean-Pierre Changeux points out:

> Mendel showed that the hereditary transmission of color in pea blossoms follows a behavior expressed by an extremely simple mathematical equation. These laws made it possible to infer the existence of stable, hereditarily transmissible determinants, but they certainly didn't predict that chromosomes, still less DNA, are the material supports of heredity.

That is, once you have heredity, you can talk about it in terms of mathematics. But starting with the laws of combinatorics and probability you would never be able to predict

DNA and chromosomes. The mathematicians and physicists gathered 'round remonstrate: Theories, once constructed, do predict new phenomena. We gave a good example earlier—the Ω^- particle. However, this argument has limited validity. If someone in the debating crowd insists that we could start with the laws of mathematics and physics and predict a human being, you have the right to denounce the idea as ludicrous.

At this, the debate adjourns for lunch and a siesta. During informal conversations with the participants you determine that positions have begun to shift, if only slightly. The mathematicians still cling to the view that mathematics is prior to all, and most of the physicists still believe that the universe is "unreasonably mathematical," but a few are now willing to concede the matter is not so obvious as they once thought.

Later in the afternoon, the discussion reconvenes at the windowless physics corpus, where amid the slaving students unregenerate academicians raise the question, Even if you attempt to explain a phenomenon mathematically, how well does the mathematics really fit?

The Four Faces of Approximation

Despite the beauty of mathematics, much of the beauty is lost when one applies mathematics to a real problem. Most calculations are in fact approximations in four distinct senses of the word. The first sort of approximation is found in the freshman physics lab, where a group of initiates is making measurements on the "simple harmonic oscillator," a technical term for a pendulum or spring without friction. A good clock pendulum that swings back and forth forever or a frictionless spring that bobs up and down indefinitely are close approximations to simple harmonic oscillators.

The simple harmonic oscillator is a problem that can be solved exactly. By that we mean we first write down Newton's famous second law of motion, which gives the pendulum's acceleration. Then we solve the equation to get the pendulum's position for any time in the future. If you can say, "Give me a time and I'll give you a position," you have found an exact solution.

However, to the consternation of everyone in the room, the springs and pendulums gradually slow and come to a halt. The reason is that no clock pendulum is perfect; all are subject to friction. So we modify the equation for the pendulum by adding a friction term. The physicists in the room try several friction terms on the blackboard, which they seem to choose for no particular reason, solve the equations and predict how the pendulum will slow down according to each. The students repeat their measurements, testing the slowdown against the predictions. Some friction terms seem to work better than others but none is terrific. The students ask what is going on and the physicists shrug.

In a deep sense, they do not know what they are doing. There is no prescription for adding friction to a system. We do not understand on an atomic level the physics of friction. The friction term is thus a fudge factor. The only way to decide which type is best is to compare the results with experiment, as the scholars are attempting to do.

Already one of the simplest systems—a single oscillator with friction—is an approximation in the sense that, although it can be solved exactly, we do not even know the correct equation to write down. Thus one cannot really be sure the mathematics is describing the microphysical universe; the only thing one is sure about is that if you pick a "correct" friction term, the pendulum slows down at a rate predicted by the equation.

You now perceive that the most elementary problems often require an Approximation of the First Kind: we pick the equations we choose to solve, which may or may not correspond closely to the physical situation. Approximations of the First Kind correspond to modeling the phenomenon. With the oscillator we at least know what equation to write down, even if the friction term must be fudged. In more complicated situations, such as the weather, so many variables are involved—temperature, air pressure, density, the rotation of the earth—that one has little idea of what the correct equations are. One can of course write down *some* equations, but whether they have anything to do with the real weather rapidly becomes debatable.

Furthermore, in all but the simplest models, the equations cannot be solved exactly. Solutions must be approximated. The debate moves into the Center for Astrophysics, where dozens of research students ruin their health at computer terminals. They are currently studying the popular problem of "large-scale structure formation." Astronomers observe billions of galaxies in the universe. As a rule, galaxies are concentrated into clusters, which may contain several dozen to several thousand galaxies, and clusters themselves are aggregated into "superclusters," consisting of hundreds or thousands of clusters. It is thought that shortly after the big bang, the primordial gas and dust began to clump under its own gravity and eventually formed the pattern of clusters and superclusters seen today.

The members of the Center for Astrophysics, along with their colleagues worldwide, want to understand the details of the structure formation process. Nevertheless, to solve the problem exactly is impossible. The researchers, who are computer literate if verbally challenged, therefore resort to approximations. Some of them use the machine to make numerical approximations to the full-fledged equations; call this an Approximation of the Second Kind. A few members of the Center who were born before mainframes take a different approach: "If I restrict my attention to the very beginning of the clumping, I can simplify the equations themselves and solve them by hand." Such a procedure represents a further abstraction away from reality, an Approximation of the Third Kind. Having solved a simplified equation in this way, the researcher exclaims, "Aha! I recognize the answer—it is a Bessel function!" Because she has attached a name to the solution, she believes she possesses sacred knowledge. But looking up the properties of "Bessel function" in a mathematical handbook, the researcher discovers that to write it down exactly requires an infinite number of terms, which she does not have time for. At this point she says with some reluctance, "Well, if I just look at very large or very small clumps, maybe I can approximate the Bessel function." In other words, having simplified an equation, she has nevertheless achieved a solution that is too complicated to understand, and is now forced to make an Approximation of the Fourth Kind, in order to make the result comprehensible.

The moral of this tale is that an answer to a physical problem is often highly abstracted. In some sense, Wigner seems to be right: it is a miracle that after so many abstractions one ever gets a usable answer.

Nevertheless, a chorus of physicists now rises: "Look guys, you can make the approximations as good as you want. This is, after all, the point of physics." Not so. As a simple example, consider the harmonic oscillator again. Everything we have said about pendulums is strictly true only if the swing is small. Suppose, however, that you swing the pendulum somewhat harder. In this case, the usual freshman physics equations no longer hold. If you now try to calculate the swing rate by making successive approximations, you can calculate from now until the crayfish whistles on the hill and you will *not* get the correct answer.

A more exciting example concerns charged particles, say electrons. As you probably know, like charges repel each other; the repulsive force increases furiously as the particles approach each other and, if the electrons are considered pointlike objects, the force actually becomes infinite when they touch.* To get rid of infinite forces was one reason that quantum field theory was invented: instead of treating particles as billiard balls of zero radius, you think of them as smeared-out fields. Smearing them out should cause the problem of infinite forces arising when particles touch to vanish.

Unfortunately, this problem does not vanish in QED, the theory of quantum electrodynamics from the First Debates. No straightforward approximation procedure will get rid of them. For this reason the process of renormalization was introduced, but as we said, renormalization is a bit of magic and generally considered a sleaze. As the debating crowd moves away from the physics corpus, someone probably quotes the famous aphorism of Einstein:

As far as the laws of mathematics refer to reality, they are not certain, and as far as they are certain, they do not refer to reality.

The Degenerate Description of Reality

By now you might be willing to concede that the fit of mathematics to the real world may be likened to a new suit of clothes made by a bad tailor: a little too big here, too small there, a bad seam in the left pant cuff.

Most of the hard-core academicians do not support this analogy; they are confident that we know how good, or bad, our approximations are. But approximation is not the major conceptual problem in understanding the relationship of mathematics to the physical world. In many cases it is simply not clear what the mathematics is describing. At the physics corpus we visited a lab full of springs and pendulums. We skipped the electronics lab. Yet, the same equation that describes the motion of a spring describes the behavior of a simple electrical circuit. Surely in the physical universe there is a difference between an electrical circuit and a spring, but the mathematical world does not distinguish them.

---

* This is because the force obeys the "inverse-square law"; it decreases with the inverse square of the distance, and conversely becomes greater as the distance shrinks toward zero.

The difficulty is similar to the one contained in the foundation of Aristotelian logic, the syllogism. As Aristotle frequently explains to Academy audiences, a syllogism is a statement of the type

All men are mortal
Socrates is a Man
Therefore Socrates is mortal.

"All men are mortal" is the major premise, "Socrates is a man" is the minor premise, and "Socrates is mortal" is the conclusion.

However, a favorite pastime among young students is to invent nonsensical syllogisms:

Where there is smoke there is fire
This man is blowing smoke
Therefore he is on fire.

The reason it is so easy to create nonsensical syllogisms is because the logic itself—the mathematics—provides no way of anchoring the statement in reality.

The Indian syllogism, on the other hand, contains five elements: The major premise; a reference to context; the minor premise; a conclusion; then a verification. For example:

Where there is smoke there is fire.
    When there is smoke on a hilltop in the summer
    you find a forest fire.
This pile of hay is smoking.
Therefore there must be a fire.
    The hay is getting warm.

The Indian syllogism is more like physics as it is practiced: a mathematician can prove a theorem, but he or she is never sure whether it is relevant to the real world. A theoretical physicist who proves a theorem must be sure the result is relevant to some phenomenon. The experimentalist's job is to establish the relevance. The logical structure of the Indian syllogism is the same as the Aristotelian one, but the verification clause gives you a reality check, which the mathematics cannot provide.

Examples of such "mathematical degeneracy" abound in physics. Planck's blackbody, an ideal object that absorbed all radiation incident on it, figured prominently in the First Debates. Since the beginning of the twentieth century, scientists understood that a blackbody had a "specific heat" that has a certain dependence on temperature.* (Specific heat is merely the amount of heat energy required to raise one gram of a substance one degree.) The specific heat of a substance is not a fixed number; as we just said, for a blackbody it changes with temperature.

---

* Specifically, it is proportional to the cube of the temperature.

Now, scientists also observed that the specific heat of ordinary solids (not black-bodies) had exactly the same temperature dependence for very low temperatures. Eventually, in 1916, Debye succeeded in explaining this phenomenon by modeling solids as crystals subject to sound vibrations—the same phonons with which we constructed a mind model yesterday. On the other hand, a blackbody consists of the mysterious Planckian oscillators, which were eventually shown to be oscillations of the electro-magnetic field itself.

So here is something very strange: a solid crystal is displaying the same dependence on temperature as the electromagnetic field itself. The reason for this, which we will not explain, is in fact that space is three-dimensional. It has nothing to do with the fact that you are dealing with electromagnetic waves in a blackbody of sound waves in a crystal. Which system is the mathematics modeling? Both.

The degeneracy of the mathematical description of reality extends far beyond physics. Congregated on the Academy's central plaza, the debaters have gathered around the Knowledge Marker, a giant barometer indicating the world price of academic ideas. Today it is at an all-time low. A favorite Academy pastime is to place bets on the future behavior of the Knowledge Marker.

Wall Street brokers, like Thales, engage in a similar activity when trading stock options. The most popular tool of futures traders is the Black-Scholes model. The equation of the Black-Scholes model is identical to the equation in physics that governs how a small particle moves through a gas or a fluid. (This is known as a diffusion equation.) In the Black-Scholes model, however, it is not a particle that is diffusing but the price of a stock. (This behavior of stock prices was noticed on the French Bourse in 1900 by Louis Bachelier, and his paper ushered in the mathematical study of Brownian motion.) Presumably a stock price is not the same as a particle, but the mathematical world does not distinguish them.

There is a more insidious problem in the degenerate category. Many concepts in science are not well defined. When an Academy chemist asks you what a molecule is, you reply, repeating what you were taught in grade school, that a molecule is the smallest naturally occurring group of atoms. For example, a salt molecule is sodium chloride (NaCl). But as you know, salt comes in small crystals, as well as extremely large ones. Is a salt molecule one atom of sodium and one of chlorine, or ten of each, or thousands? Despite the propaganda, chemists do not really have a precise description of a molecule. Similarly, in physics we sometimes think of an electron as a point particle, like a tiny billiard ball, but in chemistry courses it is often described as a cloud. Electrons, as we mentioned, are also said to have spin, in that they behave like small tops, but none of us can really picture in the mind's eye a point particle, of no length and no width, that is spinning. As an exercise you should approach some of the biologists for their definition of a species. Be prepared to walk away, quickly.

What does this have to do with mathematics? Mathematics can describe a salt molecule, regardless of whether it contains one hundred atoms or one thousand, just as it

Figure 2.4
Precision is not
accuracy.

can describe an electron whether it behaves like a top or a cloud. (Perhaps mathematics cannot describe a species.) And yet, if these concepts are not precise, what is the mathematics describing? Because we can make a precise mathematical description of a phenomenon, we fool ourselves into thinking we have described the physical world. But in this case we have made a precise description of an imprecise concept. The problem isn't with the mathematics; the problem is we don't always know what we are talking about.

At the Academy, beginning science students are admonished not to give answers to meaningless precision—too many decimal places. In the very lab we visited a little while ago, there once hung a cartoon something like the one shown in Figure 2.4. In science, the bull's-eye's location is often not known. Scientists should take to heart the lesson of the perplexed archer.

Are There Physical Things That Mathematics Cannot Describe?

Once again you may be feeling that it is time to find a way out of the Academy. With the Knowledge Marker at an all-time low, however, the exchange rate is not bad, and Academy thought (as well as meals) cheap. In any case, a question should be bothering you. We just asked, Are there mathematical descriptions that do not correspond to any physical thing? Is not the more intriguing question the reverse: Are there physical phenomena that have no mathematical description?

There are certainly phenomena whose description lies outside the mathematical domain attributed to them. Quantum mechanics is in fact incapable of describing, within its own mathematics, one of the phenomena it is most successful in explaining: radioactive decay. We often want to know, for instance, how fast a radioactive element will decay into another element, and quantum mechanics is quite good at making such predictions. Strangely, however, the mathematical functions you need for these calculations don't have the properties John von Neumann declared they must have when he originally systematized the mathematics of quantum mechanics.* Hence physicists are in the peculiar position of routinely using a certain kind of mathematics in quantum mechanics that unquestionably works but does not have the properties they claim.

"Well," you might reply with as much bravado as you can muster. "This is all a matter of nomenclature. *Some* mathematics is operating."

Perhaps, but it is not describing something within quantum mechanics as it is defined mathematically.

At this a patch of fog descends.

A more universal example where mathematics appears to fail is the phenomenon of choice. There are those who believe that all our actions were determined the moment the universe began, but if one accepts the idea of some free will, there does not appear to be any mathematical description of how cockroaches or we make a decision.

"One can easily put a cockroach in a maze," several experimental psychologists object, "and watch it one hundred times, thereby making a statistical description of its behavior." This is not what we mean. If we ask why the cockroach turns right or left each time, or why you decide to go to a movie, there simply does not seem to be any way to describe this situation mathematically.

The claim angers the reductionists present, who reply with heated voices that choice does not belong to the natural universe. We answer gently, you can decide to hold your breath for five seconds. This seems to be a physical phenomenon. How is it described? We don't talk about choice in the natural sciences because we can't.

Most of the discussion has been about the relationship between mathematics and physics, which does not include all the natural sciences. Indeed, as dusk descended on the Second Debates, a few onlookers detached themselves to seek out biologists and chemists in order to ask, "Does mathematics describe your world?" The stares of surprise with which they greeted the question showed they had never considered it before. Generally, mathematics is less applicable to chemistry, biology and psychology than to physics, if for no other reason than that chemical and biological systems are far too complicated to describe mathematically. One should certainly ask whether biological concepts such as species, territoriality, display and so on have any mathematical equiv-

* According to von Neumann, quantum mechanics should be described by vector functions with well-defined norms; however, the functions that describe radioactive decay do not have well-defined norms.

alent, but the biologists present are still laughing at the question.

Chemists, on the other hand, do use computer programs to synthesize larger molecules from smaller ones. You might call this mathematics at work. But the rules they use for synthesizing these molecules are largely heuristic—derived from practical experience. These programs are basically "expert systems"; chemists have put into them all the practical knowledge about how molecules combine, and the code merely tries one thing after another until it finds something that works.

Academicians with reductionist leanings, such as Linus Pauling, who put chemistry on a quantum-mechanical footing, would object that this difficulty is merely a practical one: you don't know the rules, but the rules exist. As we said, such a position appears to operate in one direction only: If you hand Pauling a list of quantum-mechanical rules, would he be able to predict hemoglobin?

At this point one should probably say a few words about Gödel's celebrated incompleteness theorem, in all likelihood the most profound mathematical result of the twentieth century. Briefly put, Kurt Gödel proved mathematically that any consistent mathematical system must contain statements that are impossible to prove within that system. In that Gödel's theorem puts an absolute limit on the effectiveness of mathematics, it is one of the most powerful statements of epistemology.

Scientists have often used Gödel's theorem to assert that there exist limits to the possibilities of science. To the extent that physics employs mathematics as a language, this seems a reasonable statement—all physical theories, to your delight, must have holes in them, must eventually make assertions that do not correspond to provable mathematical statements.

Not only physicists use the Gödel argument. The eminent biologist François Jacob asserted a quarter century ago: "Undoubtedly, it will soon become possible to analyse the molecular mechanism of synapses, the articulation of the nervous cells. . . . We can be sure that to the biochemist the characteristic reactions of brain activity will appear as ordinary as digestive reactions. But it is quite another matter to describe a feeling, a decision, a memory, a guilty conscience, in terms of physical chemistry. There is nothing to show that it will ever become possible, not only because of the complexity, but also because since Gödel we know that a logical system is not sufficient for its own description."

Mathematician Connes stands to object that Gödel's theorem does not limit our understanding of the brain, it merely rules out the possibility of understanding the brain in terms of mathematical axioms.

Which is more or less our point.

Practical Exercise: Is Music Unreasonably Mathematical?

At the palazzo you stumble across academician Dirac, who is eating dinner, and in order to clarify some of today's issues for yourself you decide to push things to extremes.

"Professor Dirac, " you say forthrightly, "I am told you believe in the unreasonable effectiveness of mathematics to describe music."

Dirac remains silent, at which you commit a regrettable faux pax. "Excuse me, Professor Dirac, didn't I just ask you a question?"

After another silence, Dirac answers. "I didn't hear any question. You made a statement."

With things off to such a good start, you vent your frustration by exclaiming, "How can you believe that mathematics describes the emotional content of music?"

Dirac, somewhat puzzled, replies, "What emotion? A compact disc player merely reads a sequence of ones and zeros off a compact disc. The complete content of the music is encoded on the disc."

"Surely you can't believe that!"

"There is no ghost on the disc."

"If nothing else, the recording is not completely true to the music. The ones and zeros only approximate the music."

"Approximations can be made as close as necessary, better than the limits of human hearing. Emotion does not lie in the twenty-fifth bit."

"Then you do assert that mathematics is unreasonably effective in describing music."

"Everything that can be described about music can be analyzed mathematically. Any emotion must be an association generated on the part of the listener."

"But if I gave you a string of ones and zeros, could you tell me even whether it represents a sound wave or a light wave?"

"No."

"Yet there is a difference between sound and light."

"Yes."

A long pause while you stare at each other.

"How would you describe the difference?"

"Light waves are shorter than sound waves. Sound waves are longitudinal oscillations of a solid, liquid or gas. Light waves are transverse oscillations of the electromagnetic field. All these things can be described mathematically."

"But the description is made only once you have decided whether this is a sound wave or a light wave. The mathematics does not tell you what you are describing."

"This is not always true. Mathematics predicts new phenomena. My equation, which unified special relativity with quantum mechanics, predicted a new particle—the antielectron, or positron, although originally I thought it was the proton."

"Then you were wrong."

"I was. Not my equation. My equation was smarter than I was. Anything a composer discovers can be described mathematically: volume, timbre, tone color, harmony."

"Is the emotion describable mathematically?"

"As I have said, the emotion is not in the music."

"Where is it?"

"Probably in the chemistry of the listener?"

"In that case, music is a branch of mathematics?"

"It may be."

"If so, does the world of music exist 'out there' ?"

"These are difficult questions."

Dirac walks away.

In Which the Second Debates Are Summarized

Night has fallen and by now you have again realized you aren't going anywhere. A few hours ago, however, a fellow visitor lent you a cell phone and you managed to reach your date from last night, who found the excuse for your absence so implausible that you were abruptly left holding a dead line. A nearby academician slapped you on the shoulder, admonishing you to take heart: at least you should be able to participate in tomorrow's debates, for which today's discussion will have proved useful. With that in mind, you wandered out toward the beach, reviewing the day's findings.

Spectators discovered that mathematicians, with perhaps the exception of Henri Poincaré, believe that the mathematical world exists externally to the human mind. Although a few people maintain that mathematics is a purely cultural phenomenon, they rarely pass the Academy security check.

More contentious was Wigner's claim that mathematics is "unreasonably effective" in describing the physical world. Mathematically based theories have been extraordinarily successful in their respective domains. On the other hand, there seem to be many mathematical concepts that do not correspond to anything physical, so in that sense there are more mathematical concepts than physical concepts. But in many cases, one mathematical concept corresponds to several physical concepts, and there may be physical concepts that have no mathematical counterpart. In that sense there are more physical concepts than mathematical concepts. Moreover, almost all calculations in the physical sciences end up being approximations, so if God made the physical world to fit the mathematical world, He could have used a few lessons in tailoring.

The stars have come out now, and you are struck with the circularity of the discussion: physical scientists tend to define science only as something amenable to mathematical description. If it is not amenable to mathematical description, it is not science.

All is dark.

IS THE WORLD SYMMETRICAL?

Friezes, Particles and Groups

You boil it in sawdust;
You salt it in glue;
You condense it with locusts in tape;
Still keeping one principal object in view—
To preserve its symmetrical shape.

    —Lewis Carroll

L'univers est dissymétrique.

    —Louis Pasteur

The Question

In Moscow, Russia, a stone's throw from Red Square, stands the hotel Moscow. Visitors to the city can hardly avoid walking by it once or twice, on the way to the square or the metro. To the casual tourist the Moscow appears to be nothing more than a drab, Soviet building. To the focused mind, however, a stroll around the hotel reveals a peculiar structure whose windows on the left half are not exactly of the same design or at the same height as the windows on the right half. Neither do the facades nor the ornamentation match. All in all, a walk around the Moscow is a disturbing experience. The story goes that Stalin's architects presented him with two distinct designs for the hotel and Stalin approved both. Having no intention of crossing the Master, the architects crossed the designs instead. Regardless of the truth of the story, the Moscow does not leave one with a sense of peace and harmony. Rather it leaves one with a sense of downright unease.

How much more pleasing is the sight of the sun rising over the Academy's most celebrated structure, the Parthenon, with its seventeen identical Doric columns along each flank, eight along each end, the proportions according to tradition determined by the golden mean.[†] An even more glorious spectacle is the golden rays striking the marble of what many consider to be the world's most beautiful building—the Taj Mahal, surrounded by its four slender and identical minarets.

Such visions, long cherished after the return home, inevitably force tourists to confront the daily facts of their existence: the mad rush to the train station, a never-ending traffic jam at the construction of a new overpass, the necessity of mastering today's software upgrade, the impossibility of deciphering the VCR display. Nevertheless, there are some similarities between the Academy and the Real World. In the Real World you have undoubtedly stared up at the skyscrapers towering above and asked, What is the meaning of life? And the skyscrapers have not replied, presenting instead four identical, mute faces to the world. If you have passed through suburbia, particularly Levittown, you have almost certainly wondered whether the houses have all been stamped from the same cookie cutter. Observations of this sort may well prompt you to wonder why you were so anxious to leave the Academy. They also raise the question, Why do we surround ourselves with symmetrical constructs?

Coincidentally, as you watched the sun rise between the columns of the Parthenon on this, your third day at the Academy, you discovered that yesterday's topic of mathematics had transmuted overnight into a new debate: Why is symmetry so common? Or is it?

Is There Symmetry in Art?

The academicians today do not expect much opposition to symmetry. Lesser institutions may live by the dictum "the lower the stakes, the worse the backstabbing," but at the Academy debates are carried out in genteel fashion. In any case, anyone opposing symmetry will be torn limb from limb.

There is nothing arcane about the concept of symmetry. At the most basic level, symmetry corresponds to the idea that different parts of an object look the same. The sphere is an object of perfect symmetry since rotating it through any angle leaves it unchanged. An equilateral triangle intuitively strikes us as less symmetric than a sphere; nevertheless an equilateral triangle is highly symmetric.

With the morning calisthenics in progress, you circle the arena asking, Why is there so much symmetry around us? and find the responses depend on profession. Academy architects answer, "Form follows function": practical limitations are the overwhelming reality of design, and it is far easier to build symmetrical structures than highly asymmetrical ones. Sometimes, the architect concedes, if the technical constraints leave any freedom, aesthetic considerations play a role.

To an enraptured group of Californian tourists, turbaned philosophers extol the virtues of a "balanced life" and "moderation in all things." Both imply value in maintaining symmetry among the various aspects of daily existence. The purpose of yoga is to achieve harmony in action, by balancing the qualities that make up the world, or *gunas*. When the *gunas* are in balance, the external world disappears and *kaivalya*, or complete detachment, is achieved. With the disturbances of the external world quelled, *kaivalya* is often said to represent the state of maximum symmetry. A sense of symmetry is contained in the very title *The Doctrine of the Mean*, one of the four great books of Confucianism, whose concerns echo those of yoga. "Being without inclination to either side," writes a commentator on *The Doctrine of the Mean*, "is called *chung*; admitting no change is called *yung*. By *chung* is denoted the correct course to be pursued by all under heaven; by *yung* is denoted the fixed principle regulating all under heaven." The word *Tao* itself, the Way, not by coincidence, connotes the path to spiritual harmony.

Strangely, Academy artists rarely mention symmetry. Mostly they talk about themselves, often of inspiration, they sometimes speak of integrity, earnest ones Truth, occasionally beauty. When you enter one of their enormous lofts, filled with canvases and paint and ask, Why is symmetry so common? they reply, "Nuts." Even biologists rarely talk about symmetry. Chemists, sometimes.

Theoretical physicists, on the other hand, are always talking about symmetry. Tuning in to their conversation, you quickly conclude they have all read *The Doctrine of the Mean*. Hermann Weyl, known for introducing mathematical concepts of symmetry into quantum mechanics, is lecturing on art:

$$\text{symmetry} = \text{harmony in proportion}$$

So, even art must contain an equation. Weyl adds, "Beauty is bound up with symmetry." Because equalities go in both directions, we can then read his declaration backward: "Symmetry is bound up with beauty."

Weyl is speaking for virtually all theoretical physicists, and their answer to the question, Why is there so much symmetry around us? is unequivocal: beauty is symmetry and humans prefer beauty. As vague as it sounds, the idea *beauty* = *symmetry* has become a credo to theoretical physicists (even when they are traveling in the guise of art critics). The credo is so powerful that Weyl maintains before the public that "all asymmetries are of a secondary character. . . . Seldom is asymmetry merely the absence of symmetry. Even in asymmetric designs one feels symmetry is the norm from which one deviates under the influences of forces of non-formal character."

The crowd finds it cannot dismiss Weyl's position out of hand. It is impossible not to be struck by the symmetries found in design of all sorts, from Islamic tilework to Sumerian frescoes to Bauhaus design, and the feeling of "deviation" Weyl mentions is no better demonstrated than by your out-of-kilter stroll around the hotel Moscow .

One or two physicists, however, have been seen defecting from the Weylite camp. Fritjof Capra, standing apart from the conventionalists, regards the physicists' fascination with symmetry as a relic of the West's Hellenistic past, the Pythagorean and Platonic emphasis on patterns of numbers. In the East, Capra argues, the emphasis on symmetry is far less pronounced. But Capra has overgeneralized. It is true that Japanese art leans toward the asymmetrical. Here Daisetz Suzuki speaks about Zen and Japanese culture:

> . . . asymmetry is certainly characteristic of Japanese art, which is one of the reasons informality or approachability also marks to a certain degree Japanese objects of art. Symmetry inspires a notion of grace, solemnity, and impressiveness, which is again the case with logical formalism or the piling up of abstract ideas. The Japanese are often

thought not to be intellectual and philosophical, because their general culture is not thoroughly impregnated with intellectuality. This criticism, I think, results somewhat from the Japanese love of asymmetry. The intellectual primarily aspires to balance, while the Japanese are apt to ignore it and incline strongly to imbalance.

And is this not the point? The conflict between symmetry and asymmetry is not so much between East and West but between the rational and irrational. When Capra says, "The concept of symmetry does not seem to have played an major role in [Eastern] philosophy," one should distinguish cultures. Indian tradition is more intellectual than Japanese tradition with its emphasis on the Vedic literature* and achievement of spiritual symmetry, *kaivalya*. There is the Chinese *Doctrine of the Mean,* in which balance, equilibrium, is paramount. And what is Yin and Yang if not a kind of reciprocal symmetry? Moreover, the Japanese *sangaku* display a high degree of symmetry; they have been introduced at mathematics conferences as a method by which to teach symmetry.

On the other hand, beauty—East or West—requires asymmetry. Joining the warmup, Roald Hoffmann points out that the word "interesting" comes from the Latin *interesse,* to be in between, neither one thing nor the other. And it is true: here an artist is bringing out a perfectly white canvas and you are bored before she carries it five paces. The monotony of an endlessly repetitive piece of music quickly fills you with irritability and wrath. In conformity with their love of imbalance, Japanese artists cluster objects to one corner and a Japanese garden is a wonder of irregularity. In India, a pumpkin or cactus branch is hung to one side of a newly constructed house. Not far from the hotel Moscow stands the famous St. Basil's Cathedral. It is one of the world's most asymmetrical buildings, and one of the most beautiful. The irregular New York skyline is famous for its beauty.

Capra's association of symmetry with West and asymmetry with East has the academicians grumbling. Nevertheless, everyone can agree that the New York skyline is more indicative of the world as we find it than a pure white canvas. Exact symmetry is rarely encountered in the Real World; when it is, it is invariably an artificial construction—a piece of art or architecture—and it stands out as its own asymmetry against an irregular background.

Is There Truth in Beauty?

So, is symmetry or asymmetry prior? This is the question you now put to the authors. Or is it all a matter of taste? Suzuki, Capra, Weyl, the philosophers spoke almost exclusively about art, architecture, music. These are human creations. How can such questions be answered? Well, the authors reply, why not ask a simpler question: Can the aesthetic of symmetry be brought to bear on nature? Is symmetry or asymmetry prior in the natural world?

As the growing entourage passes through the Academy's forest, you are not immediately struck by symmetry, but if you look carefully, symmetry is present. Snowflakes have

* Veda refers to "revealed wisdom" and "Vedic literature" refers to the four collections of Hindu scriptures thought to have been composed before 500 B.C.

regular patterns, not to mention tulips, roses and lotus flowers. The leaves of palm trees and banana plants are arranged in spirals. Oak leaves and maple are approximately—if not exactly—symmetric. A hemoglobin molecule at first appears entirely irregular, but on closer inspection the different halves are mirror images of each other.

The symmetries of the world are hidden symmetries. More hidden than those of biology and chemistry are the symmetries of physics. Newton postulated that space itself was absolute, completely and unalterably the same in all directions and for all times—an infinite extension of the pure white canvas. "Absolute uniformity in space and time!" you exclaim. "Surely this is the most symmetric construct imaginable." In truth. It is also the most hidden symmetry. As Aristotle argued in the First Debates, one never encounters absolute uniformity.

But at this moment you and several others are shanghaied from the forest by Weyl's followers. As they drag you toward the central plaza, via the formal gardens awash in spray from the fountains' celebratory patterns, they assure you that the authors' exposition is mere diversion. Symmetry or asymmetry prior? What a foolish question! Just as the Taj Mahal is more beautiful than a rococo church, the path to Reality is found by paying attention to the rose, the lotus flower and the snowflake, and by ignoring the lush overgrowth of asymmetry overhanging them.

The Weylite boast is not idle. Since the time of Newton, if not earlier, natural philosophers have searched for hidden symmetries and found them. Newton's famous third law, "for every action there is an equal and opposite reaction," was one of the earliest enunciations that natural forces should act reciprocally, that is, symmetrically. Like Yin and Yang.

Reciprocity inspired Michael Faraday, one of the creators of the science of electrodynamics. In 1819, Hans Christian Oersted had proved that electricity produces magnetism. Faraday, who held "an opinion almost amounting to a conviction," set about proving the reverse—that magnetism should produce electricity. He succeeded, and with his demonstration that a changing magnetic field produced an electric field, Faraday discovered the law that bears his name (and which was discovered by Joseph Henry several years earlier).

It is 1928. You are there. P. A. M. Dirac is confronted with two distinct theories: Einstein's special theory of relativity and quantum mechanics. Both seem correct; neither takes account of the other. In combining the two consistently, Dirac predicts the existence of a particle with positive charge and the same mass of the electron. Yet, in 1928 only two subatomic particles are known: the electron and the proton. Dirac concludes that his equation must describe the proton. Here he perceives a severe difficulty—the proton is fully 1,800 times heavier than the electron. Despite his reservations, Dirac publishes his paper as one about electrons and protons. Weyl himself, a mathematician whose belief in symmetry is far stronger than any petty concerns about what particles actually exist, takes the bull by the horns and declares that the new particle must have the same mass as the electron.[†] Several years later, the "positron," or antielectron, becomes the first constituent of antimatter to be discovered. Today physicists take for granted that any matter particle must have an antimatter counterpart—that there exists a symmetry between matter and antimatter in the laws of physics.

By now the success of the program begun by people like Faraday and Dirac is evident. The Weylites carry you into the Great Library, with its endless tread of lost souls around the outer ramparts, and point to the physics shelves. There, before your eyes, stand hundreds of books on "symmetries in quantum mechanics," "symmetries in crystals," "supersymmetry," "hypersymmetry . . ." With such triumphs behind them, it is not surprising that theoreticians, such as Anthony Zee, can declare before the audience: "My colleagues and I . . . like to think we [like Einstein] search for beauty. . . . The beauty that nature has revealed to physicists is a beauty of design . . . with its emphasis on geometry and symmetry."

Although Zee is speaking of beauty rather than of experiment, you rightly perceive that the equation *beauty = symmetry* has evolved to the stronger belief

$$\textbf{beauty = symmetry = truth.}$$

Zee exclaims, infused with the spirit of Dirac: "Let us worry about beauty first and truth will take care of itself!" String theorist Michio Kaku puts it in the strongest possible terms, repeating his manifesto of the First Debates: ". . . nature, at the fundamental level, does not just prefer symmetry in a physical theory, nature demands it."

Aristotle adds, "Nature abhors a vacuum."

The confusion between laws and principles of two days past confronts you squarely now. One thing seems clear: Kaku's position elevates the principle that symmetry plays a role in physics to a law of nature. The visitors begin to murmur, Does nature demand symmetry, or do physicists demand symmetry?

But something is puzzling you about all this. The production of electricity by magnetism and vice versa seems to be quite different from the pairing of particles and antiparticles. True, and neither of these is evidently related to the kind of symmetry we find in crystals, which form regular geometric structures—prisms, parallelepipeds and so on. Nor do they obviously have much to do with what is perhaps the most common symmetry found in daily life: that between left and right.

However, the symmetrists reply, to scientists and mathematicians the situation is very much like being confronted with the Republicans and Democrats; they are not as different as they first appear. All symmetries, including those in art, can be described mathematically by the same tool. That this can be done is, in fact, one of the prime examples physicists cite to confirm that nature is inherently mathematical.

In order to judge this statement more intelligently and to participate more fully in today's discourse, it is important to have some idea of how the tool works—to understand how physicists quantify beauty.

A Short Discourse on the Mathematical Description of Beauty

The instrument scientists use to characterize symmetry is the *theory of groups*. The term "group" was first used in a technical sense by the French mathematician Evariste

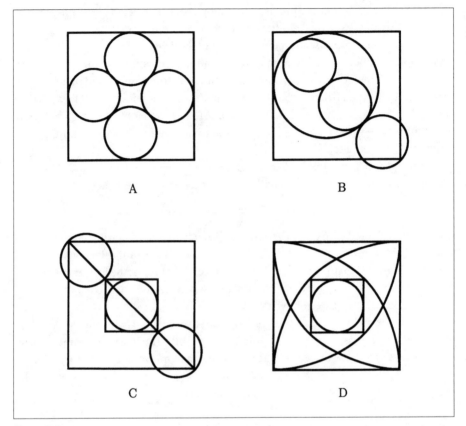

Figure 3.1
Four *sangaku* display rotation and reflection symmetries. A and D are invariant under rotations of 90 degrees, as well as under reflections across the diagonals and the centerlines that would divide the squares in half. B is invariant only under rotations of 360 degrees and under reflection across one diagonal. C is invariant under rotations of 180 degrees and under reflections across both diagonals but not under reflection across the centerlines. Which figure displays the most symmetry? (All these problems ask to find the radius of the small circles in terms of side of the square.)

Galois in the early nineteenth century before his untimely demise at the age of twenty in a duel.[†] Galois never became an academician.

You may think of a group as a collection of operations that obey a few simple mathematical rules. The most important rule is that the combination of two operations, whatever they may be, should also be an operation of the same type. The rotation of a wheel through an angle A, followed by a second rotation of the wheel through an angle B, is a rotation through an angle $A + B$.

It is easy to see how this property is related to symmetry. Some of the *sangaku* problems in Figure 3.1 are unchanged when rotated by multiples of 90 degrees. We say the figures are *invariant* under this group of rotations. Another typical example is the frieze of Persian sphinxes. If the sphinxes are displaced left or right by the proper

Figure 3.2

amount, *A*, the pattern remains unchanged. This frieze displays a very common type of symmetry termed translation symmetry: like rotations, a translation followed by a translation is merely another translation, so translations also form a group. We say the sphinxes are *invariant* under a particular group of translations, meaning those translations that shift the pattern the required distance.

Now contemplate *sangaku* A and D in Figure 3.1. Note that although the figures look very different, they are both unchanged under the rotations and four reflections described in the caption. Those operations are the only operations that leave the figures invariant. That the Persian sphinxes are invariant under translations whereas the two *sangaku* are invariant under rotations and reflections is the mathematician's way of stating that the first figure is translationally symmetric and the second displays rotational and reflection symmetry. The terms *group* and *invariant* occur in any discussion of symmetry. Indeed, to a physicist, the symmetries of any system are manifested in the operations that leave it invariant.

The symmetry between electricity and magnetism may still strike you as far more abstract than the symmetry of a Persian frieze, yet group theory handles even abstract symmetries in the same way. To understand the idea of an abstract symmetry, meditate on the simple expressions

$$x + y \qquad x - y \qquad x + xy + y$$

Note that interchanging x and y in the first and last expressions makes no difference to their values, but that the second expression is changed from $x - y$ to $y - x$, which are unequal. You cannot call the interchange of x and y translations or rotations in any physical sense. They are more akin to the transformations found in fairy tales where frogs become princes and vice versa. Nevertheless, the two outer expressions do manifest an obvious symmetry in x and y, and for that reason scientists routinely talk about symmetries in equations.

It never rains at the Academy, in order to provide for uninterrupted discourse, and the sun has climbed high into the sky. The morning session failed to cheer you and, so far today, the weather forecast has proved depressingly accurate: not much of a debate, mere skirmishing over the role of symmetry in art and science, 40 percent traditional exposition. In any case, you have now been missing from the Real World for the better part of three days and, borrowing a cell phone again, you check your answering machine to find you have lost your job.

Take heart, the academicians assure you as the crowd congregates on the central plaza in the shadow of the Knowledge Marker (which has remained unchanged since yesterday), education is its own reward. A few moments ago, the pro-symmetry majority seized the plaza, intending, now that the preliminaries are done, to monopolize the day by recounting symmetry's continued success story. The discussion promises to be long, but given that you are presently unemployed, time is on your hands and you take solace in the fact that knowledge is cheap.

The Theoreticians Reveal the Symmetries of Particles

As discourse resumes, it becomes apparent that the symmetries of concern to the physicist are closer to the abstract type of symmetry in which frogs and princes are mutually transmuted than to the simple symmetries of the Persian frieze. A symmetry by which one object is transformed into another is usually called an internal symmetry.

The Weylites begin by describing one of the first and most important examples of an internal symmetry in physics, one concerning the neutron and the proton. It is an interesting fact that the neutron and the proton have *almost* identical masses, which long led scientists to suspect the two particles were related.* On the other hand, the neutron is neutral, whereas the proton is positively charged, and the masses *are* different; consequently they are distinct particles. Nevertheless, in nuclear reactions involving the "strong" nuclear force (the force that holds the nucleus of the atom together) the neutron and the proton behave identically. For instance, a "strong" collision between two protons is indistinguishable from a "strong" collision between two neutrons. This is a bit strange because you might expect the electrical repulsion between two protons to affect the reaction. True, but as far as the strong force itself goes, the electric charge makes no difference. Strong interactions are said to be "charge blind."

As early as 1932, the year the neutron was discovered, Heisenberg proposed that the neutron and the proton are really the same. In other words, the two particles displayed a certain symmetry, which Heisenberg termed "isospin" invariance.

"Isospin?" you stop us, perplexed by the peculiar term. "What is *that?*"

It's an unfortunate historical relic is what it is, and we assure you it has nothing to do with real spin. Nevertheless, you can imagine that just as we abstractly rotated x into y above, we can rotate a neutron through an abstract angle and it becomes a proton.[†] Isospin invariance is termed an internal symmetry because the "isospin angle" through which you rotate a neutron to get a proton is not a real angle. It is something else.

Shortly afterward, Nicholas Kemmer, using isospin symmetry and Hideki Yukawa's theory of the strong nuclear force, predicted the existence of a triplet of particles known

---

* The neutron is just .1 percent heavier than the proton.

as pions, which were subsequently discovered. This was the first instance that symmetry arguments were used to predict a new particle.

At this point, the majority reminds everyone that chiseled into the cornice above the Academy physics building is the important mantra:

Use it until it breaks.

Paying heed to the immortal truth, physicists have extended the concept of isospin to other particles and interactions. Just as the neutron and the proton have nearly the same masses, other particles, less familiar to nonphysicists, are also clustered into families of narrow mass range. For example, one group of eight heavy particles, which includes the neutron, have mass differences that amount to only 10 percent of their individual masses. Once again, this similarity led physicists to assume there was some underlying symmetry connecting them. Indeed, in so-called medium-strong interactions, the eight particles behave identically.

In analogy to isospin, the octet particles are said to be identical under a larger group of transformations, which include a property called "strangeness."[†] Like isospin, strangeness is an internal symmetry. It does not correspond to electric charge or to a rotation in physical space. It is another basic property of certain elementary particles. Anticipating your question, we can say that, like "isospin," the name is best accepted without too many inquiries.

By plotting strangeness versus isospin for various families of particles, one can make various pleasing diagrams, like the baryon decouplet that came up in the Debates on Mathematics. These diagrams display symmetries in a very geometric way. Historically, in these diagrams, particles were sometimes missing. Yet the belief that the symmetry should be complete led to predictions of new particles—the prediction of the Ω^- particle mentioned yesterday. This reliance on symmetry was identical to J. Lothar Meyer and Dmitri Mendeleev's use of the periodic table of the elements, which contained gaps, to predict new elements.

Since the early 1960s, academicians have recognized that the heavier subatomic particles behave as if they are composed of more elementary particles termed quarks. Here, at long last, you screw up your courage and interrupt with a genuine question: did we not maintain a mere two days ago that no one has successfully calculated how three quarks form a proton? True, and this is why we now say the heavier particles behave "as if" they are composed of quarks. In particular, they display the symmetries required by quarks.

However, there is a second reason for the "as if." The fact is that no one has ever directly observed a quark and, according to the "theory," it is actually impossible to do so. All evidence of quarks is by indirect means. For this reason Fritjof Capra says, "Although our present understanding of the atomic world precludes the existence of quarks as physical particles, there can be no doubt that hadrons [heavy particles] exhibit quark symmetries."

At this a great cry goes up from the hard core. How can anyone maintain that quarks have no physical reality? They exist—inside the nucleus. A delicate point has arisen here, but after some negotiation, the participants agree with the authors to postpone debate until evening.

Like the heavier particles, quarks—whether they exist or not—obey rules of symmetry. Quarks come in three pairs, or "families": up and down, charm and strange, top and bottom, for a total of six "flavors." Each of the six "flavors" comes in three "colors" (red, green and blue) for a total of eighteen, and each of the eighteen has an antiquark partner for a total of thirty-six. Along with the electron and some associated particles, the thirty-six quarks are considered to be the elementary particles out of which all others are constructed. Fifty or so elementary particles strikes Academy tourists as inconsistent with the term "elementary," but it is unarguably an improvement over the hundreds that existed before quarks made their appearance. It is much like the transition from polytheism to monotheism, or perhaps to multi-theism.

The important point is that the different-colored quarks corresponding to a flavor can be transformed into one another by internal rotations in the same way as the neutron and the proton or the octet particles can be transformed into one another.[†]

The Symmetrists Show That the Symmetry of Forces Is No Different

Isospin, strangeness, octets, quarks, rotational symmetries . . . It reminds visitors of some witch doctor's incantation, "GDP, CPI-U, CPI-W, IMF, ILEI . . ." On the plaza, vendors begin hawking vocabulary sheets.

This is useful, for the symmetrists have only begun. As we have said, their boast that symmetry is at the bottom of everything is far from idle. The Weylites now remind the crowd that, not only do particles obey certain rules of symmetry, but the same symmetries can be extended to forces. In nature, physicists once recognized four distinct forces: the electromagnetic, the "strong" nuclear force (the force responsible for holding the nucleus of the atom together), the "weak" force (responsible for certain types of radioactive decay), and gravity. The name "electromagnetic force" is itself suggestive. Two hundred years ago, electricity and magnetism were thought of as separate forces. However, the work of people like Henry and Faraday showed they were connected, and Maxwell finally showed they were two aspects of the same electromagnetic force.

With the twentieth century behind us, it is not surprising that forces manifest symmetries; physicists have learned that every force is transmitted by a particle, and so the symmetries of forces are essentially the symmetries between the force-transmitting particles. To give things a name, particles of matter—like the neutron and the proton—are referred to as *fermions,* while force particles are termed *bosons.* A familiar boson is the light particle, the photon, which is responsible for transmitting the electromagnetic force.

At hearing the last terms, you grudgingly purchase a vocabulary sheet. Nevertheless, we suspect that you have begun to be swayed by the simplicity and beauty of the symmetry concept. If so, you are in good company. Symmetry principles are, by their very nature, principles of simplicity; they introduce order and design into what before seemed unstructured and chaotic. The idea that there must be unity in diversity has haunted physicists since Maxwell unified electricity and magnetism. By now the general procedure of particle physicists, we hope, is clear: To connect the seemingly dissimilar, find a symmetry group. The more dissimilar the things that are to be connected, the more complicated the symmetries must be.

The next step in the program toward unification, the symmetrists continue, was to connect the weak nuclear force with the electromagnetic force into one *electroweak* force. The weak force, about 100 billion times weaker than the strong nuclear force, is unfamiliar to most people; nevertheless it is important on a subatomic level. For example, the neutron is actually a radioactive particle that tends to decay into a proton plus an electron plus a particle known as an antineutrino. The weak force governs the rate of this decay. Examined closely, the weak force turns out to be a force between electrons and neutrinos. The electron and the neutrino are governed by the same internal symmetry group as the proton and the neutron. By the rule that all forces are transmitted by particles, the weak force must have bosons associated with it, which indeed were found. (They are called the W and Z bosons.)

To combine the weak and the electromagnetic force, the groups governing their behavior are merely spliced together into one bigger group, which has all the symmetries that were inherent in the electromagnetic force and the weak force separately.[†] (Much as some *sangaku* contain both rotations and reflections.) The unification of the electromagnetic and the weak force was carried out by Sheldon Glashow, Abdus Salam, and Steven Weinberg.

As is said, "Superior men know how to combine elegance with simplicity." Not content with half-measures, physicists have added into the mix the strong force. The strong force holds the nucleus of the atom together and is the strongest force in nature. Because the protons and the neutrons of the atomic nucleus are thought to be composed of quarks, the strong force should then be a force between quarks; the bosons responsible for transmitting these forces are termed gluons. The "standard model" of particle physics splices the strong and the electroweak forces into one big symmetry.[†]

But now, as the Weyl camp continues its recitation, an anonymous academician raises a timid voice in protest, pointing out that the standard model is much like a college fraternity that has limited its membership to students above a certain income level. Even as an amazed hush falls over the crowd we say: True. The standard model has been quite successful in describing all known elementary-particle interactions but it does not include the force of gravity. Gravity is by far the weakest force in nature and is insignificant at the level of elementary particle interactions. But the strong and the weak nuclear forces do not generally operate over distances greater than the atomic nucleus, and the electromagnetic force operates only between charged particles; the universe, in the large, is neutral. Thus, gravity is the single force responsible for the large-scale evolution of the universe itself. Nevertheless, it does not participate in the standard model.

The symmetrists shrug off the objection. Progress in science, they rightly maintain, exists. Since Einstein, academicians have endeavored to wed gravity to the other forces into a "theory of everything." The latest attempt along these lines is known as superstring theory. It is true that string theory is far from complete, but recently there has been much excitement in the string community, which is slated for discussion in the Eighth Debates, on theories of everything.

The debaters adjourn.

Unification Without Unification

The sun passed the zenith a few hours ago and with the crossing, a number of spectators vanished from sight, you among them. We sympathize with your disappointment; only one feeble objection was raised (that gravity has not been incorporated into the standard model) and that was quickly crushed. With everyone chanting the same litany, no wonder you attempted to flee. A little while ago a rumor was circulating that you planned to join a Buddhist monastery. "Nam myoho renge kyo. . . ." This is not a bad plan, but as you have already discovered, escape from the Academy is difficult.

Back at the plaza, the situation appears unchanged. Someone is quoting an old Persian saying, "If a hair from a sage's beard falls into a bowl of milk, the flaw is easily observed." To the physicists on the plaza, the proverb's meaning seems clear: The perfect symmetry of the surface is now marred. *Beauty* = *symmetry* and asymmetry is a mark of imperfection. Reviewing the day's remarks, it strikes you that the word "asymmetry" has hardly appeared. A cogent observation. So imperfect is asymmetry that theoretical physicists in fact rarely use the term. Instead the phrase is "broken symmetry." The very choice of words is instructive. It underscores the physicist's belief that symmetry is primary, asymmetry secondary. Weyl said it, the Persian said it.

But is the hair from the sage's beard really an accident? This is the question troubling the skeptical minority. Asymmetry not only exists in the macroscopic world, where it is manifested in the New York skyline, but it persists in the microscopic world. "After all," one of the participants now asks forthrightly, "where is the antimatter on Academy grounds?"

The academician has a point. Virtually the only antimatter known to exist is the tiny amounts produced in accelerators and cosmic ray showers. If antimatter and ordinary matter existed in equal amounts, the two would annihilate and that would be the end of us. The fact that we are here argues for the lack of substantial amounts of antimatter in the universe. But if nature demands symmetry, should there not be as much antimatter as matter? Should we not have long ago been converted into photons? Confronted with this stark reality, the symmetrists do not capitulate easily.

They answer instead that, at the beginning of the universe, matter and antimatter existed in equal amounts but that a tiny excess (one part in a billion) of matter quickly developed. That tiny excess was left over after the annihilations and constitutes the matter in the universe today.

The authors prompt you to complain that this is not really an explanation. Although the theory explains how matter and antimatter annihilated, the appearance of the matter excess is put in by hand in order to agree with observations.* No theory at present gives a more fundamental explanation.

The symmetrists are unmoved. Very well. In that case we have no recourse but to point to the more severe asymmetry of gravity. Gravity is only an attractive force; "antigravity" does not exist in nature. Unlike electric charge, gravitational mass is always "positive"; two masses always attract, never repel. This seems to be a fundamental asymmetry of nature. No unified theory, to our knowledge, predicts at any energy a symmetry between positive and negative mass.

While the Weyl camp ponders this undeniable asymmetry, you realize the central question of the Third Debates has returned. In the case of matter–antimatter, physicists have assumed that the universe was initially symmetric. Then, as if dropping a hair into a bowl of milk, they "break the symmetry" by putting in a small surplus of matter over antimatter, producing the universe we see today. Symmetry is prior, all asymmetry is merely "broken symmetry," as the Persian tells us. Almost all fundamental theories of particle physics incorporate symmetry breaking. But then we are left with the question, Is the asymmetry of gravity a result of symmetry breaking, or is this a fundamental asymmetry of the universe? Is symmetry or asymmetry prior?

Now that a few weaknesses have been exposed, the authors feel more confident in pointing out further ambiguities in the symmetrist position. Maxwell unified electricity and magnetism into the electromagnetic field; Sudarshan and Marshak (followed by Feynman and Gell-Mann) unified all the weak interactions; since then other researchers have unified the electromagnetic and the weak nuclear forces into the electroweak force, and the electroweak force and the strong nuclear force into the standard model of particle physics.

In what sense are these unifications?

A provocative question. Although Maxwell unified the electric and the magnetic fields into a single electromagnetic field, Maxwell's equations display an equally important asymmetry, both in form and function, between the electric and the magnetic fields. The equations describe the behavior of electric charges. They do not describe the behavior of magnetic charges for the simple and profound reason that magnetic charges do not exist. If you break a magnet in half, two magnets each with a north and south pole result, not isolated north and south poles. Here is an analogy with the lack of negative gravitational mass in the universe. Maxwell's unification of electricity and magnetism did not produce magnetic monopoles. How, then, in practice are the fields unified?

Physicists often say that unification is achieved by going to a higher dimension (or by flying off to a comet). For example, Maxwell's four equations contain two for the electric field and two for the magnetic field. The differences in the equations are apparent to the eye—a visual display of the asymmetry between the electric and magnetic fields. Yet, if one treats time as a fourth dimension—as Lorentz and Einstein did—one can rewrite the equations in a way that visually makes no distinction between electricity and magnetism. This is, in fact, what academicians commonly mean by unifying the two fields.

---

* Each annihilation produces two photons; after a billion annihilations, two billion photons are left for each remaining matter particle, about what is observed.

Figure 3.3

However, the situation is similar to NAFTA. NAFTA facilitates trade between the United States, Canada and Mexico, but it does not mean that the three countries are identical. Crucial in the four-dimensional version of Maxwell's equations is a thing called F, the "field tensor," which is a matrix consisting of rows and columns like a spreadsheet or a calendar. This spreadsheet contains rows and columns for the electric and magnetic fields. But if you put the fields in the wrong places, you get the wrong equations. Thus, while F "unifies" electricity and magnetism, each has its allotted place; one cannot replace the other.

Much the same can be said about the unification of the electromagnetic force with the weak nuclear force. One constructs a "field tensor" that contains both the electromagnetic field and the weak field. However, they are kept distinct within this object. You can imagine a checkerboard with a single white checker at one corner. Surrounding it are black and red checkers. The rules for moving the black and red checkers are such that they can be shuffled only among themselves. According to these rules then, the red and black checkers can be transformed into one another but not into a white checker. It is a reasonable question as to whether such a scheme represents true unification among checkers.

In a somewhat more complicated way, the same holds true for the electromagnetic and the weak force: the constituents of the weak interaction cannot be interchanged with the constituents of the electromagnetic interaction. Exactly the same ambiguities concern the standard model, the model of particle physics that unifies the electromagnetic, weak and strong forces.

Apart from the question, Is this unification? another issue lurks here, one that an Academy artist now raises. The motivation of symmetry was to produce a beautiful theory. The splicing together of the standard model has produced a theory that indisputably works, but which in other respects resembles a Frankenstein's monster. In addition to the fifty-odd "elementary particles" there are at least nineteen arbitrary parameters that are not accounted for by the theory and must be put in by hand.

"You call this beautiful?" the artist exclaims, then walks off laughing.

Well, as Roald Hoffmann says, "Beauty has nothing to do with simplicity."

Physicists have not been blind to the artist's point. Because the standard model is so jury-rigged, and does not truly unify the forces, theorists have sought to find a more natural symmetry that combines all the interactions in one group rather than keeping them distinct as the splicing standard model does. In the 1970s and 1980s this endeavor produced Grand Unified Theories, or GUTs, which the academicians will undoubtedly take up in a few days. For now we note only that no one ever found any evidence for GUTs and that more recently they have been replaced by superstring theories, for which there is also no direct evidence.

Countless Manifestations

The previous arguments have restored some balance to the debates and seem to have rekindled your interest, which had flagged, causing you to temporarily vanish. For this reason you pay attention when the litany of ambiguities in the symmetry program continues. Earlier the authors mentioned symmetries among the three quark colors as well as among certain "octets"—families of eight heavy particles called *baryons*. It turns out that exactly the same group describes both the quark symmetries and the baryon symmetries. But that means the same group describes the symmetry among three particles and among eight particles. Therefore, group symmetries alone are insufficient to explain all the properties of the system, in this case the number of particles.

"But then what is the difference between the two types of particles?" you ask.

The frieze of Persian sphinxes manifests the difference. The pattern is invariant under shifts of the appropriate distance. This remains true whether there are three sphinxes or eight. The symmetry properties say nothing about how many sphinxes actually exist.

The missing element in the description is related to what mathematicians call a "representation." It is not enough that particle properties satisfy the symmetries of a certain group. That group must also contain an "eight-dimensional representation," corresponding to the eight particles of an octet. As the astrologers say, "The stars impel, they do not compel."

To better interpret the astrological advice, contrast the frieze of the Persian sphinxes with a second frieze of Persian sphinxes alternating with Greek sphinxes. Both friezes are translationally invariant. Yet you must shift the second frieze twice as far as the first to make the figures match up. Loosely, we might say that the translation group with shift 1 describes the Persian sphinxes and the translation group with shift 2 describes the Persian sphinxes and Greek sphinxes. The two shifts distinguish two representations of the same group.

Suppose you spent the afternoon at the Academy zoo considering whether the translation group describes known frieze species. You know that the shift-1 representation describes the Persian sphinxes alone. If you then realized that the same group contained the shift-2 representation as well, you would predict that another particle—the Greek sphinx—existed. Alternatively, if you discovered the Greek sphinx, you would examine the group to determine whether it contained a shift-2 representation. If it did not, you would be forced to throw out the group as a description of nature.

Notice that the Persian-Greek frieze does not move back into itself with a shift of three units. Therefore this frieze cannot represent three animal species. The same applies to the groups of physics. On the one hand, it is remarkable that the particles discovered in nature conform to groups with the appropriate representations. On the other hand, it does show that symmetry is not the entire manifestation of reality.

But Do Quarks Exist?

A great grumbling has arisen from the visitors on the plaza. Why are we debating the subtleties of particle physics if we aren't even sure whether the fundamental constituents of the standard model—the quarks—exist? Earlier in the day, Fritjof Capra cast serious doubt on their reality.

Capra's position does not sit well with particle physicists, who counter it by describing the hunting of the quark. Recall, they tell the crowd, that if you bombard an avocado with BBs, most pass straight through, but occasionally one hits the hard pit in the center and gets stuck, or it bounces off. This is precisely how scientists discovered the atomic nucleus; they bombarded atoms with high-energy particles and observed how they scattered. Indeed, when one bombards an atom with particles of sufficiently high energy, the atom fragments, ejecting neutrons and protons. Hence "atom smasher."

Now, when we shoot high-energy particles at the neutrons and protons themselves, they behave as if three objects are rattling around inside—the quarks. It is that simple.

The crowd seems convinced by the explanation.

But wait.

It is precisely here that the analogy with the avocado or atomic nucleus breaks down. According to current beliefs, regardless of how hard you hit the proton or neutron, you can never break it into quarks. So, despite all the hissing from the right, Capra does have a point: there is no *direct* evidence for quarks, and one could choose to regard them as useful fictions to explain the symmetries of all the known particles. In many ways quarks are like the ether of the nineteenth century; it was a useful mental crutch but ultimately, as Einstein showed, unnecessary.

Yet the motivation behind Capra's statement is equally weak. Capra intends to draw parallels between particle physics and "Eastern" philosophy. As he maintained earlier, symmetry is Western and asymmetry is Eastern. But he is then faced with the difficulty of explaining all these particle symmetries, which the quarks do so well.

Capra's solution is to account for the particle symmetries in terms of another model, known as the bootstrap model. The bootstrap model is a perfect democracy; in contrast to the quark model there are no fundamental particles. All particles are equally fundamental. Capra is attracted to the bootstrap model because, whereas quark symmetries connote static, if beautiful, patterns, the bootstrap model is based on constant, dynamic interactions. Daisetz Suzuki puts it this way: "Buddhists have conceived an object as an event and not as a thing." The bootstrap theory is concerned with events; the quark model is concerned with things. Just as the Chinese Book of Changes, the *I Ching*, shows how cosmic events are organized into patterns of hexagrams, the bootstrap model dynamically organizes particle interactions into quark symmetries.

Unfortunately, there is simply no reason to believe the bootstrap model is correct. Most physicists abandoned it twenty years ago. The bootstrap model is not only incapable of explaining the particle symmetries, but it is also unable to explain why the neutron and the proton behave as if there are three particles rattling around inside. And it does no better than QED in accounting for the existence of the hydrogen atom.

Analogies and metaphors have often proved pivotal in expanding our thoughts both within and without science, and so one should not discourage the attempt to synthesize apparent opposites. However, citizens of the New Age often forget that, when they involve science, analogies should be tempered by experiment and calculation.

Your head is swimming. Do physicists really base their most fundamental theories on entities that cannot in principle be observed? Yes. If so, what does it mean to exist? Is this what was meant in the Upanishads: "The eye cannot see it; mind cannot grasp it"? What about the following syllogism: If art, as André Malraux maintained, is antidestiny, and symmetry, as Michio Kaku maintained, is destiny, then art must be antisymmetric? QED.

The strain of the day is taking its toll. Many visitors before entering Academy grounds underestimate the rigors involved in continued disputation. (For this reason the Academy maintains extensive athletic facilities and spas.) Nevertheless, the debate goes on; although the sun has set, the pro-symmetry forces have not capitulated.

Far from it.

Are the Laws of Nature Symmetric?

Regardless of how you may feel this evening about symmetry's success in describing the properties of particles and fields, everyone agrees that there are yet more important areas in which the tool of symmetry has been brought to bear. One of the past century's most remarkable discoveries has been that the laws of nature themselves manifest symmetries.

This is not an obvious statement. To come to terms with it, remember that for several hundred years all natural philosophers believed that Newton's laws, which describe how objects move and interact, governed the behavior of the entire universe. The laws themselves were not the whole story. Of greater importance was a strongly held principle that if a given law or equation held in London in 1688, then it should also hold in New York in 1998. Furthermore, if it held on the ground, then it should hold on any *inertial frame*, which is any system—say a train—moving at constant velocity (e.g., 100 kilometers an hour, north). Daily experience seems to verify the assumption that Newton's laws are valid in inertial frames: When you toss a ball on a train it falls back into your hands as if you were not moving at all; indeed with respect to the inertial frame of the train, you are not moving.

In addition scientists assumed that if you were riding a horse at 20 kilometers per hour with respect to the ground, then to a person on the train you would be moving at 120 kph or 80 kph, depending on which direction you were heading. Velocities add.

With the assumptions that the laws hold in all inertial frames and that velocities add, it is quite easy to show mathematically that Newton's laws are essentially like the Persian sphinxes or the *sangaku*—unchanged under certain shifts or rotations. These transformations may merely be a hop to a different place or time or to a system moving with a different velocity (as in the case of the train). The transformations that leave Newton's laws invariant are known as the Galilean group in honor of Galileo, who first gave the rules for going from one inertial frame to another. The Galilean group describes the symmetries inherent in the laws of nature themselves.

In the 1860s Maxwell wrote down his famous equations that govern the behavior of the electromagnetic field. Maxwell's equations contain the speed of light. When one subjected Maxwell's equations to the same symmetry operations—the Galilean transformations—that left Newton's laws invariant, Maxwell's equations were *not* left invariant. This led scientists to suspect Maxwell's equations were incorrect (Maxwell himself had doubts).

The dilemma was resolved only by the work of Lorentz, Poincaré, and finally, in 1905, by Einstein. They accepted the idea that the laws of nature should be the same in all inertial frames. But in acting on Maxwell's equations by the Galilean group, one is assuming that the speed of light gets added the way other velocities do. Einstein threw out this assumption and made a new one: that the speed of light remained constant to all observers. With this single change to the old theory he showed that Maxwell's equations remained invariant under a different set of operations, today known as the Poincaré group.

In the Galilean group the shifts that take you from one space location to another are separate from the shifts in time, much as the rotations and reflections in the *sangaku* are distinct. On the other hand, in the Poincaré group, the transformations combine both space and time into spacetime. One can think of them as rotations through *both* space and time. For this reason, since Einstein, we speak not of space and time separately, but of spacetime. Einstein's discovery implied that Newton's laws were not exactly correct and that the Galilean group did not correctly describe the symmetries of nature. In discovering the set of symmetries inherent in Maxwell's equations, Einstein created the special theory of relativity.

You have no choice but to be impressed.

Is Motion Symmetry?

Symmetry has again gained momentum and appears headed for victory. Even so, the symmetrist forces on the plaza have another point to make, perhaps the most sublime of all.

Symmetries involve patterns, be they the geometric designs of a Persian frieze or the abstract invariants found in equations. Newton's laws, Maxwell's laws, govern the flight of bodies under the action of forces. Symmetry connotes the static, motion the dynamic, to some the West and the East. But we have just seen that Newton's and Maxwell's equations, which describe how objects move and interact, embody certain symmetries. This is a profound fact of nature. It suggests that perhaps the dichotomy between symmetry on the one hand and dynamism on the other is not sharp. If the embodiment of symmetry by dynamical theories is a profound fact of nature, yet more profound still, physicists argue, is that the symmetries of space itself correspond to the laws of nature.

To see what they mean by this, recall that one of Deepak Chopra's seven spiritual laws of success is the law of "least effort." It is "based on the fact that nature's intelligence functions with effortless ease and abandoned carefreeness. This is the principle of least action, of no resistance. This is, therefore, the principle of harmony and love."

Whether or not love figures into it, Chopra did not originate the principle of least action. The *action* was introduced into science in 1747 by Pierre-Louis-Moreau de Maupertuis, who believed that action is minimized through the "wisdom of God." Maupertuis, like Chopra, was evidently convinced that nature acts with minimum effort but, unlike Chopra, he assigned a definite mathematical meaning to the word. In the scientific sense, the action is a mathematical function[†] that you can in principle write down for any system under consideration, be it a swinging pendulum or a planet in orbit around the sun. Having written down the action, you minimize it and—presto! There spring forth Newton's equations of motion, or Maxwell's or Einstein's. Hence the principle of least action. To the extent that the equations of motion entirely describe the behavior of the system, and that the laws are coiled within the action, then the action, miraculously, contains all the information about the system, both static and dynamic.

Scientists have often been struck by the universality of the action, to the extent that many declare: "In the beginning was the action."

What is the basis for such exultation?

Imagine yourself far from the Academy, in a universe that is completely uniform. All the matter is distributed evenly, everything is the same color and temperature. This is the most symmetric situation one can imagine. One can translate an arbitrary distance and everything remains exactly the same. This includes forces. Any forces present must be the same everywhere. And so in a completely uniform universe you can never tell you have moved from one place to another. This means you must be standing still or, what is effectively the same, moving at constant velocity.

Now recall the law of inertia: any body not acted on by an outside force moves at a constant velocity. However, if a body is moving at a constant velocity, we say its momentum is conserved. Here is one of the great laws of nature: the momentum of an object not affected by outside forces never changes.

Yet, we see that the law of conservation of momentum results from the fact that space is uniform! In a similar way, if a space is rotationally symmetric (isotropic, to use the word from the First Debates), such that it is impossible to discern which direction you are pointing, then an analysis of the action leads to another great conservation law: the conservation of angular momentum. If space is such that the action remains unchanged in the course of time, then it yields the conservation of energy.

These results are special cases of a celebrated theorem proved by Emmy Noether in 1918, that every symmetry in the action is related to a conservation law. Many consider Noether's theorem to be one of the most profound results in physics. Symmetry and laws of motion cannot be separated.

Symmetry Without Symmetry

It seems that symmetry has won the day, after all. On the other hand, Aristotle in the First Debates cautioned against claiming that momentum is conserved *because* space is uniform. They are just two ways of saying the same thing. In any case, asymmetry has not quite capitulated. Her adherents, gathering a last bit of strength, point out yet another important way in which the symmetry picture of nature is incomplete: symmetry deals only with the laws of nature themselves. But laws are themselves insufficient to describe the initial state of the universe.

Until approximately the time of Newton, scientists did not distinguish among teleology, dynamical laws and initial conditions. The concept of teleology came up the day before yesterday when Aristotle exclaimed, "Nature abhors a vacuum," and today when Michio Kaku declared, "Nature demands symmetry." Systems evolve with a purpose. The principle of least action, which says that nature minimizes the mysterious action, is also apparently a teleological statement. Why should nature minimize the action? We advise you to keep the question in mind.

Regardless of the answer, all physical theories require one to supply the initial conditions by hand. Although Newtonian mechanics describes the trajectory of a baseball in a gravitational field, it does this only after the ball's initial position and velocity are specified. Because the symmetry description of nature corresponds to the description of *laws,* nothing in the *truth = beauty = symmetry* equation explains how things began. Furthermore, although the baseball obeys the laws of conservation of energy and momentum, which can be regarded as manifestations of the symmetries of space and time, nothing in these symmetries tells you in which direction the baseball is moving.

A Contemplation: Symmetry Is Asymmetry

The debate among academicians has gone far into the night and the visitors, those who remain standing, are at a loss to declare a winner. The question is moot, for in the noise and shouts the participants have forgotten that "when duality comes to an end, O Arjuna, they and I live in one place, with one name." For the fact is, symmetry requires asymmetry. We recognize symmetry only once it has been broken, and thus there can be no description of the world based on symmetry without asymmetry.

To see this, suppose the world was entirely one color—red. In that case we would never invent the concept of "red," because "red" is needed only to distinguish it from other colors. In fact we would never invent the concept of color unless there was more than one. Or suppose you saw a spinning sphere. Because a sphere is absolutely symmetric, you cannot tell that a spinning sphere is spinning. You must mark it. You must destroy the symmetry in order to observe the symmetry.

The same is true in the physical world. In a completely uniform universe, one whose properties are identical everywhere, you cannot tell you have moved. To do so, there has to be some marker in the universe, informing you that one place is different from another.

The symmetries of subatomic particles do not escape this fact. Due to isospin symmetry, the neutron and the proton are indistinguishable to the strong force. A neutron can be distinguished from a proton only by measuring their electric charges, and we do this by observing how the neutron and the proton behave in an external electromagnetic field. (A uniform magnetic field will deflect protons but not neutrons.) Until you turn on the external field, isospin symmetry makes the two particles identical. Without the external field to break this symmetry, we would never have devised isospin symmetry to explain why the neutron and the proton are nearly—but not exactly—the same. The neutron and the proton would have in fact been indistinguishable, identical particles.

Quarks also suffer from symmetry blindness. Although we refer to quarks by their "colors," we cannot see them. There is no external "color field" in the world to break color symmetry. Quarks of different colors cannot be distinguished and the external world is "colorless."

The expanse of milk is not observed to be flawless until the hair from the sage's beard has fallen into it. Only by observing the asymmetries inherent in the New York skyline do we realize that many of the buildings are symmetrical.

Practical Exercise: Is Symmetry in the Eye of the Beholder?

Even as the sun comes up, a few diehards have persisted. Some, fighting off exhaustion, repeat the observation made in the First Debates: most physicists are experimental physicists, but most physicists who write books are theoretical physicists. Roald Hoffmann, in his unflagging assault on the physics hegemony, describes the phenomenon this way: "Physics has a real psychological problem, a kind of schizophrenia, for it is an experimental field but (because of its buying whole hog reductionist logic) one which has opted to have its heroes theoreticians. . . . The theories experimentalists use are often so simple they're ashamed to put them into print, so they invent grander reasons than those that initially moved them."

The remark raises the question whether the physicists' emphasis on symmetry might not have arisen from the experimental-theoretical split. The two communities really do often seem to have nothing in common. Theoretical physicists write books. Experimental physicists spend their days building and fixing apparatus and, when the stars are in the right position, take data. Data may consist of strings of numbers, or of particle tracks in a detector.

"That being the case," ventures an experimentalist, "shouldn't one point out that experimental physicists don't measure symmetry directly? We measure particle properties: mass, energy, momentum, angle of emission."

"Nevertheless," a theoretician responds, "symmetry is a powerful scheme for correlating these properties, even a language."

"A powerful language, fine. But isn't it a rule in science that the same phenomena can be described by different languages? You know better than I, but when Julian Schwinger and Richard Feynman independently invented their versions of quantum electrodynamics, weren't their languages so different that it took a third physicist, Freeman Dyson, to show they had created the same theory?"

"True, but the point is, they are the same theories, and exhibit the same symmetries."

The experimentalist thinks before answering. "No, that ain't the point. Yes, the value of a good, simple formalism in physics cannot be overstated. The mysterious action does not give you anything more than Newton's laws themselves—"

"—but to derive the equations necessary for a problem from the action is almost always easier than from Newton's laws directly."

"Agreed—"

"So much so that progress in theoretical physics would probably have halted a century ago if the action formalism were unavailable."

"All true. But if you'd allow me to finish my thought, a formalism is a language, not the phenomenon. Einstein did not discover relativity by thinking in terms of the symmetries; this was a construct applied afterward. Is there any difference between saying that momentum is conserved because the universe is uniform and saying that momentum is conserved in the absence of forces?"

"Hmm . . ."

"It is far from clear to me that the Omega-minus particle could not have been predicted by other, more clumsy, means, or that it could not have been discovered by us experimentalists before you guys predicted it."

"That never would have happened."

"Typical theoretician's arrogance. You call yourselves artists. Fine. Just as the idea that artistic beauty derives from symmetry is an aesthetic, so the contention that natural beauty derives from group-theoretical principles is an aesthetic."

"It has been extraordinarily successful, a validated principle."

"It is a subjective judgment, a metaphysical principle, unverifiable by experiment."

"Physics doesn't use metaphysical principles. As Maxwell said, a metaphysician is merely a physicist disarmed of all his weapons."

"Armed I would say. The success of symmetry in predicting particles demonstrates above all else, it seems to me, the importance of metaphysical principles for physics. Why deny it? You believe in the action—"

"Of course."

"Why should God require the action to be minimized?"

The theorist thinks as the Fourth Debates begin.

In Which the Third Debates Are Summarized

Once again you failed to escape the Academy and hours ago you became too tired to entertain any such thoughts. As you lay down in the palazzo for a nap, symmetric visions danced in your head and the Third Debates replayed themselves, instantly.

The first question was whether art is symmetric; all but the physicists thought it was a ridiculous question. The main topic of the day concerned the physicists' claim that physical truth follows from mathematical beauty. The successful exploitation of symmetry in the hunt for natural laws is the most striking affirmation of such a belief. Proponents pointed out that the application of symmetry has led to the discovery of new particles, new relationships among particles, the unification of forces and a unification between laws and the very properties of space itself. Particularly crucial in this unification was the *action,* a mathematical quantity that gives the equations of motion and whose symmetries describe the symmetries of space.

Despite these successes, the opposition could point to some obvious asymmetries in the universe: the lack of substantial amounts of antimatter, the lack of magnetic monopoles and the lack of negative gravitational mass. No unified theory explains why gravity is only attractive. An Academy artist also pointed out that the standard model of particle physics is not beautiful, and the authors reminded the crowd that symmetry does not deal with the initial conditions of a system, and that the very recognition of symmetry requires asymmetry.

Finally, most of the discussion took place at the level of the microscopic world. No one would currently claim to have an explanation of the symmetries of a banana plant or of a lotus flower, which take place on a scale much larger than that of elementary particles. Although the Directors have slated a debate on this topic in a few days' time, we are still much in the dark about the origin of familiar natural symmetries. Those who enjoy basking in the warm glow of the mysterious can point to the rose and say, without fear of contradiction, "Here is something we do not understand."

WHY DO THINGS HAPPEN?
Causality, Synchronicity and All That

The discovery of causal laws is the essence of science, and therefore there can be no doubt that scientific men do right to look for them. . . . The maxim that men of science should seek causal laws is as obvious as the maxim that mushroom-gatherers should seek mushrooms.

—Bertrand Russell

The law of causality . . . is a relic of a bygone age.

—Bertrand Russell

The Question

At the top of Mt. Royal in Montreal, Quebec, tourists and vacationing academicians can relax in the visitors' pavilion and gaze at the large murals that line the walls. One of these shows a man, bearded, garbed in leather and furs—the archetypal seventeenth-century trapper or hunter. He stands on the shore of a northern lake, surrounded by fir trees. In the background his fellows ready the canoes for some expedition. The legend to the painting reads: "LaSalle about to discover the Mississippi."

The caption elicits a laugh, at least from academicians, because of its paradoxical nature: How could LaSalle know he was about to discover the Mississippi? One imagines the great leader waving his crew on, "Come on boys, let's discover the Mississippi!" LaSalle's prescience is a good example of one of Aristotle's final, or teleological, causes. Things happen for a purpose. The cause of LaSalle's expedition was to discover the Mississippi.

Such a statement sounds very strange to modern ears. This was confirmed on the morning of the Fourth Debates when, arising from your nap, you spied the Academy's reproduction of the famous mural hanging in the palazzo and chuckled, like everyone else. Your reaction demonstrated that, excluding instances of precognition, we tend to believe that events tend to occur in certain orders, that discoveries are made after expeditions set forth; that effects follow causes. The effect was the discovery of the Mississippi; the cause was LaSalle's expedition, not vice versa.

Strangely, the sight of the mural seems to have violated causality, because as the noise around you gradually resolved into words, it became apparent that during your nap the academicians had turned their attention to that very topic.

Figure 4.1

The notion of causality—that events have causes and that a cause is something that precedes an effect—is so ingrained in our existence that we could not live without it. When you deposited fifty cents into the Academy vending machine for an aspirin, you expected something to happen; if nothing did you would have assumed that something had gone wrong, that the natural course of events had been violated. Society as a whole accepts the principle of causality. Legally, all of us (except insurance companies) are responsible for our actions. People travel across the continent, not because astrologers tell them it is a good day to fly, but because they believe aircraft engines have the ability to lift planes off the ground. Generally, people live by the maxim

An event without a cause is like chickenpox without a virus.

On the other hand, as you watched the morning fistfights break out, you probably began to suspect that the matter is not straightforward. This is true. If during your tour of Academy farmland someone had asked, "Why did the chicken cross the road?" you would, intending not to be hoodwinked, have first given the age-old teleological reply, "To get to the other side." (The final goal—the other side—was the cause.) Here you recognized that the chicken and LaSalle have much in common. But you then noticed the farmer had forgotten to lock the chicken coop, allowing the chicken to escape, and that a motorist had dropped some food on the other side, which perhaps the chicken smelled. Thus, the event (the fabled crossing) evidently had multiple causes, causes that in retrospect proved to be difficult to isolate.

But even here we assumed that the many causes conspired to produce a single effect; we have isolated the effect. However, if the riddle had asked, "What happened as the chicken crossed the road?"—in other words, if we termed the effect the entire situation along the road during the crossing—you would have been confronted with a much more complicated picture. The farmer might have been chasing the chicken. A road crew might have been cleaning up the chicken's meal, as well as other litter. Happy campers could have been eating lunch at the picnic area. In looking back on the situation, it rapidly becomes less clear whether all these events are connected and, if so, how.

In short, what is a cause and what is an effect?

And does the universe operate by them?

Are There Causes? In Which Philosophers Befog One's Mind

As the sun climbed over the transformed quarrels and you shook off the lethargy due to an acute lack of sleep, your first thought was, once again, to escape. Rumors reached us that you circled the Academy confines, crying "I want information!" only to be told by each person you met, "You won't get it from me." They were, however, only too happy to talk. Specifically about cause. And effect. It gradually became apparent that, like the notion of symmetry, many people object to the idea of causality but nonetheless accept it as a matter of course.

Lawyers, doctors, politicians and judges, you discovered, are basically in favor of causality. They understand that, as mentioned just moments ago, society could not operate without the principle of causality; the result would be pure anarchy. The scientists you encountered were more cautious, but a straw poll showed that, like the lawyers and doctors, most are registered causalists: if all events took place randomly, if anarchy reigned, to describe the universe would be impossible. Except for a few abstract painters everyone agreed that only an apparent order to affairs allows any description whatsoever. The laws of nature are one way of describing such order, but virtually all laws of nature assume causality. Newton's third law, "for every action there is an equal and opposite reaction," implicitly accepts the fact that the reaction does not take place before the action. It also accepts the fact that there is such a thing as an action—a describable cause. In this sense the principle of causality is more general and sacred than most of the laws of nature.

So fundamental is the principle of causality that even those who decry Western modes of thinking cannot live without it. Many visitors this morning at the Academy spa have been caught citing Deepak Chopra's third spiritual law of success, the law of *karma:*

> Karma is both action and the consequence of that action; it is cause and effect simultaneously, because every action generates a force of energy that returns to us in like kind. There is nothing unfamiliar about the law of karma. Everyone has heard the expression, "What you sow is what you reap." Obviously, if we want to create happiness

in our lives, we must learn to sow the seeds of happiness. Therefore, Karma implies the action of conscious choice-making.

So the New Age is not so new that it is able to escape causality. Also fashionable among the left is the belief that Jung's concept of synchronicity is a demonstration of acausality. Synchronicity appeared briefly the day before yesterday. Jung's idea is that all events in the universe are connected in an acausal way—a way that does not require "causes" to be before "effects"—and that this explains those meaningful coincidences in our lives. Jungians have traditionally turned to quantum mechanics for support with the claim that, like synchronicity, quantum mechanics is also acausal. Even as the mists lift over the Academy farmlands, it is apparent that visitors and academicians are spoiling for a fight on this issue. To be sure, the prospect of a slugfest on synchronicity and quantum mechanics is so intriguing that you temporarily put aside the search for an exit.

But quantum mechanics is not necessary for one to be puzzled about cause. Indeed, on this fair morning the loudest have been the philosophers, who have wrestled with the ambiguities of causality since Aristotle confused everything. One thing on which all factions have agreed is that the notion of cause and effect requires change. If nothing whatsoever changed, there would be no sense in speaking of causes, as the immortal chicken demonstrated. It is far easier to talk of causes of specific events.

Recognizing this or not, Aristotle himself distinguished four different types of causes. If you walked into the Academy with an Uzi (something that has crossed your mind more than once in the past few days) and massacred all the philosophers, you would be considered the "efficient" cause of their deaths. The victims themselves (the substance in which a change is wrought) would be regarded as the "material cause," and the "formal cause" would be the dead bodies (that into which something is changed). The reason for your actions (to rid yourself of academicians) would be the already encountered final cause.

Aristotle evidently agreed with the National Rifle Association, which maintains that "bullets don't kill people, people kill people," for bullets do not figure in Aristotle's scheme. To rectify the oversight, Thomas Reid and John Stuart Mill distinguished "physical causes" (bullets) from efficient causes (you), Mill arguing that efficient causes are difficult to confirm—you are the only one who knows whether you were out of your right mind or knowingly decreed, "Death to philosophers!"

Here you protest that as long as we are considering bullets, we should take into account the Uzi, and once again causes begin multiplying. For this reason, David Hume, two hundred years ago, advanced a more contemporary idea: one merely observes from experience that one event follows another; we never observe exactly what connects them. However, even Hume's reasoning runs into difficulty. If, as he said, similar events A (terrorist runs into Academy with Uzi) are always followed by similar events B (massacre of philosophers), the question arises: how similar is similar? The only event exactly similar to a terrorist running into the Academy with an Uzi is a terrorist running into the Academy with an Uzi. Add a lost tourist to the scenario and who knows what will happen. This point was made by Bertrand Russell, who eventually decided that the whole notion of cause should be discarded. Yet, even after Russell, few of us doubt that a bullet can kill you.

Thus.

As the Fourth Debates crystallize, you have begun to wonder whether philosophers have said anything useful about causality in the past two hundred years. It is a reasonable question. Most introductory philosophy texts, even recent ones, carry out the discussion without any reference to twentieth-century physics, which makes as much sense as writing a book on the philosophy of space and time without any reference to relativity. The day promises to be long.

In Which Physicists Befog One's Mind

Physicists, as you probably already suspect, regard themselves as causality's arbiters. There is some justification for their hubris: if they have not solved the problem of causality, they have at least clarified it more successfully than philosophers. The clarification has come about through two of the most sacred principles of science: the much-maligned reductionism and the equally notorious mechanism. Reductionism is hardly more than a manifestation of causality; all events are ultimately based on a small number of similar causes. Mechanism in physicists' minds is hardly separable from reductionism, and for good reason. It is history's greatest manifestation of reductionism.

In the seventeenth century, Descartes divided the universe into the two separate realms of mind and matter. It is often said that in making this division—the celebrated Cartesian dualism—Descartes set Western science on its present course by allowing scientists to treat matter as dead, inert and independent of themselves. But it was Kepler who in 1605 first clearly enunciated the concept of mechanism, the idea that the universe operated like a giant machine, a clockwork, subject to mechanical forces. Newton clarified the nature of these forces, and with his work, scientists and philosophers accepted that the clockwork operated according to universal laws, unalterable by human or divine intervention. Mechanism received its clearest expression in the words of the Marquis Pierre-Simon de Laplace, who, when asked by Napoleon why he had not mentioned the Creator in his famous treatise on the operation of the solar system, supposedly made the definitive reply: "I have no need for that hypothesis."

The Newtonian universe is rigidly causal, or deterministic. The equations allow no room for error. Once the Creator started things off, Newton's laws took care of the rest. Everything from the big bang—or whatever beginning one chooses—is determined. As Laplace's famous, if apocryphal, remark shows, by the seventeenth century science recognized God as an encumbrance. There is no point at which He needed to interfere. The idea that Krishna could prevent the sun from setting is no longer considered possible. Nor is Joshua's similar feat of staying the moon and the sun in his battle against the Amorites. Physics took care of itself. God should not be dragged into such minor affairs. God cannot be dragged into such minor affairs because each event in the deterministic, Newtonian universe is caused by all the preceding events.

Newtonian mechanics therefore provided a definitive answer to the question Pao Ching-Yen posed in the prologue to the debates when he asked, "Why should illustrious Heaven be brought into the matter and why should it have given such precise instructions?" In Newtonian mechanics God is not permitted to intervene, once He let Newton be.

The morning is already half done and you walk up to one of the New Age visitors and ask innocently, "Why do you rely on karma to explain causality, when according to the authors it would be far more efficient to cite the Newtonian universe for this purpose?"

The New Ager hits you over the head saying, "What goes around, comes around."

Thus dazed, you approach one of the Newtonians and ask, "If we cannot actually measure causes, how can you maintain that a Newtonian universe is strictly deterministic?"

The Newtonian bangs you over the head, saying, "You don't believe in determinism? Take that, heretic!"

Having produced in you a nearly unconscious state, this Newtonian evidently believes that "deterministic" is synonymous with "predictable." But is it? Does determinism imply predictability? Is determinism a verifiable concept? Your excellent question will be discussed tomorrow—if you survive.

Apart from reductionism, mechanism and arrogance, what were the crucial steps in the physicists' simplification of the notion of causality?

First, they isolated certain occurrences. Because a search for causes makes sense only if things change, physicists relegated ordinary circumstances to a quiet background, which is assumed to be unchanging. You don't ask why the day advances, or why ice cream flows downhill or why apples fall to the ground. This isn't news. If you saw an apple flying sideways, or the Red Sea parting, this would be news. Physicists focus attention on exceptional events. Once you isolate an event, it is easier to speak about causes, as the chicken demonstrated.

A particularly crucial step was taken by Galileo, who decided that uniform motion (motion at a constant speed and direction) should be relegated to the class of unchanging things. Aristotle had been known to wander Academy corridors insisting that the "natural state of motion is rest" and that to keep objects in motion required causes. In a celebrated coup, Galileo toppled Aristotle from the directorship by declaring uniform motion to be equivalent to rest—uncaused. Newton, who succeeded Galileo in the directorship, appropriated the statement as his first law of motion, the law of inertia: any object travels at a constant velocity unless acted upon by an outside force.

In the law of inertia is implied the next great simplification: all physical causes are identified with forces. When a Newtonian physicist speaks of causes, she has no inkling of "efficient causes" or "formal causes." She has only one thing in mind: a force. Newton himself provided the modern definition of a force and what it amounts to is merely a push or a pull. All changes in Newtonian physics take place because of pushes and pulls, which can be measured. In that sense Newtonian mechanics gave a clear answer to the title question of these Fourth Debates, *Why Do Things Happen?* Things happen because of forces.

The twin steps of relegating uniform motion to the realm of the uncaused and limiting the discussion of causes to measurable forces provided an inestimable simplification over "efficient," "formal," "material" and "final" causes, most of which required consideration of the participant's state of mind. With the mechanization of the external world, physics split off the mental from the physical universe and stopped dealing with the mind. Teleological thinking fell into disrepute, and Aristotelians were sent across the road to tend Academy farmlands. Alchemy, which intertwined notions of the soul's purity with the ability to transmute elements, was eventually replaced by chemistry, which eliminated the soul from consideration. Astrology, which supposes some unnamed influence of the planets on human behavior, was also banished to the outer darkness.

Here a great cry goes up from today's Academy visitors, who with horoscopes in hand, point out that astrologers to this day outnumber astronomers. There is no 900 hotline for astronomy. So the mechanists clearly got it wrong and of course synchronicity provides a basis for astrology. We'll see, the authors answer, but point out that one result of unimaginative, mechanistic thinking was that societies eventually ceased to burn people at the stake for witchcraft.

During the above exposition a great cry also went up from the more moderate philosophic quarter. Surely we were not claiming that physics resolved the problem of causality. Not at all. Causes became associated with measurable forces but there remained the problem, as in the expanded chicken riddle, of multiple causes and of deciding what constituted an effect. Moreover, there had always been the suspicion that an effect should "immediately" follow a cause, but academicians had never resolved whether "immediately" meant instantaneously or after a short interval of time. In which case, how short?

Two other problems, deeper ones, dwell within the Newtonian cosmos. The first is that if the universe is strictly deterministic, then there is no essential difference between cause and effect. Given the initial state of a system, one can, à la Laplace, predict anything that happens thereafter, including the final state of the system. By the same token, given the final state of the system, one can equally well retrodict the initial state of the system. Which conditions one calls "causes" and which one calls "effects" would be entirely arbitrary were there no direction of time. And as academicians preparing for tomorrow's debates will aver, nothing in Newtonian physics provides a direction of time.

Finally, we might agree with Hamlet: "Find out the cause of this effect / Or rather, say, the cause of this defect." What the Dane had in mind was the paradox of the Newtonian universe: in a strictly causal world, the existence of causality cannot be established. If every event is determined by every preceding event, then the concept of free will is meaningless, as is the notion of running an experiment, which presupposes that conditions can be varied. Yet, if the experimenter's very actions are predetermined, then nothing has been varied and no "experiment" has been carried out. To put it another way, one needs a defect in causality to verify causality. That is to say ...

Apparently, Newtonian physics did not clarify all issues.

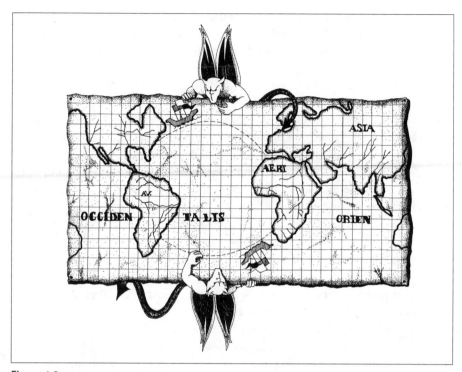

Figure 4.2
Arctic demons sucking on ships bend their routes to north and south.

Some Fraudulent Causes Are Unmasked

As if to underscore the state of affairs, one recalcitrant academician points out, ringing her glass for our attention, that despite everything the Newtonians have led us to believe, causes do not exist.

Huh? you ask yourself, wondering if the Newtonians have suddenly decided that the blows they gave you to the head this morning were imaginary.

Well . . . . Imagine a satellite in a circular orbit around the Earth, with an orbit inclined to the northwest. Although such an orbit traces out a circle on an ordinary globe, on a flat map the orbit appears to oscillate, first north, then south. Similarly, on a typical wall map, any great-circle shipping route in the Northern Hemisphere is found bent toward the North Pole and any great-circle route in the Southern Hemisphere is found bent toward the South Pole.

The bending of shipping routes, the story goes, caused a great deal of consternation among ancient seafarers. What, they asked, is causing the shipping lanes to be pulled toward the poles? Is there an Arctic demon sucking on the ships?

"Come now," you reply, perceiving a fraud, "Nothing is causing anything." The apparent bending of the routes toward the poles is an illusion due to the fact that we have used the wrong type of map to display the satellite orbit. This is a problem with language, and another example of how a single phenomenon can be described by different mathematical constructs. Using the wrong mathematics produces a fictitious cause. In fact, to explain the bending, a physicist would need to introduce a "fictitious force," a force that is purely the result of using a flat map instead of a globe.

You remain unimpressed. As you said just a moment ago, "Nothing is causing anything." Why are we making an issue of nothing? Because one may ask whether all causes in the sense of the bad map projection are fake. Suppose, for example, an archer shoots a container of frozen yogurt into the air. Normally, physicists would say that the gravitational force of the Earth causes the yogurt to follow a parabolic trajectory until it hits the ground. However, there are other ways of looking at this problem in which forces, and therefore causes, vanish.

The theory of general relativity, for example, dispenses with forces and describes gravity as a curvature of space and time. From the standpoint of relativity, the yogurt is merely traveling along the straightest path allowed by the curvature of space in the Earth's vicinity and experiences no forces. In this picture gravity is the same as the fictitious force acting on the ships caused by the Arctic demon. It arises because you have made the bad choice of plotting the yogurt's trajectory in flat space. In curved space, the trajectory appears straight—like the great circles on a globe. Just as we do not ask for the cause of straight-line motion in Newtonian physics, we do not ask for the cause of "great-circle motion" (more accurately "geodesic motion") in curved space.

Einstein has persuaded you that even so well-defined a concept as a force—the basic cause in Newtonian mechanics—can be made to exit the scenario. In which case the notion of force as cause is ill-defined.

Loud harumphing lofts over the Academy fields; one member declares that the whole thing is an insult and stomps out. Yet the disgust is ill-placed. One does not have to resort to general relativity to make the same point. During the Third Debates the concept of the *action* figured prominently.

The action is a function from which one can derive the equations of motion for a particular problem. To be more concrete, in Newtonian physics the action is the energy of a system* multiplied by the time over which you observe the system.

Newton's laws are a "local" statement; the action formulation is a "global" statement. Newton's laws tell a particle how it will behave at the current instant of time and guide its motion step by step. Like an adolescent, there is no concern about events in the distant future. The action principle, by contrast, is more like a wise village elder, who is able to see the entire course of events. To write down the action we need to know the system's energy over the entire history of observation. This is the first key point.

The second key point is that there is no mention of forces in the definition of the action; there is only energy and time. Given that the action contains all the information about the system, forces are unnecessary. And now, listening attentively to this exegesis, you have caught an apparent contradiction. Did we not state that from the action one derives the usual equations of motion for a problem? Those equations describe motion under forces.

---

* Actually, the kinetic minus the potential energy.

Correct. Hidden in the action, like a genie in a bottle, are the equations of motion, which involve forces, so the two descriptions are equivalent. But we can write down an action for yogurt traveling in curved space and bypass the idea of forces altogether.

There is more. In yesterday's Practical Exercise, the experimentalist said to the theorist, "The success of symmetry in predicting particles demonstrates, above all else, the importance of metaphysical principles for physics. Why deny it? You believe in the action. Why should God require the action to be minimized?" And the theorist thought until the next debate.

The theorist prepares to speak.

An Argument on the Teleological Nature of the Action

However, the debate is interrupted as it adjourns to the Academy fishpond, where the topic can be discussed more vividly. Laplace has begun lecturing on flying yogurt. Normally in freshman physics, he explains, one would inform the students at what velocity the yogurt is launched, at what height above the ground, and at what angle (the initial conditions of the problem), and Newton's laws would do the rest. They tell us that the yogurt's path is a parabola, they would tell us how high the yogurt flew and at what time it would hit the ground. "As I have demonstrated in my *Méchanique céleste,*" Pierre-Simon de Laplace remarks, "once the universe is started off, all that is required is Newton's laws."

Now he contrasts it with the action. We do not give the yogurt's initial conditions. Instead we gave the initial and final state of the yogurt and its energy. We then "minimize" the action and—voilà—out pops the parabolic path.

"My good fellows," says Heron of Alexandria, the first to employ minimum principles in science, "do not enter an action against me for impiety, but I find Laplace's explanation too abstract for the common person. Let me illustrate with my own discovery." He does so, crouching by the pond. "The sunlight reflects off the pond, into one's eyes. It is easily shown that the shortest path from the sun to the eye is that one that gives the universal law of reflection: the angle of reflection from the pond is equal to the light ray's angle of incidence." To paraphrase Heron, one writes down the action for the light path, minimizes it, and gets the law of motion for the light ray.

However, as you watch this exchange, something strikes you as a bit funny. "Surely," you say, "the shortest path is the one from the sun directly to the eye."

"True enough," intervenes the modern physicist. "Sometimes there are cases when one maximizes the action. It should not really be called the principle of least action but the 'minimax' principle. Nevertheless, what Heron says is basically correct. Assuming the light ray is reflected from the pond, the ray takes the shortest path from the sun to the eye."

"But do you not grasp it?" exclaims Aristotle, who has refrained from experimental activity. "Heron's demonstration forces us to accept what is apparent: The ray understands that the goal is the eye and from all possible paths selects the shortest one to it. In the motion of the light is knowledge of its destination. Such discoveries prove to

Figure 4.3
Heron demonstrates his law of reflection. Which ray follows the shortest path from the sun to the eye?

Figure 4.4
Which is the shortest path from the sun to the pond to the eye? Simple geometry can be used to show that the shortest path (solid line) is the one in which the angles of incidence and reflection are equal.

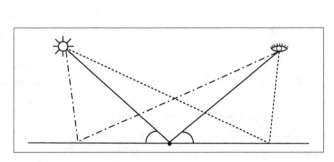

anyone who may doubt it that final causes are as important as primary causes in natural philosophy. I shall sacrifice a bull on the altar."

To everyone's surprise, the Aristotelians who were banished to Academy farmlands return from internal exile and the discussion, once thought dead, continues.

Philosophers and scientists (and discoverers of the Mississippi) throughout history have in fact often agreed with Aristotle, though we know of no sacrifices having been made. On the basis of Pierre de Fermat's work (in which he showed that minimizing the *time* of flight results in Heron's law and Snell's well-known law of refraction), Leibniz argued that final causes were essential to science. Pierre-Louis-Moreau de Maupertuis, who actually introduced the action as such to physics, maintains to the crowd that final causes imply a

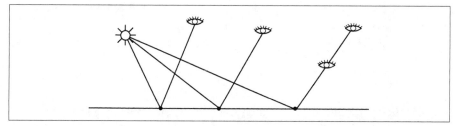

Figure 4.5
Does the sun's ray know beforehand the shortest path to the eye? No. Heron's law works for many different eye positions. That the position of the eye is arbitrary is made most clear by the final path, in which a new eye is positioned between the pond and the old eye. Obviously the law of reflection does not distinguish between these two eyes. (Strictly speaking, the rays from the sun should be drawn parallel in these diagrams.)

Supreme Being and that the action is minimized "through the wisdom of God." The great mathematician Leonhard Euler casts a fishing line and agrees, stating that, "since the fabric of the universe is most perfect, and the work of a most wise Creator, nothing whatsoever takes place in the universe in which some relation of maximum and minimum does not appear," and that final causes are as satisfactory an explanation of effects as forces.

Most twentieth-century physicists do not pause to consider whether the action principle is an example of teleology; however, this is one perennial question that appears to have an answer, the answer the modern theorist has been patiently waiting to give:

On the one hand, the action contains more than the equations of motion. Working with Newton's laws, one must put in by hand the rule for reflection (angle of reflection equals angle of incidence) to get the correct path for the light ray. The action formulation, remarkably, gives you this rule for free. On the other hand, it is not true that the light ray needs to know the position of the eye to select a path, or in general that the action principle requires knowledge of the final state of the system. As Figure 4.5 shows, one could position another eye somewhere between the pond and Heron and the same result would obtain. The same is true for the sun. Any intermediate points will do to describe the action. As a matter of fact, you can choose the two intermediate points arbitrarily close together, so close that they are really the same point, in which the global, teleological action principle becomes a local principle, Newton's laws.

The relation between Newton's laws and the action might be summarized by considering the giant multinational World Wide Widgets. The CEO is paid to say, "I want to increase profits by forty percent within five years." This is a global, teleological statement. The question is how to get there. To do this, the CEO sets middle management the task of guiding WWW to the goal. Middle management converts the global problem to a step-by-step solution. In this they are like Newton's laws—they work locally. However, if the CEO lacks vision, there is not much difference between his action and the action of middle management.

We also point out that just as the action principle has some advantages over Newton's laws, it also has some disadvantages. In order for the CEO of WWW to make a forecast, he must assume the external situation is not changing, that a meteor will not come in and pulverize the WWW headquarters. Teleology and freedom of choice are in this sense incompatible.

With that the Aristotelians again cross the road to exile and the debate adjourns for lunch.

Throughout the morning the impression followed you that something had gone astray. Were we not going to talk about synchronicity and quantum mechanics? Once again your attention drifted to thoughts of escape to the Real World and you boarded one of today's tour buses that had come through. However, to your frustration, the other visitors decided to remain in order to witness the promised discussion. In the meantime, the unfathomable academic appetite for discourse made itself manifest again and the debate continued through lunch, uninterrupted. One of the academicians returned to an issue briefly raised earlier.

The Academicians Contemplate Action at a Distance

The question was, Does an effect immediately follow a cause, or is there a time interval between them? More generally, how are forces transmitted?

The answer is not obvious. Ancient Greeks distinguished between "action by contact" and "action at a distance." When Hercules clobbered the Nemean lion with a club—as well as your own drubbing this morning—that was action by contact. Action by contact is a daily experience and readily accepted. But the Greeks also recognized that magnetism works across space, hence "action at a distance." Since that time philosophers have taken "action at a distance" to mean the ability to influence distant objects without any intervening medium. This proved to be an indigestible concept; so opposed were the Greeks to the idea that an action can take place across a void that they classified magnetism as a property of animate, rather than inanimate, bodies. The association of action at a distance with living things survives today: telepathy, if it exists, must surely be called action at a distance, since it cannot be explained as the transmission of any known forces.

Gravity in particular was a vexing issue. (The Academy's monument to vexation is marked on the standard tourist map.) Newton had successfully described the action of gravity with his famous law that gave the magnitude of the gravitational force between two objects in terms of their masses and the distance between them. Yet the formula said absolutely nothing about how the gravitational force was transmitted between the Earth and the moon. Is the force transmitted by something? Is it transmitted instantaneously? Or does gravity take time to pass from the Earth to the moon?

Like the Greeks before them, the philosophers of the seventeenth century were extremely reluctant to abandon the idea that a force needed a medium to transmit it, so reluctant that they invented all sorts of unseen mechanisms to do the job: vortices, pendulums, slings. . . Newton, amidst the swirling controversy, attempted to remain noncommittal. The intractability of gravity elicited his celebrated remark "*Hypotheses non fingo*" ("I frame no hypotheses") meaning he did not know what the causal mechanism involved in gravitational attraction was and, as far as his formula went, it didn't matter. Later, on the other hand, he called action at a distance "so great an absurdity that no man, who has in philosophical matters a competent faculty of thinking, can ever fall into it."

Most philosophers agreed and continued to search for a reasonable agent to transmit a force. In the 1840s the agent took more or less concrete form when Michael Faraday, elaborating from the work of Ruggerio Boscovich, postulated the *field*. The field. The field of battle. The field of consciousness. The karmic field. The force field. At the time there was no evidence for fields, but for those who believed that forces should be transmitted with finite velocity and not act instantaneously the concept was a godsend.

Here, the field having pulled you off the bus, you lean to the nearest academician and ask the reasonable question: Why?

During the 1840s, a number of scientists independently established the law of conservation of energy: energy could be neither created nor destroyed. But suppose body A loses some energy, which is eventually absorbed by body B. Suppose also you accept action at a distance—the energy is transmitted across the void; nothing intervenes. Then you are in a fix. During the time of transit, body A has lost energy, but body B has not yet received it. Energy has vanished into thin nothingness. Conservation of energy is violated.

"Ah!" you exclaim, perceiving the solution. "The field saves you. During the crossing time, the missing energy is stored in the field. Conservation of energy is restored!"

Now you have second thoughts: What is the field other than an imaginary construct invented to transmit forces and save conservation of energy? That is exactly what it was. Furthermore, if one believes forces are transmitted instantaneously, one needn't put oneself through such elaborate contortions.

All true, and with experimental evidence lacking in favor of either fields or instantaneous transmission, the situation for much of the nineteenth century was extremely confused, even, as the philosopher Ernst Mach said, "unintelligible." Nevertheless, in the usual way theoretical science operates, the field concept bolstered the idea that forces were propagated with finite velocity. By the mid-nineteenth century the phrase "action at a distance" came to refer to a situation in which forces were propagated without an intervening medium *instantaneously*. Such ideas fell largely out of favor when Maxwell showed that electromagnetic radiation travels with the large, but quite finite, velocity of light. After Maxwell, electromagnetism would be forever thought of as being associated with a field.

The situation became more extreme. Because Maxwell's equations allowed for the transmission of electromagnetic radiation without the presence of electric charges, the field became an entity in and of itself, independent of objects like charges or gravitational masses. The existence of independent electromagnetic fields was verified first by Joseph Henry and then by Heinrich Hertz. Today such freely propagating electromagnetic fields are known as radio waves.

The final blow to action-at-a-distance theories was delivered by Einstein in 1905 with the creation of special relativity, according to which no signal can be transmitted faster than light. And so, with the advent of relativity, physicists came to believe that all causes propagate at finite speeds. If a new planet suddenly came into existence between Earth and Mars, its gravitational attraction would not be felt until it had time to reach Earth at the speed of light.

Certainty at last! the assembled visitors sigh.

Well . . . First, what relativity really shows is that *if* causes could propagate at superluminal velocities, *then* effects could precede causes. However, we never receive letters before they are sent (given the state of the postal service, this goes without saying). Consequently it is our daily experience that causes take place before effects—our belief in causality—that leads us to reject superluminal signaling. Relativity itself does not prove causality.

Second, it turns out that electromagnetism and all other field theories mathematically show no distinction between past and future. This means they allow effects to precede causes. (A debate on the indifference of field theories to past and future is brewing for the morrow.) A radio transmitter can be turned on to broadcast radio waves, but according to standard electromagnetic theory, an equally valid situation is for all the radio waves to converge at the transmitter at the moment it is switched on; the radio waves, like LaSalle, seem to know to converge at the transmitter in order for it to be switched on.

But we never observe such a situation in nature. Is this not, you might wonder, recalling the debates of two days past, another case where mathematics seems to inadequately describe the real world? Perhaps, and it is another case where the world appears to be asymmetrical. Physicists term solutions in which effects precede causes "advanced," and call the usual situation in which causes precede effects "retarded" (meaning only "delayed"). We reject advanced solutions on the basis of our notions of causality, not for any reason built into the theory.

At this moment, a lone academician interrupts the exposition to insist that the authors have lied: relativity failed to kill off action-at-a-distance theories. We confess. Since 1905, a few Academy diehards, notably Hugo Tetrode and A. D. Fokker, and John Wheeler and Richard Feynman, attempted to make action at a distance consistent with relativity. We do not go into these efforts, pointing out only that, like the older action-at-a-distance theories, they implied a violation of energy conservation. Furthermore, in such theories action and reaction between bodies A and B are assumed to take place at the same *time.* If Fred sends out his aura, which later acts on Ethel, then for her aura to react on Fred at the same time, she must emit it *before* receiving his. Thus effects precede causes, and Fred and Ethel are very much like LaSalle about to discover the Mississippi.

For these reasons and others, even "modern" action-at-a-distance theories have found eternal rest in the lost theory graveyard of the Congo.

The Attraction of Advanced Effects Leads the Academicians to Synchronicity

As the evening set in, and the lone academician retired, you perceived that another bit of certainty had been achieved: fields exist in nature, causes in the physical world do not propagate with superluminal velocities. This is true. You also perceived that physicists do not lightly entertain the notion of causality violation and their lack of imagination disappointed you.

Lack of imagination or no, a similar situation exists in other walks of life. Generally we like to assume we can interfere with things, improve our lot. Most people prefer not to think that something in the future caused our present behavior. It is a bit like the fable in which the jackal says it will eat the lamb. "What mischief!" exclaims the lamb. "What have I done?" "You are muddying the water I drink," answers the jackal. "How can that be?" asks the lamb., "I am downstream from your drinking spot." "Ah," replies the jackal, "you forget your parents were here before you and muddied it upstream." With such advanced effects in force, the lamb is having difficulty improving its lot.

In a more optimistic vein, from John 9:1 we learn that as Jesus leaves the temple, he passes by a man blind since birth. His disciples ask, "Master, who did sin, this man, or his parents, that he was born blind?" And Jesus answers, "Neither hath this man sinned, nor his parents, but that the works of God should be made manifest in him." In other words, he was born blind in order that Jesus could later cure him of his blindness.

All biblical prophecies share with the story of the blind man the assumption that God arranges things in advance, like the radio waves converging on the transmitter. Because such events are considered improbable, they are reserved for special occasions. Some people believe they are reserved for occasions so special they never happen.

And so, like physicists, people tend to regard advanced effects with great skepticism if not outright denial. On the other hand, from a psychological point of view, advanced effects have their attraction. Certainly there is a large element of future orientation in human behavior. Transformational counselors admonish us to choose our goals and work toward them. In which case, is the goal the cause or the effect?

It is not difficult to understand why people prefer future orientation. As today's debates have shown, causality is not an observation, but a way of describing how things happen. Usually it is a very successful way and all the doctors and lawyers agreed that science and society rely on it. But, many of the New Age tourists are pointing out, it is a reductionist approach; all things future depend on the present and on the past. Exactly. In this respect Newtonian causality does not allow for anything really new to happen; it does not allow for *creativity*. This is why people rebel against the notion of cause.

However, like their attitude toward advanced effects, people's attitude toward causality appears schizophrenic. We rebel against determinism, seeking freedom, but when we find something that appears unmotivated, we immediately search for a cause. This is true in both East and West. For example, if genuine creativity is disallowed in the Newtonian universe, a description of creativity must transcend a causal description. But when confronted with great creators like Mozart or the mathematician Ramanujan, whose methods appear inexplicable, the first thing that we do is attempt to assign a cause. Typically, people invoke divine grace.

Similarly, despite the Newtonian universe's determinism, almost any realistic situation is far too complicated to be analyzed directly in terms of causes. In those situations we abandon causality and talk about probability. Whether we are buying stock on the market, flipping a coin or making a weather prediction, we speak in terms of the odds. This is one way we attempt to circumvent the shackles of causality.

"One moment," objects Laplace. "Probability is nothing more than a measure of our ignorance. Our feeble intellects are unable to perceive all the influences operating in a particular situation, and therefore we declare the outcome to be unknown, or that one outcome is no more improbable than another. This is illusory. An intelligence vast enough to see all influences and submit them to analysis could have no doubt as to the outcome of any particular situation."

We will take up this contention tomorrow. For the moment we are merely pointing out that if people do not know how things are going to turn out, they have a sense of freedom. Yet, even when they know the astronomical odds against winning the lottery, people still buy tickets. Either they "feel lucky," which is a way of saying that there is a hidden cause at work, or they buy ticket numbers corresponding to their birthdays, which is also saying there is something mysterious at work.

"Superstition," you interject.

Well yes, superstitions are a way of assigning causes to apparently uncaused events. In the Kannada language of India, for instance, the word for luck is *adrishta*, meaning "unseen." Luck is not chance; if we were aware of other things, we would find causes behind luck. So even in the mystical East they search for causes. New Agers, though they may deny it, do as well: today a poll taken among Academy visitors showed that many of them attribute karma and predestination to Indian philosophy, although as your head experienced in the morning, it is really the Western universe that is completely deterministic.

Much the same concerns coincidences. What is a coincidence? Usually we think of coincidences as being unusual events that occasionally pop up, due to the odds turning in their favor. You walk down a street in New York and meet a friend you haven't seen in twenty years. What a coincidence! What are the odds of that happening? We won't calculate the odds, but if they are truly coincidences, they have no obvious cause.

Laplace would of course argue that coincidences are merely a measure of our ignorance. An intelligence vast enough to see all forces at work would be able to calculate the odds; indeed there would be no odds. The vast intelligence would be able to predict precisely whether something would happen or not and the notion of coincidence would not exist. (Laplace nods from the palazzo with a turkey drumstick in his hand. We have correctly represented his position.)

Most people are closet Laplacians. Suddenly a hush descends among the New Agers in the dining room. What an insult. How dare we? It is true. Most people do not like to leave things without explanations. Many of us have bought a book for no reason whatsoever, only to find years later that it contains a needed quotation. Discovering a relevant quotation in a randomly chosen book is an experience so common that Arthur Koestler called it an example of the "library angel" in action. That is one explanation. Another is perhaps the well-known phenomenon (at least to authors) that when you are writing a book, everything becomes remarkably relevant. Another might be the subconscious mind at work, guiding you to a forgotten passage. All are attempts to assign a cause to the event, *adrishta*.

But, you now interrupt, does substituting the word "library angel" or "subconscious" for "cause" explain anything? A good question. Does it? Another explanation for such coincidences is synchronicity.

At last, the cry goes up.

What Is Synchronicity?

What happened next has never been clear to you or to us. At the very question What is synchronicity? the debate spontaneously reformed at the central plaza, where the opposing camps began to jostle for position. Many tourists also took a keen interest and the crowd grew large. Everyone agrees that the first speaker was Carl Jung himself, who attempted to explain his theory of synchronicity, which he had designed in turn to account for the meaningful "coincidences" in our lives. However, no one can agree on exactly what he said, even less on what he meant, or whether he even finished his exposition.

We can say that according to one of Jung's own letters, it was a 1912 lecture by Einstein on the relativity of space and time that started him thinking about a possible connection to the psyche. But it was not until he came in contact with Eastern yoga that he formulated his idea of synchronicity. On the plaza, Jung himself seems to have had difficulty in elucidating the concept: "It seems as though time, far from being an abstraction, is a concrete continuum which possesses qualities or basic conditions capable of manifesting themselves simultaneously in different places by means of an acausal parallelism, such as we find, for instance, in the simultaneous occurrence of identical thoughts, symbols, or psychic states."

The remarks catapulted many of the day's visitors into raptures, left equally many scratching their heads. For the latter, recall that causes, according to relativity, cannot be transmitted at superluminal velocities. Hence, in relativity, you must be very careful about what you mean when you say two events took place "at the same time." If an event takes place "now" in a galaxy far, far away, we cannot know about it for millions of years. Does it make any sense to talk about an event taking place "now" when we cannot have had any knowledge of it?

Synchronicity was an outgrowth of the collective unconscious and Jung's attempt to get around the strictures of relativity in accounting for "meaningful coincidences"—"meaningful" referring to events that are attached to some powerful psychic image. The archetypal synchronistic event took place when Jung was treating a patient who "had an impressive dream the night before, in which someone had given her a golden scarab—a costly piece of jewelry." While the patient was still telling Jung of her dream, a large insect flew into the room. "It was a scarabaeid beetle or common rose-chafer, whose gold-green color most nearly resembles that of a golden scarab. I handed this beetle to my patient with the words, 'Here is your scarab.' This experience punctured the desired hole in her rationalism." The treatment then continued with satisfactory results.

This is a very famous story, but even before Jung has finished it, right-wing academicians attack. "Why is it so unusual for a common beetle to fly into a window?" someone shouts from the crowd. "During summer it happens all the time."

The vocal reactionary has of course raised the main objection to synchronicity, that such events may be dismissed as "mere coincidence." But dedicated synchronists to your left are ready with an answer: "Synchronistic coincidences are too elaborate to be explained by chance, you fascist pig!"

Arthur Koestler, for example, recounts the typical synchronistic experience of a publisher named Simmons who, after twenty-five years in the business, was forced to sell an unsuccessful book to a paper mill for pulping. He asked his manager for the name of a mill; the manager did not know of any. At that moment a boy from the warehouse walked in and suggested Phillips Mill, which was near his home. Simmons asked the receptionist to call them and she replied, "The sales representative is here." He had just walked through the door.

You, as well as most of the visitors, have experienced enough synchronistic coincidences of your own to find such arguments persuasive, and therefore above the increasing noise you agree that it would be impossible to calculate the odds of such a thing happening. Perhaps, but the right wing replies, "Our notion of coincidence is extremely one-sided. It concerns things that happen, not things that don't happen."

"Of course," you shrug. "What else could it concern?" But the right has expressed a rare point. You are walking around the upper West Side in Manhattan and don't run into anyone you know. You don't call the fact that nothing happened a coincidence. But suppose we told you that your entire high school graduating class lived in the neighborhood. Then the odds might really favor your running into someone and it would be a coincidence that you didn't. Exactly the same mathematics governs each case, but about the only time we count things that don't happen as coincidences is when a soldier survives a ferocious campaign, and then we call it a miracle. All for good reason: generally we are unaware of nonevents. If we could be more evenhanded about such things, our perception of what is likely and what isn't would undoubtedly be far different from what it is now.

The blood of the left is now boiling. Allan Combs and Mark Holland, devoted synchronists, quote from their book, *Synchronicity:* "The probabilistic interpretation of synchronicity . . . is not dealt with in the main text . . . . We wish to avoid the tiresome academic writing style that one answers one's opponents' objections on every page."

But now, as the Knowledge Marker begins to fluctuate wildly, the right howls in protest: "If they don't deal with counterarguments, they are merely assuming their own conclusions." We agree. If nothing else, it is odd that Combs and Holland attempt to bring science (in particular quantum mechanics) to bear on synchronicity when their entire approach of assuming one's conclusions is antiscientific. This is especially true of a subject such as synchronicity, for which the evidence relies entirely on highly debated coincidences.

It may have been the remark of Combs and Holland that touched off the general mêlée, but no one is certain. At about this time the right managed to get another objection heard: it is the synchronists who decide what is meaningful and what isn't. For a coincidence to count as synchronistic it must, according to Jungians, have psychic res-onance. Yet what is meaningful to one person may not be meaningful to another. It is told that a man Ajamila had a son named Narayana, which also happens to be one name of Lord Vishnu. As Ajamila lay dying and messengers of death approached him, he cried out in terror to his youngest son, "Narayana! Narayana!" Hearing the cry, Vishnu sent his retainers to Earth. They vanquished the messengers of death and took Ajamila to Vishnu's kingdom, where he lived eternally thereafter.

Now, this was a pretty silly thing for Lord Vishnu to have done, mistaking Ajamila's cry to his son for a plea for salvation. By the same token, in detective stories it is often the case that a chance remark by one of the characters sparks in the mind of the detec-tive a solution to the problem. Also, it is believed that in the mere presence of a guru, the disciples become enlightened. The guru doesn't need to *do* anything. What is mean-ingless to one person may be quite meaningful to another.

Is this synchronicity? Looking back on the evening, at that stage you were beginning to doubt you knew what synchronicity was. Truthfully, as they cited the early debates on scientific terminology, many scientists this night refused to allow synchronicity to be called a theory because of the lack of clear-cut evidence. "Where is the data? Where are reproducible experiments?"

Here is a good example in which terminology lends credence to a concept which some feel it does not deserve. Let us at least agree that if you have a dream in which a cousin dies (a common type of synchronistic dream) and the next morning you dis-cover that the cousin did die during the night, then we have one of two choices. Either it was a coincidence, or there was a deeper explanation. If, in the latter case, some psy-chic substratum of the cosmos is invoked, then it must evidently transmit information in a way that violates known physical laws. Synchronists call this "acausal," by which they mean that one can have instantaneous knowledge of an event happening at great distances, which violates the principle that no information can be transmitted faster than light.

However, "acausal" properly means "without cause." The bent trade routes on flat maps that came up for discussion earlier in the day are in the true sense acausal—without physical cause. Jung, to the contrary, is attempting to find a cause; it is simply not a cause that would be accepted by most scientists. In fact, if in the heat of the evening, visitors had not forgotten the earlier discussions, they would realize that syn-chronicity is a kind of action-at-a-distance theory: the causes, whatever they may be, are transmitted instantaneously.

What Does Synchronicity Have to Do with the East?

As some observers remember the evening, things were still relatively calm at this stage. When the hard core objected that synchronistic events are "mere coincidence," the Jungians attempted to bolster their arguments by turning Eastward, in particular to the philosophy of the *I Ching*. The *I Ching* probably evolved between the seventh and third centuries B.C. from peasant omen compilations.[†] Originally used for divination, the *I Ching* consists of sixty-four hexagrams, which, like the tarot cards of the West, have symbolic significance. The difference between the *I Ching* and Western divination texts is that centuries of commentaries gave the book an added ethical and cosmological dimension, until it came to be regarded as a repository of the basic principles governing all natural phenomena. In some sense, the *I Ching* became very much like Newton's laws in the West, sacrosanct, except that the outcome was exactly the opposite. Whether one regards the *I Ching* as the cause or the effect of Chinese thinking, faith in the book was arguably one of the most important reasons that science as we know it did not develop in China. Until even a century ago, the Chinese claimed that truths about light, heat and electricity were held in the hexagrams of the *I Ching*.

Given that the *I Ching* is supposed to be a kind of cosmic telecommunications switchboard, it is not surprising that Jung came to connect it with synchronicity. As he said, before some people remembered things going out of control:

> The manner in which the *I Ching* tends to look upon reality seems to disfavor our causalistic procedures. The moment under actual observation appears to the ancient Chinese view more of a chance hit than a clearly defined result of concurring causal chain processes. The matter of interest seems to be the configuration formed by chance events in the moment of observation, and not at all the hypothetical reasons that seemingly account for the coincidences. While the Western mind carefully sifts, weighs, isolates, selects, classifies, isolates, the Chinese picture of the moment encompasses everything down to the minutest nonsensical detail, because all of the ingredients make up the observed moment.

There is something very attractive in imagining an event as the outcome of the momentary configuration of the entire universe. This is obviously very different from the Newtonian, or even Einsteinian, view of the cosmos, and to believe it one must throw out virtually everything that we have said about causality today. But many people who reject the dualistic philosophy of the West, that science and psyche are split, nevertheless see a similarity between Western physics and Eastern philosophy. Capra and the *Tao of Physics* have a pedigree that can be traced to Jung and synchronicity, if not to Erwin Schrödinger himself.

Can Quantum Mechanics Explain Synchronicity?

Some visitors have claimed that the melee did not begin until Jung undertook to quote from his 1949 thoughts on the *I Ching* and quantum mechanics. After mentioning how curious it was that such a gifted and intelligent people as the Chinese never developed what we call science, and reiterating that Western science is based on the principle of causality, which is considered to be an axiomatic truth, he then took the fateful (and to many, fatal) step. "But a great change in our standpoint is setting in. What Kant's Critique of Pure Reason failed to do, is being accomplished by modern physics. The axioms of causality are being shaken to their foundations: we know now that what we term natural laws are merely statistical truths and thus must necessarily allow for exceptions."

At that moment, several academicians within your eyesight dropped dead of apoplexy. Their reaction was perhaps extreme. Unaware of the riot breaking out around him, Jung continued at length:

> This assumption [that the hexagram of the *I Ching* was an indicator of the moment] involves a certain curious principle that I have termed synchronicity, a concept that formulates a point of view diametrically opposed to that of causality. Since the latter is a merely statistical truth and not absolute, it is a sort of working hypothesis of how events evolve one out of another, whereas synchronicity takes the coincidence of events and space and time as meaning something more than mere chance, namely, a peculiar interdependence of objective events among themselves as well as with the subjective (psychic) states of the observer or observers.
>
> The ancient Chinese mind contemplates the cosmos in a way comparable to that of the modern physicist, who cannot deny that his model of the world is a decidedly psychophysical structure. The microphysical event includes the observer just as much as the reality underlying the *I Ching* comprises subjective, i.e. psychic conditions in the totality of the momentary situation. Just as causality describes the sequence of events, so synchronicity to the Chinese mind deals with the coincidence of events.

To interpret, Jung makes two main points regarding science: The first is that causality is "statistical." The second is that, as in ancient Chinese philosophy, to the modern physicist the world has a "psychophysical" structure. In the first case he is referring to the famous probabilistic nature of quantum mechanics; in the second case he is apparently referring to the well-known Copenhagen interpretation of quantum mechanics, in which the observer of an experiment cannot be separated from the experiment's outcome.

Regarding the former, Jung simply seems to have misunderstood quantum mechanics. Yes, quantum mechanics has a statistical aspect. Does that mean that the laws of nature must allow for exceptions? No. Regarding the latter, Jung is passing on a particular interpretation of quantum mechanics. But that is an interpretation, not quantum mechanics, and it is an interpretation that has fallen increasingly out of favor among physicists.

The authors would gladly explain all this in greater detail but by now it is impossible. On the left, descendants of Jung continue to claim that quantum mechanics is acausal, that "unlike the general theory of relativity, quantum theory does not deal with the existence of objects but with actions or events," that quantum mechanics has "conclusively established the veracity of the spiritual notions," that quantum mechanics deals with waves that "cannot be called waves in space and time at all," that particles in quantum mechanics can exchange signals "faster than the speed of light . . ."

On hearing such statements, you are mightily intrigued and beg for more, but the physics militia has retrieved its firearms and it appears that a general shootout is imminent. Nothing like this has ever been seen at the Academy before. The authors attempt to invoke Robert's Rules of Order but to no avail. Ignored by all sides, they have no choice but to step in and impose a cease-fire. Further discussion of quantum mechanics will be postponed while we enforce a cooling-off period and decide what to do next. An exceptionally mild-mannered academician suggests that today's subject matter is equally applicable to the most fundamental mystery in physics, why time itself moves forward, and that we might turn to that debate instead. This is not a bad idea. But you join in with the general protest from the visitors: all of you were certain of a quantum-spiritual-synchronistic connection and demand to see it resolved.

A Practical Exercise: Can Coincidences Be Computed?

While the authors consider the most prudent course of action, they propose a practical exercise. One of the persistent claims of synchronists is that synchronistic coincidences are so improbable that the odds cannot be calculated. The statement is not entirely obvious. The authors are going to take an event that happened to one of us not too long ago and each of us is going to independently estimate the odds. Then we will compare results. If something like this has ever happened to you, you might try it yourself.

In the winter of 1996 I was staying at my mother's house near Princeton, New Jersey. One afternoon I found a note she had taken off the answering machine and left for me. Apparently a Marsha M* had called from Vermont, asking my help in tracking down Fang Lizhi, a well-known physicist who had been China's leading dissident and who was now living somewhere in the United States. Because I had recently been working in Vermont and had been involved with Soviet dissidents in the past, I assumed that Marsha M* had called my previous employer who had passed on my name.

The next day I reached Marsha M*; the conversation went something like this:

"This is"—I gave her my name—"returning your call."

"Who?"

I repeated my name. "You phoned me yesterday."

"I did? What's this about?"

"You tell me."

"What organization are you with?"

"I'm not with any organization. You phoned regarding Fang Lizhi."

"You're not with the Institute for Higher Learning?"

"No."

"I phoned the Institute for Higher Learning."

"There's no such organization in Princeton."

"Yes, there is. I called the operator in Princeton and she gave me your number."

Now things became clearer. In attempting to track down Fang Lizhi, Marsha M* decided to call Princeton, where the famous Institute for Advanced Study is located. But she could not remember or did not know the name and asked for an Institute for Higher Learning. My mother, a psychotherapist, runs a small Institute for Experiential Learning and Development. The operator mistakenly gave Marsha M* my mother's number; she called and, without speaking to anyone, left a message on the machine, which my mother assumed was for me. I recognized Fang's name and because of my involvement with dissidents in the past, also assumed the message was for me.

"But was this a synchronistic coincidence?" you ask.

"Umm, according to Jung's original descriptions, probably not. There were no dreams involved or archetypal images. It seemed meaningful in the sense that I am a physicist, had been working in Vermont, and had a connection with political dissidents. It does seem as synchronistic as some of the stories reported by later writers, like the one about the paper mill, and there were certainly at least five places where the chain of mistakes could have gone wrong. But let's suppose it is merely a coincidence. Can we estimate the odds? If you were trying to track down a physicist, what are the odds you would call Princeton?"

"Well. . . oh, I don't know."

"All right. Princeton has a high concentration of physicists, surely at least 1 percent of all the academic physicists in the country. So let's say the odds are .01 to call Princeton. Having called Princeton, there are basically only two institutions to choose from, so let's give a probability of .5 that Marsha M* decides to call the Institute. But she doesn't know the name. What are the odds of her getting my mother's place instead?"

"Oh please, this seems impossible to guess."

"Not really. Princeton is a small town. I doubt there can be more than ten or so institutions with similar names. So shall we give it a probability of .1? Does that sound reasonable?"

"I suppose," you answer with reluctance.

"OK, having gotten my mother's place, what are the odds that a physicist would be living there?"

"There aren't many physicists. Maybe one out of a thousand?"

"It's probably a bit higher, and in the Princeton area they're entirely run-of-the-mill. I would say .1 but be willing to go as low as .01. Agreed?"

"OK. But how many would recognize the name Fang Lizhi?"

"He's pretty well known among physicists, if only for his dissident activities. I would say at least 50 percent of all physicists would recognize the name. So what does that give us?"

Quickly we multiply all the numbers together and the upper estimate gives a probability of 2.5×10^{-5}, or 1 out of 40,000, and the lower estimate is 1 out of 400,000. I certainly can't get more than 1,000 phones calls a year. So I'd expect something like this to happen maybe once in my life and although I've experienced other strange events, I don't remember ever having gotten a phone call like that."

The second author seemed to remember that Fang had been at the Institute in Princeton at one time and that Marsha M* probably knew this. So he gave her a high 1 in 5 probability of calling Princeton. He then assigned a 1 in 20 (.05) probability that she would mistake the name. That she would get the wrong Institute he gave a probability of 1 in 3. That a physicist would get the message, .1; that the physicist would recognize Fang's name, .5. Although author number two broke things down somewhat differently than author number one, the result was about the same: roughly 1 in 100,000.

As a check we phoned Princeton for a listing of institutes. There are precisely ten and only two have the word "learning" or "study" in their titles. The fact that both of us independently got the same result, that our estimates were consistent with the facts, and that our answer was large enough that such a thing *might* happen once in a lifetime indicates that perhaps one should not be too eager to dismiss coincidences.

After hearing these results you agree to cogitate on the matter but nevertheless insist that quantum mechanics be discussed tomorrow.

In Which the Fourth Debates Are Summarized

The Directors of Sports and Recreations, agreeing to consider your request, consulted the *I Ching* to determine whether the topic of tomorrow's debate would be quantum mechanics or the direction of time. After dividing the yarrow stalks, the hexagram was *Tun,* retreat. "Retreat is not to be confused with flight. Flight means saving oneself under any circumstances, whereas retreat is a sign of strength." It appears that tomorrow's topic will not be quantum mechanics.

Having no choice but to hang around for the outcome of the struggle, you reviewed what had been said during the course of this memorable day.

Everyone agreed that the notion of cause requires change. If nothing changes in the system under discussion, talking about causes and effects would be irrelevant. One of the main questions was whether the universe operates by causes. Regardless of the answer, academicians agreed that it is impossible to measure causes directly; we can speak only of "correlations" among events. One reason that events appear to have causes is that as a rule we isolate effects and speak of multiple causes producing a single outcome. But if we enlarge our view of the entire system then it becomes less clear that anything has changed and less clear that there is any real distinction between cause and effect.

Newtonian physicists simplified things by associating causes with forces; no force, no cause. The Newtonian universe was deterministic, meaning that it was rigidly causal; each set of events determined exactly the next set of events. In principle, if you knew the initial state of the universe—meaning the positions and velocities of all the particles—then the future behavior of the entire universe was fixed. However, in such a universe causality, paradoxically, cannot be determined because any "experiment" you set up, as well as its outcome, are also predetermined. Teleology made an appearance as well when natural philosophers introduced the "principle of least action," which seemed to imply that final causes operate in physics. Although many people still believe this to be true, here was one case where we reached a definite conclusion: no.

Nowadays fields have replaced forces as the medium by which causes are transmitted, but field theories admit instances in which effects precede causes. Because such "advanced" effects are not observed in nature, physicists discard them, although nothing in the mathematics requires them to do so. Here is another example when mathematics apparently does not correspond to the real world and another example of nature's asymmetry.

When situations are too complicated to follow the trajectories of individual particles, we resort to probability and statistics. Laplace argued that this is not a fundamental restriction, merely a matter of ignorance. Such ignorance results in coincidences. However, in the Indian Vaiseshika system of philosophy, *adrishta*, the unseen, is an essential element of cosmology. In sympathy with such ideas, Jung introduced his concept of synchronicity in order to explain meaningful coincidences. As the unrest began, Jungians pointed out some similarities between synchronicity and both Chinese philosophy and quantum mechanics. Jung gave some hints that the classical conception of causality is no longer valid in quantum mechanics. The authors warned about accepting Jung's opinion on these matters before we could discuss the matter more fully.

After that things went out of control.

DOES TIME GO FORWARD?
Past and Future

The law that entropy always increases—the second law of thermodynamics—holds, I think, the supreme position of the laws of Nature. If someone points out to you that your pet theory of the universe is in disagreement with Maxwell's equations—then so much the worse for Maxwell's equations. If it is found to be contradicted by observation—well, these experimentalists do bungle things sometimes. But if your theory is found to be against the second law of thermodynamics, I can give you no hope; there is nothing for it but to collapse in deepest humiliation.

—*Arthur S. Eddington*

Michele has left this strange world just before me. This is of no consequence. For us convinced physicists, the distinction between past, present and future is an illusion, although a persistent one.

—*Einstein*

The Question

When Siddhartha, in the seventh year of his quest for enlightenment, folded his legs beneath the sacred fig tree, he prepared himself for meditation and took a solemn vow: "Come what may—let my body rot, let my bones be bleached by the sun—I will not get up from here until I have found the way beyond decay and death." Thus, Siddhartha passed into deep meditation, determined to achieve nirvana. The tempter Mara sent to Siddhartha his daughters of unearthly beauty, then assaulted him with his armies of lust, doubt, desire for fame and fortune, hypocrisy and cowardice, and, finally, Mara appeared himself. But Siddhartha defeated all these temptations, and on the eighth day of his meditation, he passed into the state of enlightenment. In this state he recalled his past lives and in particular the time he met the Buddha of the previous age, who blessed him and assured him that he would be the Buddha of the age to come.

On the morning of the Fifth Debates, you awoke with yesterday's disturbances on your mind, but unlike the Buddha you had no accurate vision of what was to come in the next cycle of existence. One of the most fundamental and depressing facts of reality is that for those of us who remain unenlightened, we remember the past but not the future. This is true despite the White Queen's remark, "It's a poor memory that only works backwards."

But not only memories distinguish past from future. Academicians grow younger only in science-fiction films or cosmetic ads, and then we tend to attribute it to vitamin E or Industrial Light and Magic. Eggs can be scrambled but not unscrambled. A piece of music played backward strikes us as aural nonsense, except perhaps for the music of Hector Berlioz, which sounds pretty much the same in either direction.

The vast majority of processes, in fact, look extremely peculiar when run backward. Peculiar? No. When we see an omelet unscramble itself into an egg, it strikes us as so absurd that we immediately assume we are watching a movie being run in reverse. On this morning of the Fifth Debates, when you were jolted from fitful sleep by the dawn tolling of the clock, it probably struck you that we tell time by such *irreversible* processes—phenomena that proceed inexorably in one direction—which we term forward. If processes were *reversible*—if eggs unscrambled or we could see Buddhas past and future—we would have no right to say time goes forward.

But, in fact, according to Newtonian physics, electrodynamics, relativity and quantum mechanics, all natural processes *should* work equally well forward and backward; at a fundamental level, all phenomena are *reversible,* not irreversible. Furthermore, yesterday we discovered that the same theories admit advanced effects which precede causes—yet we never see such things in nature. And so we have a great paradox, perhaps the greatest in physics: all our fundamental theories are symmetric in time, but despite their great successes (and despite any claims of the symmetry faction two days ago), nature exhibits a clear asymmetry between past and future.

Since you wandered into the Academy, the entire debates have centered around the question, Is the universe describable? Yesterday's variant of the question was, Why do things happen? Not much would happen and not much could be described if there were no distinction between past and future. Which brings us to the question the academicians have already begun murdering each other over: Why is there an arrow of time?

Despite the fact that we have delayed quantum mechanics by imposing martial law, virtually all the visitors find the question intriguing enough to withhold protest. As usual we begin with background research.

In Which the Paradox Is Clarified

This morning you were able to find only one tourist, from Southern California, who believed that time does not move forward. A "cosmic visionary," she claimed, like the Buddha, to see all events, past, present and future, at once. Apart from her, everyone at the Academy was willing to accept the evidence of direct experience, that time has an apparent direction. The question is why. And here there is bitter disagreement.

The dispute began, as is so often the case, with Aristotle, who declared, teleologically, that "the natural state of motion is rest." Aristotle was taken in by friction. When you roll a ball on a floor, sure enough it tends to stop, due to frictional forces. This is a garden-variety irreversible process; except in *America's Funniest Home Videos* you never see balls spontaneously speed up from rest and hit where it hurts. The reason most everyday processes are irreversible is because they are governed by friction. A few days ago we pointed out the lack of a microscopic theory of friction. So you see this small, insignificant gap in the physicists' description of the world has major consequences. This gap is also not without reason.

It took the monumental achievement of Galileo, Kepler and Newton to banish friction to the realm of illusion. Galileo rolled objects on inclined planes and concluded that if friction didn't interfere, an object would roll indefinitely. He generalized this conclusion to what is today called the law of inertia: any object unaffected by outside forces will continue to move at a constant velocity. In yesterday's debates the law of inertia figured as the proclamation that constant velocity is to be relegated to the class of uncaused things, of things that require no explanation. Aristotle would be quite astounded at this news, and protest feverishly, but no one would listen.

Newton confiscated Galileo's law of inertia as his first law. Newton's laws, in particular the second, may be regarded as the mathematical rules that describe an object's path, or trajectory, in space. Yesterday, academicians agreed that Newton's laws were deterministic—once God started the universe off with a bang, Newton's equations determined everything that was to come thereafter. However, the laws manifest a stranger property. The second law is perfectly *reversible,* or in the language of the debates on symmetry, time-symmetric. Mathematically, this simply means that the second law does not distinguish time forward from time backward. If you shoot a movie of the planets in orbit around the sun, someone watching the movie later on cannot tell whether it is being run forward or backward—the planetary trajectories look as sensible in one direction as the other. The same time symmetry is embodied in all the modern theories we have discussed: electrodynamics, relativity, and so on. Their equations contain no arrow of time.

Having heard out this exposition, a few modern academicians object to the statement that Newton's laws and the others allow prediction with absolute precision. We beg that this argument be deferred until later in the day.

But at this hour, in the morning, you need to remember that the solar system, which begat modern physics, is frictionless and to a high approximation reversible. Like the enlightened Buddha it is indifferent to past and future. Most processes, such as aging or eggs scrambling, are quite irreversible and a movie of them run backward appears ridiculous.

Once more we arrive at the central paradox: since Newton proclaimed them, scientists have believed his laws are at the bottom of everything. But if the individual particles in an egg obey Newtonian dynamics, then at a microscopic level friction does not exist, and their trajectories should be time-symmetric. Why do eggs only scramble and not unscramble? Why was Sophocles correct when he wrote, "The immortal gods alone have neither age nor death. All other things almighty Time disquiets."

"Hmm," you say.

You are not alone in being stopped in your tracks. For one hundred years scientists have debated this question without resolution.

Will today be different?

Is Entropy Possible?

The debate caught fire in 1872 when the Austrian physicist Ludwig Boltzmann announced that he had derived the second law of thermodynamics from Newtonian mechanics. The second law is the infamous law of entropy increase, which says specifically that the entropy in an isolated system (a system that exchanges neither matter nor energy with the outside world) never decreases.

Very likely you have heard the word "entropy" so often that you either are convinced you know what it means or have given up the attempt to understand it. Deepak Chopra indirectly refers to entropy when he says that if you know how to expend energy in an efficient way you can create any amount of wealth, and that when you seek power and control over others you spend energy in a wasteful way. When you harness the power of love you can use energy creatively. In this Chopra is supported by prophet Jeremy Rifkin, who once declared, "Love is not antientropic, as some would like to believe. If love were antientropic, it would be a force in opposition to becoming, for the entropic flow and becoming go hand in hand."

Already several academicians are being carried off to the medical center, and we recall the discussion of language barriers in the First Debates. In several days we will need to come to grips with the vexing issue of scientific metaphors. This morning we can say that the idea of entropy originally arose not from the study of love or creation of wealth, but from the study of steam engines. By the mid-nineteenth century scientists realized that regardless of how well built or how efficient a steam engine was, during its operation some energy was always lost as waste heat. For this reason perfect engines—perpetual motion machines—are impossible. Rudolph Clausius quantified the observation by introducing a quantity, which he termed entropy, to measure the energy wasted in any conversion process. Clausius intentionally chose the word "entropy" to resemble "energy" because the two quantities were similar. Perhaps too similar; scientists often speak loosely of entropy as waste energy and to most people the word "entropy" has become synonymous with waste and decay.

However, although energy and entropy are similar, there is a crucial difference. Energy is conserved. Entropy is not. The second law of thermodynamics enshrines the observation that in any energy-conversion process entropy increases. Only frictionless systems, like the solar system, which do not dissipate heat, show no entropy increase. Such systems are referred to as "reversible." Systems like steam engines, or virtually anything else in the real world, generate entropy and are termed "irreversible."

You may be puzzled here by the use of the terms "reversible" and "irreversible." A moment ago we used the latter to describe systems whose time-reversed counterparts could be distinguished from their time-forward counterparts. This is not coincidental. The rolled ball is slowed down by friction; in the process entropy increases. That is why the film run backward looks different. To be sure, by the 1860s it was clear to everyone that the second law was quite different from all the other laws of nature. The other laws are time-symmetric. The second law gave a direction of time.

One further lexigraphical clarification. Physicists often use the term "thermodynamic systems" to refer to macroscopic systems, like containers of gas or steam engines, which contain many particles, as opposed to microscopic systems, which contain few particles. Thermodynamic systems are those that manifest irreversible behavior. In that case, the two main pillars of physics—thermodynamics and Newtonian mechanics—rest on apparently incompatible assumptions. According to reversible Newtonian mechanics, if it can even be defined, it cannot increase.

Boltzmann's Solution Is Presented and Viciously Attacked

It was precisely because the second law of thermodynamics alone of all the laws of nature contained an arrow of time, resulting in the complete incompatibility between thermodynamics and Newtonian mechanics, that Boltzmann attempted to link the two. More precisely, he set out to prove that the second law could be derived from Newton's laws.

Boltzmann began by considering a container of gas. Because even an ordinary bottle of gas contains a huge number of particles, of order 10^{23} molecules, it is difficult to follow the particles individually. You tend to treat them in bulk and deal with averages. Nevertheless, Boltzmann did his best to peer into this gas, as if with a mental microscope, and examine its behavior due to the motion of individual particles moving and colliding according to Newton's laws. His original approach was to define a quantity H, which was related to the position and velocity of a typical particle in a sample of gas. He then asked how H would change as particles collided, and he managed to "show" mathematically that H always decreased. Hence the famous "H-theorem" of 1872. Finally, Boltzmann found that he could identify H with the negative of the entropy, which therefore always increased. Voilà. Boltzmann had explained the great riddle of why time goes forward.

Now wait a minute, you protest mightily. Even amateurs can detect a swindle. How could Boltzmann derive time-irreversible behavior from time-reversible laws?

Very good. This is precisely the question that was raised by academician Joseph Loschmidt, one of Boltzmann's colleagues, and this is precisely the question that has divided physics ever since. Already the Academy physicists are arming themselves and taking up positions on the opposite side of the central plaza. The Knowledge Marker has suddenly plunged to zero.

As you perceived, it is a mathematical impossibility to derive time-asymmetric behavior from time-symmetric laws. This is much like setting a positive number equal to a negative number or equating apples and avocados. What did Boltzmann do?

Boltzmann's H-theorem confused people then and confuses people now, but the increase in entropy it predicts is due to collisions of individual molecules. In the course of these collisions, if one believes the theorem, the particles become more uniformly distributed—just as smoke from a cigarette gradually fills a room—and this shows up as an increase in entropy. Entropy increases until the particles are uniformly distributed—until the smoke has spread throughout the room. At that point the entropy has reached a maximum and no further change takes place. Such a state is called *equilibrium* and Boltzmann might draw the graph shown in Figure 5.1. Nevertheless, as Loschmidt pointed out, this simply doesn't seem possible. What is going on?

The key to Boltzmann's theorem was an assumption, known as the *Stosszahlansatz*, or "molecular chaos" assumption. According to this assumption, the particles are always "uncorrelated" before they collide. Now irretrievably sucked into the debate, you press for an explanation. "Uncorrelated" merely indicates that the motion of the particles in the gas is taken to be random: at any moment roughly the same number of particles are moving north as south, east as west, and they are all traveling with more or less the same speed.

Now two particles engaged in such random motion collide. They each have a certain speed and direction. Unfortunately, their motions are no longer uncorrelated. Once the collision takes place, the laws of conservation of energy and conservation of momentum tell you their relative speeds and positions afterward. So if you know where one is, you know where the other is. Correlated. Particles retain memories of their collisions.

"Hence," concludes Loschmidt, "Herr Doktor Professor Boltzmann has, in the basic assumption of the theorem, introduced a distinction between past and future. He has, unfortunately, assumed what he set out to prove. Because Newton's laws are time-reversible, for each collision that increases the entropy, the reverse collision *must decrease* the entropy. The entropy graph should look like this." He adds the line marked "L" shown in the illustration.

After some thought, Boltzmann is ready with an answer. "Herr Doktor Loschmidt's objections are ingenious but ultimately sophistic. After only two particles have collided, the reversed motion of the entire gas would again be random except for the velocity of the two particles under consideration. Hence, entropy would increase in even these circumstances."

Boltzmann's reply has struck everyone as reasonable and the majority are prepared to believe there is something in his proof. The academicians on the plaza are also nodding approval and prepare to disperse for coffee.

But Loschmidt replies, "No, Herr Doktor Professor Boltzmann is mistaken. If one is considering the collision of two particles only, then reversing the gas causes these two particles alone to collide and the entropy must go down."

For a moment the crowd hesitates, not being able to make up its mind. During this silence Ernst Zermelo, a student of Max Planck, unexpectedly mounts the platform. "I beg your attention," he says. "There is yet a further objection to Herr Dr. Professor Boltzmann's theorem. The number of particles in the gas is finite. Let us assume particles 1 and 2 collide. Afterward their motions are correlated, as we have agreed. Then particle 1 will collide with particle 17 and particle 2 will collide with particle 318. Now all four velocities are correlated. Eventually, all the particles in the gas will have collided and all their motions will be correlated. At that moment the motion of the gas can no longer be considered random and the molecular chaos assumption is violated. To the contrary, the particles are now highly organized; their positions and velocities are specified.

"From this moment on, if we reverse all the velocities, the particles 'know' how they are predestined to collide; in accordance with the time-reversibility of Newton's laws they must retrace their steps and entropy must decrease. Alternatively, we may simply continue

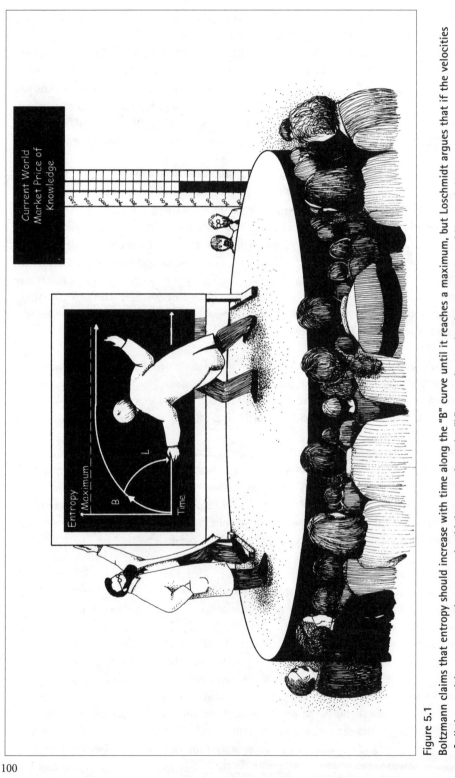

Figure 5.1

Boltzmann claims that entropy should increase with time along the "B" curve until it reaches a maximum, but Loschmidt argues that if the velocities of all the particles are reversed, entropy should decrease along the "L" curve due to the time-symmetry of Newton's Laws.

to run the clock forward. Because Professor Boltzmann's assumptions are no longer valid, it is not the case that entropy will necessarily increase. In this regard, Henri Poincaré has proven that any closed system must sooner or later return to the state from which it started. The conclusion is inescapable: the entropy must eventually decrease again."

Much of the crowd is very impressed by Zermelo's argument and begins to applaud. Oblivious, he continues: "In any case Boltzmann's theorem is reversible. It is of practical value only because he has assumed the universe began in a low-entropy state. But the universe will sooner or later return to that state. Unless he can show why the system started off like this, he has assumed what he set out to prove." Zermelo draws a new diagram (Figure 5.2).

Throughout Zermelo's rebuttal, Boltzmann has been extremely depressed. Nevertheless he answers that the time required for the universe to return to its initial state according to Poincaré's "recurrence" theorem would be so inconceivably long that it is pointless to talk about it.

At this moment Poincaré himself steps onto the platform and weighs in. "Are the hypotheses legitimate? Are they self-consistent? I do not believe they are. I do not wish to discuss them here; but there is no need for a long discussion in order to challenge an argument of which the premises are apparently in contradiction with the conclusion, where one finds in effect reversibility in the premises and irreversibility in the conclusions."

After a long silence Boltzmann acknowledges that one can prove the H-theorem only by assuming the system starts in a low-entropy state. However, he adds, "But surely, if our region of the universe began in a high-entropy or equilibrium state, for which there was no direction of time, then no physical processes could take place and life would not evolve. This seems to me the only way one can understand the validity of the second law: that because life exists, our part of the universe must have begun in a low-entropy state, after which the H-theorem became operative." Furthermore, the theorem would only operate until the time limit imposed by Poincaré's theorem.

In this defense, Boltzmann was making an early use of the famous *anthropic principle,* the idea that the reason the universe is observed to be as it is, is because we are here to observe it. Were it otherwise, life would never arise. This subject will undoubtedly recur.

The day is not done. Boltzmann and others devised further versions of the H-theorem in an attempt to circumvent the objections raised by Loschmidt and Zermelo. It is these later versions, in fact, that most scientists have in mind when they speak of the H-theorem, although over the years the distinctions have blurred.

The modified H-theorems dealt with the same bottle of gas, only described it differently. The basic idea is, again, to treat the huge number of molecules with statistics. But instead of dealing with the typical behavior of one molecule in a gas sample, we take a larger picture: we first pretend to tag the positions and velocities of all the particles in the gas. You recall from previous debates that knowing the position and velocity of all particles in a Newtonian system determines *everything;* hence when each molecule is thus tagged physicists say one knows the "state" of the system.

Figure 5.2
After Zermelo declares that entropy must eventually decrease along the "Z" curve, Gibbs points out that Boltzmann's entropy should also increase if time is run backward, and he adds the "GA" and "GB" curves.

Next we declare, "This is not kosher. With so many molecules it is impossible to know the exact state of the gas." Therefore we retreat and say it is too difficult to separate two states in which a few positions and velocities differ only slightly. Let's instead lump nearby states together. It's very much as if you were observing a crowd without your glasses. Instead of seeing distinct people they get blurred together; there are fewer states to worry about. Although you might be tempted to call such a blurring of states "averaging," bear in mind we are not computing a mean. For that reason physicists instead term the procedure "coarse-graining," which is another way of saying that we are ignorant of the gas's exact state.

The downfall of Boltzmann's H-theorem lay in the correlations among molecules; molecules remembered they had collided. But when one "coarse-grains" the state of the gas, one loses information, including information about correlations, and Loschmidt's objections vanish. It's like sex. If you observed the same crowd again, you would find many men and women together in pairs—they are correlated. But without your glasses, you might hesitate to say who is a man and who is a woman. Under ignorance, correlations vanish. Since the correlations vanish, the motions of the gas molecules may be considered random and Boltzmann's idea goes through—entropy increases.

Moreover, we know from observations of real gases that quantities like entropy and energy actually fluctuate a bit around mean values. Boltzmann's original H, on the other hand, didn't fluctuate; it always decreased. Since we have decided we do not know the precise state of the gas, why not make a virtue out of necessity and choose to regard H as a statistical quantity that can fluctuate. Then it makes sense to ask for only its most likely, or probable, behavior. It turns out that entropy no longer *always* increases, as it did in Boltzmann's original theorem; if the system starts off in a low-entropy state, entropy merely increases most of the time, in fact an *overwhelming* amount of the time.

Boltzmann is relieved.

An analogy will make the discussion more concrete. Suppose you found a deck of cards that was automatically reshuffling itself every second. Suppose also at the moment you found it, the deck was perfectly ordered in suits and cards. Call this the state of lowest entropy—perfect order. Now, if you had to bet on whether the deck would remain in the same order after the next two or three shuffles, you would be willing to bet never. The odds are overwhelming that the shuffling will make the deck more *disordered* in the future. Entropy increases! Of course, after a few more shuffles, the deck might become slightly more ordered again. Entropy decreases, a little. Nevertheless, the trend introduced by shuffling is to disorder the deck. This corresponds to the idea that the second law of thermodynamics is a probabilistic law. In almost all circumstances entropy increases, but very occasionally it decreases. This is the Boltzmann picture of entropy.

"Now, wait a minute," you protest, having listened carefully to the entire argument. "Just because you maintain you don't have a powerful enough microscope to see the troublesome correlations doesn't mean they don't exist."

Yes, you have hit the nail on the head. In the first place, as long as you can see the values of all the cards, no sequence is any more "disordered" than any other—all sequences of fifty-two distinct cards have the same probability. So it doesn't really make sense to call the initial sequence, with all the suits and values lined up, the most ordered. Nor does it make sense to call a highly shuffled deck disordered. Having perfect knowledge of the cards corresponds to the Newtonian picture: you are able to distinguish and track individual molecules. But if you can do that, then no one state of the gas is any more "random" than any other and the whole concept of entropy ceases to have meaning.

Boltzmann listens attentively.

So you must find a way to make some states more probable than others. How do we do this? We "coarse-grain." Suppose you begin to play poker with the magic deck of cards. To get a royal flush is highly unlikely. But let's say you are suddenly afflicted with a strange sort of nearsightedness that makes the jack of spades and the jack of clubs indistinguishable If you are working on a royal flush in clubs, the odds of getting it have conveniently doubled! Your grandfather, looking over your shoulder, might be even more nearsighted, so that all four jacks appear identical. In that case, the odds of getting a royal flush have just quadrupled.

We see that averaging (in the sense of coarse-graining) can indeed make certain configurations more probable than others. High-entropy states are like those in which all the jacks appear identical—more probable than the other states. But who decides how the averaging be done? You or your grandfather? Averaging is a subjective procedure. There is nothing in Boltzmann's theory that tells you how to make an average. If one believes the probabilistic interpretation of entropy, one is forced to conclude that it is at least in part an anthropomorphic concept.

Boltzmann and the rest of the crowd continue to focus their attention, but the physicists have already lost interest.

"This is not the final difficulty," interjects J. Willard Gibbs, the American physical chemist, who now takes his stand before the Knowledge Marker. "It should not be forgotten that the very use of probability assumes a direction in time. We often use probabilities to predict, but it is rarely the case that we use probabilities to retrodict."

To translate, imagine once more stumbling on the deck of cards at the moment it is completely ordered. It is true that after the next shuffles the odds overwhelmingly favor the deck becoming disordered. But if you found it at the moment of complete order, it is equally likely that the previous shuffle ordered it; the odds are equally high that the deck was highly disordered in the *past*. One can show, in fact, that for a closed system the probability that the entropy will increase in the future is exactly the same as the probability that the entropy would increase if time were run backward. Even the probabilistic interpretation of entropy shows no arrow of time. Gibbs adds two branches marked Gibbs A and Gibbs B to Zermelo's diagram, showing that entropy rises in both time directions.

The Knowledge Marker also rises, slightly, but several academicians wave their hands in disgust.

Throughout, you found your head swimming. The only questions that passed through your consciousness were, Who is right? Where is certainty? At lunch the conversation remained animated but consisted entirely of dismissing the other point of view as foolish and unworthy of attention. In that activity the physicists revealed no uncertainty of their own correctness. Soon they ceased altogether to listen to each other. Afterward, you were saddened to learn that Boltzmann, perhaps as the result of the controversy surrounding his work, committed suicide.

The Never-Ending Debate

Within a few years after Boltzmann's death in 1906, physicists accepted his statistical interpretation of entropy, that in some way entropy arises from our ignorance of the exact state of affairs, and this is the view that the majority of physicists hold today. For instance, academician Murray Gell-Mann maintains, "There are more ways for peanut butter and jelly to contaminate each other's containers than to remain completely pure." Only Max Planck and a few others held out against the general trend. Planck, in his *Treatise on Thermodynamics,* wrote:

It would be absurd to assume that the validity of the second law depends in any way on the skill of the physicist or chemist in observing or experimenting. The gist of the second law has nothing to do with experiment; the law asserts briefly that *there exists in nature a quantity which changes always in the same sense in all natural processes.* The proposition stated in this general form may be correct or incorrect; but whichever it may be, it will remain so, irrespective of whether thinking and measuring beings exist on the earth. . . . The limitations to the law, if any, must lie in the same province as its essential idea, in the observed Nature, and not in the Observer. . . . The law once discovered must receive recognition of its independence, at least in so far as Natural Law can be said to exist independent of Mind. Whoever denies this must deny the possibility of natural science.

The powerful defense of the second law as a fundamental law of nature in its own right went basically ignored. But Planck also pointed out an interesting contradiction in the psychology of physicists. The second law is equivalent to the statement that perpetual motion machines are impossible. Bearing that in mind, try the following experiment: Ask any Academy physicists you meet whether they believe perpetual motion machines are possible. We guarantee that every one will declare, "Impossible!" Ask them whether they believe this declaration to be a matter of opinion. Of course not. Ask them whether they believe that the increase of entropy is only a statistical law, due to "coarse-graining," a subjective procedure.

Given the difficulties inherent in the statistical interpretation of entropy, you might be wondering, why do scientists continue to accept it? This is a difficult question. Partly, it is due to the "coarse-graining" of history. Most textbooks no longer discuss the objections of Zermelo, Loschmidt and Gibbs, and the current generation of authors has either forgotten them or merely assumed that the issues were resolved decades ago. But perhaps the main reason is given by Steven Weinberg:

> For a while during the 1880s and 1890s a battle was fought between the supporters of the new statistical mechanics and those like Planck and the chemist Wilhelm Ostwald who continued to maintain the logical independence of thermodynamics [i.e., that thermodynamics could not be derived from Newtonian physics]. Ernst Zermelo went even further and argued that . . . the assumptions about molecules on which statistical mechanics is based must be wrong. This battle was won by statistical mechanics, after the reality of atoms became generally accepted early in this century. Nevertheless, even though thermodynamics has been explained in terms of particles and forces, it continues to deal with emergent concepts like temperature and entropy, that lose all meaning on the level of individual particles.
>
> Thermodynamics is more like a mode of reasoning than a body of universal physical law; wherever it applies it always allows us to justify the use of the same principles . . . and [the use of these principles] inevitably leads us down to the level of the elementary particles . . . .

Weinberg has declared the problem solved, as have many before him, but we see that the reason is one of faith, faith in reductionism. Physicists today are almost without exception the intellectual descendants of Boltzmann. Their worldview lies rooted in time-symmetric theories and they believe that thermodynamic systems, as Weinberg says, can be boiled down to a collection of particles obeying Newtonian dynamics. Weinberg is as aware as anyone that "when we say that one truth explains another . . . *we do not necessarily mean that we can actually deduce the truths we claim to have explained.* Sometimes we can complete the deductions, as for the chemistry of the very simple hydrogen molecule. But sometimes the problem is just too complicated for us" (emphasis ours).

Surely thermodynamics is one of those problems that have until now proved "just too complicated." Weinberg may believe that thermodynamics and with it the second law have been "explained in terms of particles and forces" but that is a leap of faith, which may ultimately prove to be right or wrong; it is not an uncontested fact. Nothing in the century since the modified H-theorems appeared has altered the fact that coarse-graining is a subjective feature of the theory and nothing has altered the fact that, as Gibbs noted, the use of probabilities is time-symmetric.

So one must search further. Nevertheless, Weinberg is far from alone in believing the problem solved. As the afternoon wears on, the cosmologists take over.

The Cosmological Solution Is Disputed

Just as the day before yesterday the symmetrist forces held overwhelming battle-field superiority, today the Boltzmannites have marshaled vast numbers. Only a few Planckians are to be found. Yet, because the issue of the arrow of time is so controversial, even the Boltzmannites are divided into subfactions. The three problems anyone claiming to have resolved the paradox of time must remove are: that for entropy to increase, the system must have started in a low-entropy state; that probabilities are time-symmetric (the Gibbs problem); and that coarse-graining is a subjective procedure.

Boltzmannites tend to solve the first problem by fiat: our world began in a low-entropy state. "We have to assume the system began in *some* way," they maintain. "Everyone agreed yesterday that to determine the trajectory of a Newtonian particle one must supply its initial position and velocity. All our most fundamental theories require the input of initial conditions. Therefore we are justified in choosing the initial state of the system to be one of low entropy."

This point of view is generally termed the "solution by boundary conditions." Earlier Boltzmann himself resorted to it when he declared that our sector of the universe must have begun in a low-entropy state, otherwise life could not evolve. The boundary-conditions solution remains popular today with theorists, in particular cosmologists. Roger Penrose is probably its most famous exponent.

Having listened to this argument, you ask why the Earth should have begun in a low-entropy state. Here Penrose provides the cosmological answer. The second law of thermodynamics says, strictly speaking, only that entropy increases in isolated systems—systems that exchange neither energy nor matter with the outside world. With the help of a suction pump one can move the cigarette smoke filling a room (a high-entropy situation) into one corner (a low-entropy situation). The energy for the pump must be provided from an external source, such as an electric outlet, and so the room is not an isolated system.

In the same way, neither is the Earth isolated; it receives energy from the sun in the form of highly organized sunlight. Photosynthesis and like processes transform the sunlight into low-entropy food and other energy sources, which eventually power both body and industry. In the course of transforming the energy from sunlight to food or fuel, plants, animals and factories produce heat, which increases the entropy around them. Global warming is a prominent result.

This answer, of course, merely pushes the boundary conditions back a notch: how did the sun's low entropy originate? The sun was formed when a cloud of diffuse gas condensed under gravitational attraction into a star. This cloud was part of the material that formed the galaxy, and the galaxy was formed from a cloud that fragmented into the local cluster of galaxies—and so on back to the big bang. All astronomical evidence indicates that at the big bang the matter in the universe was distributed with remarkable uniformity. Ultimately, in Penrose's view, the low-entropy situation on Earth is due to the vast entropy decrease that took place when the primordial gas cloud collapsed through gravitation into stars and galaxies. If the primordial gas cloud had been clumped from the beginning, no such collapse would take place. In a word, we attribute the second law to the state of the universe at the big bang. The big bang is the ultimate boundary condition. We cannot talk about anything earlier. Therefore we are entitled to cut off the Gibbs A branch from the diagram.

Penrose has actually taken a further step and proposed a kind of H-theorem for the entire universe. He calls it the Weyl-tensor hypothesis. The Weyl tensor, W for short, is one of the quantities in general relativity that describe the curvature of space-time. It has the property that if the universe began with perfect uniformity W would be zero. Penrose therefore conjectures that W was zero at the big bang, when the matter was smoothly distributed, and continues to grow as the matter clumps under gravitational attraction. By the Big Crunch, when most of the matter in the universe will have collapsed into black holes, W will approach infinity. If Penrose is correct, W is analogous to Boltzmann's H and provides a measure of the entropy for the universe.

Having heard Penrose out, most of those present nod in agreement, but Penrose's colleague Stephen Hawking dissents. He dislikes the Weyl-tensor hypothesis for two reasons. First, he says, if the universe began with perfect uniformity, W would have been zero and remain so. This appears to be a rather Aristotelian statement. It is true that if the universe began with *perfect* uniformity, W would remain zero, but perfect uniformity is an extremely unstable situation. It is much like a marble perched atop an infinitely thin needle. Such situations are never encountered in the real world; the marble inevitably falls in one direction or another. In the same way, matter at the big bang would inevitably begin to climb and W would move away from zero.

But Hawking's main objection is that he believes, like Weinberg, that the fundamental laws of nature should be time-symmetric (precisely, CPT invariant, which we discuss later), and he dislikes any proposal that distinguishes past from future on a microscopic level, which Penrose's would do.* Hawking's second objection is also unconvincing. Radioactive atoms decay, heavy subatomic particles decay into lighter ones. We never see such processes taking place in reverse, so they can be called irreversible, but they *are* fully time-symmetric (in the sense of being CPT invariant, shortly).

Independent of Hawking's claims, there are some more down-to-earth objections to the Weyl-tensor hypothesis. Although Penrose has championed it for several decades, he has never actually calculated how W behaves in any cosmological model. Several investigators have, including one of us (T. R.), and it simply doesn't grow in the way Penrose requires. The proposal does have some interesting features but it may be time to throw it out.

To no one's surprise, Hawking has his own solution to the arrow-of-time problem, which he developed with James Hartle in 1983. They call it the "no-boundary" proposal. At the mention of the proposal, certain academicians call for it to be discussed in the Ninth Debates along with quantum cosmology. We agree and so for now say only that the proposal really does make a choice of a time direction; it is neither more nor less reversible than Penrose's W hypothesis. Furthermore, there is certainly no direct evidence for the conjecture, and so the question is raised that will be raised again, Has science become metaphysical?

Although unbridled cosmological speculation is popular among the visitors, subterranean rumblings have begun. "Enough of this already!" cries one of the few Planckians present. "What does the big bang have to do with the increase of entropy in my teacup?" The academician, taking everyone by surprise, has expressed the view that it seems implausible to have to invoke the big bang to explain today's entropy increase, which may indeed take place in a teacup. The Boltzmannites reply, no, to prepare tea in a cup requires an external energy source in order to lower the entropy. Ultimately, the required energy can be traced back to the gravitational field of the universe.

Most of the spectators are convinced by this argument and prepare to give the day to the Boltzmannites. Yet the Planckians are not finished. "But if the direction of entropy depends on the initial configuration of the universe," they remonstrate, "that seems to imply that if two regions of the universe have different initial conditions, then entropy in one could increase while entropy in the other could decrease."

Some of the opposite faction remain silent, but a few others are quick to respond: "Only if entropy were increasing could life evolve."

Ah, they have retreated to Boltzmann's anthropic explanation for the second law. Many scientists are enthusiastic about the anthropic principle and during the brief exchange numerous heads around you began to nod. However, you have perhaps already detected something akin to teleology in the anthropic principle—one more step and

---

* Actually, any change in W is perfectly reversible, since it is a quantity from general relativity, a time-symmetric theory. The only thing that makes it irreversible is that Penrose chooses W to be zero at the big bang, and something very large near the big crunch. Penrose does speculate that a future theory of quantum gravity would be intrinsically time-asymmetric and provide the boundary conditions on W. Conceivably this speculation is what Hawking is objecting to, though he does not say so.

the universe would be here for our benefit. And you thought we had dispensed with teleology yesterday. The Aristotelians again return from exile across the road.

The Planckians, meanwhile, have not allowed their tactical wedge to go unpursued. "You Boltzmannites," one of them insists, "like most people have tacitly been equating disorder with high entropy and order with low entropy." Disorder, like the papers on a messy desk or smoke filling a room, is, in Boltzmann's view, high entropy, whereas an ordered desk or smoke confined to one corner of a room is low entropy. "But this is in general true only for isolated systems near equilibrium."

The Planckians have a serious point. In far-from-equilibrium situations, where the state of the system is usually dramatically changing, the more *ordered* state can be one of high entropy. The gravitational field of the universe itself represents a situation with no equilibrium whatsoever—matter always collapses under gravity, unless a pressure is present to oppose it. It may be true that the gravitational collapse of matter after the big bang produces a low-entropy situation but this statement is not based on any calculations. A consistent theory of the entropy of the gravitational field does not yet exist. "Thus Penrose's claim that the entropy on Earth can be ultimately traced to the entropy of the gravitational field at the big bang is pure speculation."

Hearing all this, the Boltzmannites remain firm. "The issue is one of entropy, not of order. The second law is due to a low-entropy big bang."

"Prove it."

The dispute is interrupted. Let us bring our microphones in closer. Someone in the crowd—surely a Planckian—is demanding that the Boltzmannites explain away the subjective nature of the second law in the statistical interpretation. Penrose himself answers the objection. The odds of entropy increasing rather than decreasing are so astronomical that it doesn't matter how you carry out your coarse-graining—for any "reasonable differences in viewpoint" the answer will always be the same.

Most of the Planckians are silenced by Penrose's response, but an idolator calls forth, "Odds? Odds botkins! By that reasoning the odds that your grandparents met each other were so remote you would never have been born!"

A moment passes in silence, then a general grumbling arises. Although neither side raises further objections, it is clear that everyone is a bit dissatisfied by Penrose's response, which merely says "live with it."

The Gibbs Branch Reveals an Insidious Assumption

For the past hour, it appears, something has been gnawing at the back of your mind. Finally you find the courage to turn to the nearest academician, who happens to be a Boltzmannite, and ask, "You've been talking a lot about entropy increase, but what about the direction of *time?*"

At which the academician stares at you with incomprehension. "You have understood nothing. The direction of time *is* the direction of entropy increase."

If you recover from the insult, however, you should persist, referring to Gibbs's drawing (Figure 5.3). "I understand why you want to chop off the left side of the

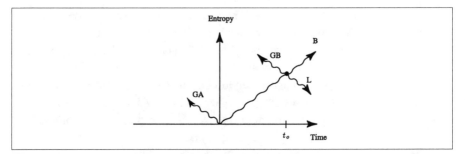

Figure 5.3
Because entropy is a statistical quantity and fluctuates, Boltzmann, Loschmidt, Zermelo, and Gibbs should have drawn their diagrams with small squiggles. Nevertheless, Gibbs's point can be seen on either diagram. Suppose the clock is started at time equals zero. In the forward direction entropy climbs along the B branch, but because probabilities are time-symmetric, if the clock is run backward, entropy should increase along the GA branch. The same holds true at any moment in time. If the clock is reversed at zero, entropy should cease increasing along B and continue increasing along GB. The question then becomes, Why can we ignore GA and GB?

diagram, the Gibbs A branch. This is the big bang boundary and we can't talk about anything earlier. But," you continue, pointing to the Gibbs B branch on the diagram, "why can't you go up here? In which case, it seems to me time would zigzag, depending on which way entropy increased."

You are in fact restating Gibbs's objection: to get an increase of entropy into the future, you must use probability in the forward direction only. Reverse time and entropy goes up in the backward direction too.

Meanwhile, the academician has been thinking. "My dear tourist, if the direction of entropy increase *is* the direction of time, then you wouldn't know whether it is zigzagging or not."

"But you have marked time increasing to the right on the x-axis. This seems to assume a prior direction in time. Is there?"

In a huff, the academician abruptly walks away.

At that moment physicist Rudolf Peierls appears, agreeing with you and Gibbs. *Even* if you start off the system in a low-entropy state, entropy should increase in either the positive or the negative time direction (following the "Boltzmann" branch in the positive direction and the "Gibbs B" branch in the negative direction). In that case there is no time asymmetry and Boltzmann's program is incorrect. But then Peierls declares, "The extension to negative times is, however, not of practical interest, because it does not describe a possible situation. In the laboratory this is due to the fact that we can remember the past and make plans for the future, but not vice versa."

"Wait a minute!" you exclaim, mightily perplexed. "Why can't we remember the future? You seem to be saying that our brains work by one set of laws and thermodynamics by another. *Of course* if you allow only one direction of time then entropy will increase in only one direction. Has this entire argument been based on a prior direction of time?"

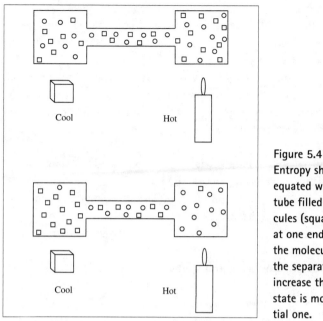

Figure 5.4
Entropy should not always be equated with disorder. If a metal tube filled with two types of molecules (squares and circles) is heated at one end and cooled at the other, the molecules will separate. During the separation, molecular collisions increase the entropy, but the final state is more ordered than the initial one.

Evidently. In lopping off the Gibbs B branch, Sir Rudolf (and apparently most of the Boltzmannites) have tacitly assumed, if not a prior, at least another arrow of time. One such arrow might be the "memory," or psychological, arrow of time. Another might be the expansion of the universe. Let us agree that time increasing to the right means the universe appears to be expanding. In that case if you crossed to Gibbs B, moving to the left, the universe would appear to be contracting. That we don't see the universe suddenly contracting suggests that the time arrows always point in the same direction, in other words that the world doesn't go up the Gibbs branches.

So there is more than one arrow of time?

Yes.

This intelligence catches many visitors off guard and they suddenly realize that the matter is more involved than they had previously suspected. Once more you experience the overwhelming temptation to find a cell phone and call for a rescue helicopter. Nevertheless, you take pleasure in the fact that you stumped at least one academician and consider applying for tenure.

But now, under cover of darkness, the Planckians have gathered their strength and decide to counterattack. Gathering under the Knowledge Marker, they put forth

The Thermodynamic Solution, Which Also Comes under Fire

The thermodynamic solution to the problem of entropy is most forcefully advocated by Ilya Prigogine, the eminent Belgian physical chemist and theoretical physicist. Since about 1995 he has claimed that his group in Brussels and Austin, Texas, has resolved the paradox of time.

Like Planck before him, Prigogine rejects the idea that the second law is due to ignorance or coarse-graining; the second law must be considered fundamental. It cannot be derived from Newtonian mechanics. The second law in and of itself is time-asymmetric. Entropy increases because entropy increases.

In that case, of course, the entire Boltzmann program is unnecessary and incorrect. Because one no longer attempts to derive the second law by imposing statistics on the behavior of Newtonian particles, coarse-graining and subjectivity never enter the picture. Neither does Gibbs's objection figure, since entropy's increase is no longer based on the use of probabilities.

Such a view is completely incompatible with the one expressed by Weinberg and indicates how deep a schism the arrow-of-time question has left in the world of physics. Nevertheless, if Prigogine is correct, all the problems under discussion today vanish. In that case, however, we are evidently left with the original paradox: two incompatible branches of physics, time-reversible mechanics and field theories, and time-irreversible thermodynamics.

To leave physics so divided would amount to an excommunicable heresy, but Prigogine's intent is otherwise. He too would like to unify physics, but rather than begin with Boltzmann's "bottom-up" approach, based on Newton's laws and particles, Prigogine finds his inspiration in Planck and an extraordinarily prescient statement of Poincaré made before the assembled Academy: "Perhaps the kinetic theory of gases will serve as a model. . . . Physical laws will then take on a completely new form, they will take on a statistical character." Prigogine's group goes from the top down.

The first step of the Brussels school is to declare along with Planck that the thermodynamic properties of a gas such as density, temperature, pressure and entropy (properties Boltzmann would treat statistically) are fundamental. Assume thermodynamics is primary, not secondary. But a cry immediately goes up from Loschmidt and Zermelo: it is no less impossible to pass from irreversible thermodynamics to reversible Newtonian mechanics than vice versa. True enough, and so any theory encompassing both must include an escape hatch.

The Brussels group finds their escape hatch in "unstable dynamical systems." In stark contrast to Weinberg's belief that our most fundamental description of the universe is found in time-symmetric theories, Prigogine declares, "Thanks to Poincaré's work, we know that the fundamental description of the universe is based on unstable systems."

What are they? Let us say only that unstable dynamical systems include the chaotic systems by now widely familiar to Academy visitors. To dedicated Newtonians a trajectory is more than just a path in space and time. Laplace demonstrates by hitting a baseball over the central plaza, thrice, the second time with just slightly more force than the first time, the third time in a slightly different direction. "As expected," he says, "when the initial conditions are changed slightly, the trajectory changes in only the slightest degree." For nearly two hundred years physicists tacitly assumed Laplace was correct. "Am I not?" asks Laplace with umbrage. No, the crowd roars, for by now all tourists know the answer. In chaotic systems the slightest change in initial conditions—the proverbial flap of a butterfly's wings—makes the system entirely unpredictable.

Academicians Michael Berry and David Ruelle take on Laplace. Consider an air molecule above the Academy plaza, they tell the assemblage, one that has followed a hypothetical trajectory, colliding with other molecules. Now remove a single electron from the far edge of the observable universe—ten billion light-years away. The gravitational force acting on the air molecule will now be ever so slightly different. The difference is impossible to imagine, really. Nevertheless, after only fifty-six collisions with other air molecules, the path of the test molecule will have diverged so much from its original trajectory that it misses one of the molecules it hit in the first place. After that, the new trajectory has nothing in common with the original one!

Clearly, removing an electron from the far edge of the universe represents a *slight* perturbation, but chaotic systems are infinitely sensitive. Strictly speaking, one needs *infinite* precision to follow particle trajectories in chaotic systems. Here a side debate on determinism and causality suddenly flares up among the academicians. We tune in later. Whatever, infinite precision is not found in the real world. "In such situations, Newtonian trajectories are a mathematical idealization," says Prigogine. "They do not exist." Since in chaotic systems one cannot follow trajectories, forward or backward, such systems are obviously irreversible.

Next, Prigogine declares that real-world thermodynamic systems, for instance gases, belong to the class of unstable dynamical systems. "After all," he says, "the existence of irreversible systems is an experimental fact. Their interpretation in terms of dynamics is a different story."

Having listened attentively, you perhaps see where this is going. Boltzmann would have us decompose the bulk properties into properties of individual particles following Newtonian trajectories. But if in an unstable system Newtonian trajectories do not exist, a decomposition into trajectories does not either.

Now, as was mentioned in the Second Debates, quantum mechanics relies on a particularly simple class of mathematical functions that have vector properties. Newtonian mechanics can also be described by such functions. A vector is the mathematical equivalent of an arrow (we don't mean an arrow of time); it has a length and a direction. The most important property of vectorlike functions is that a "length"—called a norm—can be simply defined. But mathematically speaking this is a rather small class of functions and most functions do not have well-defined norms.

What the Brussels-Austin group did in the mid-1990s was toss out the assumption of "normed" functions. Then the group found that stable systems, like the solar system, decompose into Newtonian trajectories, as everyone would expect. "It could not have been otherwise," affirms Laplace.

But unstable systems behave differently. When they are decomposed, the result is functions without well-defined lengths; these "non-vectorlike" functions are associated with irreversible effects such as friction and entropy. In other words, in unstable dynamical systems, Prigogine assures us, an arrow of time shows up naturally. Normless functions may sound mysterious but they have been known for decades and are actually quite common. The everyday exponential function that describes the radioactive decay of atoms is a function without a well-defined norm.

Despite the fact that we've left the idea of non-vectorlike functions a little vague, you suspect a swindle. It appears that the Brussels group merely changed the rules of the game to get the answer they wanted, much as Captain Kirk did at Star Fleet Academy to survive the Mitsubishi-Subaru maneuver.

In one sense you are correct, but every new theory does the same. To invent general relativity, Einstein had to reject the usual Euclidean mathematics as a description of spacetime and go to the geometry of curved spaces. Quantum mechanics broke several major rules. The question is not whether you break the old rules, the question is whether the new ones work.

However, you are far from alone in your protests. Several hours ago the sun set, but the debates show no sign of abating. Among the torches lit on the plaza, many of those present have objected that the Brussels approach is no more than an explanation without predictions. Perhaps television trial lawyers like theories that have nothing to do with the facts, but the hard core on the right calls that metaphysics.

Not so, the Brusselians reply. They have modeled numerous real-world systems, such as dense gases, and computer simulations verify the results; they have predicted new effects and have begun devising experiments.

The claim does little to pacify their opponents. Truly, the most extreme Boltzmannites will have nothing of it. "I think they've set back physics one hundred years," one of them exclaims and stomps away. This Boltzmannite speaks for the majority. The idea that statistical properties are fundamental is terrifically difficult for most physicists to swallow. Stephen Hawking says, "Physics is time-symmetric." Period. End of story.

But is this a fact or a prejudice? If we imagine a cloud-covered world, perhaps like Venus, where optical astronomical observations would be impossible, it is conceivable that historically thermodynamics would have developed before Newtonian mechanics. On such a world it might be the reversible systems that strike us in need of a more fundamental explanation and the Planckians, not the Boltzmannites, would be in the majority.

Such arguments do not please the academicians among the crowd. Neither are we going to claim that Prigogine is correct. There do appear to be technical difficulties in Prigogine's work, which limit the results. The theory does predict a microscopic origin of entropy increase, rather than a statistical one, but because the systems are chaotic and unpredictable in the future, they must be chaotic and unpredictable in the past as well. In that case, entropy will increase in the backward-time direction, just as Gibbs foretold. This point is currently under dispute, but the worst one can say at the moment is that the Brussels-Austin group might yet be proven wrong.

Does the Brussels Approach Explain Quantum Mechanics?

The debate shows every sign of going on longer than the one on symmetry two nights ago. By midnight neither the Planckians nor the Boltzmannites had retreated an inch, despite efforts by the authors. Evidently all factions were well aware of the Darwinian Principle of Science:

Without Publicity There Is No Prosperity

also known as the "Zel'dovich Principle," after the famous Russian astrophysicist who adopted it as his motto.

A brief lull in the fighting took place late in the evening when someone brought news of a civil war in Africa, but at the words "quantum mechanics" everyone's attention was once again riveted on the subject at hand. For according to Prigogine, the Brussels-Austin theory resolves a long-standing "dualism" in quantum mechanics—the measurement problem. Those in the crowd who insisted that today's debate be on quantum mechanics are suddenly pacified by the prospect of a sneak preview.

In quantum mechanics, a system—say an electron—is represented by an object called a wave function that gives the probability of finding the electron, say, here or there. The evolution of the wave function is determined by the basic equation of quantum mechanics, the Schrödinger equation, which is as reversible as Newton's second law. However, when we make a measurement on the electron, we do not measure the probability of its being here or there, we measure it definitely in one place or another. This measurement is irreversible; we never see electrons dissolve backward into probabilities.

But since Schrödinger's equation is perfectly reversible, it cannot describe the transition from probability to actuality, and so quantum mechanics doesn't either. This has led to seventy years of debate about how the act of observation turns a reversible system into an irreversible system. The usual words are that an observation causes the wave function to "collapse" into a definite state, but how this comes about has never been satisfactorily explained. Weinberg himself is forced to concede that this duality has prevented a unified view of nature because the observer seems to lie outside the rules of quantum mechanics, and to be sure, as will become apparent in the Quantum Debates, physicists have often invoked the observer—or even consciousness—as the mechanism to explain wave-function collapse.

But according to Prigogine, such metaphysics has no place in the new Brussels formulation. In quantum mechanics there is also a thing called the density matrix, which can correspond to the bulk, statistical properties of Newtonian physics. The Brussels group finds that if they treat the density matrix as the fundamental quantity instead of the wave function, then just as trajectories could not be recovered in the classical case, the wave function cannot be recovered in the quantum case. Furthermore, a measurement introduces an irreversibility that automatically causes the system to relax to one of the definite states traditionally found after a measurement. No extra assumptions, such as wave-function collapse or observers, are required.

Visitors and tourists, and even some academicians of the cosmological faction, frown at the news. Without far-out speculation, what is the point of science? One shouldn't worry; even when scientists don't speculate, journalists do it for them. According to writer John Horgan, Prigogine's statistical approach means that "science in the future will be more probabilistic and speculative," and hence have even less to do with reality than it does now. This statement must be rated decidedly weird. Both statistical mechanics and quantum mechanics are probabilistic, and though many have called quantum mechanics mystical, no one has called it speculative. Quantum mechanics, as we have said elsewhere, is as close to true as science gets.

The Other Arrows of Time Are Listed and Briefly Contemplated

We now abandon the Planckians and Boltzmannites to their struggles, which will continue until no one is left standing, and pick up an issue that earlier jarred everyone. Yes, there is more than one arrow of time. Roger Penrose has, in fact, compiled a list of seven arguably independent arrows, some of which have already surfaced in the debates. The first arrow, the subject of today's battle, is the increase of entropy.

The second is the psychological arrow of time. As Rudolf Peierls suggested, time seems to go forward; we hold memories of the past, not of the future. Penrose sees this as independent of the thermodynamic arrow, but Hawking has argued that to recover a memory in the brain requires the expenditure of a small amount of energy with a concomitant increase in entropy. In that case the psychological arrow would be linked to the thermodynamic arrow.

The third arrow has already been mentioned several times. During the debates on causality we revealed that the theory of electrodynamics admits solutions in which effects precede causes, such as radio waves converging on an antenna, or LaSalle discovering the Mississippi. The same holds true in Newtonian physics, relativity, quantum mechanics and field theory. Yet we never observe "advanced" solutions, as we called them, in nature and we throw out advanced solutions, not for any reason in the theory but because radio waves converging on an antenna from all over space, or ripples in a pond converging on the center just as a stone is thrown in, are never observed.

Because electrodynamics is a perfectly reversible theory, it is not obvious that advanced solutions are connected with the thermodynamic arrow of time. Most physicists discard them on the grounds of boundary conditions: to arrange for radio waves to converge simultaneously from all directions on an antenna, or to arrange for ripples to converge from a pond's edge at its center, would take an implausible amount of arranging. Nobody believes this can really happen. The third arrow is connected with the entropy arrow in that Boltzmannites use boundary conditions to explain both of them. If one could show that low-entropy boundary conditions disallowed advanced solutions, one might kill two birds with one stone.

We have also mentioned the fourth arrow, the expansion of the universe. Astronomers observe the universe to be expanding, but an equally valid solution to Einstein's equations of general relativity is for it to be contracting. The fact that astronomers have not observed the universe to suddenly reverse direction suggests that the various arrows remain pointed in the same direction. But general relativity, like Newtonian mechanics, does not distinguish between past and future, so it is difficult to see any link with thermodynamics.

The fifth arrow is also of cosmological origin. Evidence from the Hubble Space Telescope has convinced virtually all astronomers that black holes exist. Anything can fall into a black hole, but (classically) nothing can get out. However, if you imagine the time-reverse of a black hole, you get a white hole—an object that spews forth everything and captures nothing. Although a television might spring forth from a white hole, which strikes many as implausible (if not distasteful), there is no obvious reason that such an object could not exist. And so we would expect to see as many white holes in the universe as black holes. That we don't has led Penrose to propose the "Cosmic Censorship Hypothesis," which states that nature abhors white holes. At present there is no more fundamental explanation of nature's preference for black holes over white.

We alluded to the sixth arrow not long ago: the quantum arrow of time. Because the Schrödinger equation is reversible, it cannot explain how the quantum mechanical wave function "collapses" into a definite measurement. To put it another way, measurements never dissolve backward into probabilities. Academicians are already warming up to discuss this subject in detail during the quantum debates.

The seventh and last arrow is usually said to be given by the puzzling decay of the subatomic particle known as the neutral kaon. Like many other unstable particles, the kaon decays with a characteristic half-life into a few lighter particles. Until 1964 scientists assumed that all particles would decay at the same rate whether time ran forward or backward. In that year, a celebrated experiment by Val Fitch and James Cronin at Princeton shattered many illusions.

The experimenters could not change the direction of time. Rather, they relied on one of the fundamental theorems of quantum mechanics, which Hawking mentioned earlier, the CPT theorem. CPT for "charge, parity, time." This is another symmetry theorem, which states that all processes involving particles remain invariant if the particles are changed into antiparticles ("charge conjugation"), the system is reflected in a mirror ("parity change," or left goes to right), and time is run backward ("time inversion"). As an example of CPT invariance, a negatively charged electron spinning counterclockwise traveling forward in time must be indistinguishable from an antielectron (positron) spinning clockwise traveling backward in time.

Although they could not reverse time, the Princeton experimenters could change the charge (C) and parity (P) of the particles in the kaon system. To everyone's amazement, the kaon's decay rate changed under those operations; CP was violated. That meant in order for CPT to remain invariant, the system would have to change if time were run backward.*

Thus, one is faced with a choice: either the CPT theorem is incorrect, or the kaon decays in a different way in a universe where time runs backward. The CPT theorem is so general that it is hard to see how it could be wrong, and so most physicists accept the fact that the kaon decay rate distinguishes past from future.

Nature is being very subtle here; only in the neutral kaon decay does nature reveal CP violation. Uranium, thorium, radon . . . all decay in ways independent of time's direction (CP is conserved). Nevertheless, and here we depart from conventional wisdom, all radioactive decays are *irreversible*, in the sense that we never see a particle undecay.

---

* Imagine you have three numbers whose product is fixed, say $2 \times 3 \times 4 = 24$. If you change 2 and 3 to 1 and 2, then you must change 4 to 12 in order not to change 24.

This is true whether CP is violated or not. Any particle decay provides the seventh arrow of time and this is why Hawking's statement that microscopic physics doesn't distinguish past from future seems incorrect. The particle decay arrow seems completely unconnected with all the other arrows of time.

The existence of multiple arrows of time raises one further point. As the belligerent academician suggested in his argument with you, one arrow defines the direction of time, so it doesn't matter which way it points. This suggests that we should talk about only six independent arrows of time.

Practical Exercise: Does Chaos Exist?

The sun rose hours ago on the second day of the debates about time, but still the struggle showed little sign of ending. You may have caught a brief glimpse of the Planckians and Boltzmannites shouting and gesticulating as you fought off exhaustion and searched for something to eat, but your attention would have been distracted by an argument held over from the day before yesterday and which flared up earlier. The question under review was whether chaos had implications for the notion of causality. Everyone agreed, yes.

With the advent of chaos theory, scientists were forced to accept the fact that Laplace's clockwork universe was no longer tenable. Contrary to his famous remark that both the past and the future would be present in the eyes of an intelligence who knew at a given instance the positions, velocities and forces on every particle in the universe, this is far from the case. We now know that chaotic systems rule out such universal knowledge. Complex systems may be deterministic—they are governed by Newton's laws—but they are no longer predictable. Before chaos, one used the word "chance" to describe situations (like the roll of dice) that were presumably governed by Newton's laws but whose outcome we could not predict only for lack of sufficient information. Now we know that in many cases we can never have enough information. "Chance" or unpredictability has become a fundamental feature of the universe.

This is a genuine conceptual advance. Among the most powerful statements in science are those that mark absolute limits to knowledge. Gödel's incompleteness theorem is one of these, as is Heisenberg's uncertainty principle. The existence of deterministic chaos also falls in this class. Perhaps for that reason, in 1986, over Laplace's protests, Sir James Lighthill remarked for all physicists: "We collectively wish to apologize for having misled the general educated public by spreading ideas about the determinism of systems satisfying Newton's laws of motion that, after 1960, were proved incorrect."

On hearing Lighthill's *mea culpa*, spectators have launched into a sub side argument about the old problem of free will. If the universe is deterministic, do we make choices? Two days ago we saw that it was difficult to isolate causes in the best of circumstances; chaos theory now seems to remove the question forever beyond the range of decidability. Even if free will does not exist in some absolute sense because the universe is deterministic, the behavior of complicated systems will contain enough unpredictability (for instance in the firing of neurons in the brain) that it will behave as if free will is operative. In other words, one will never be able to predict the future with enough accuracy to rule out the concept of free will. The circumstance is closely analogous to Gödel's theorem, which states that all mathematical systems contain unprovable statements. By the same token, the very concept of determinism may well be undecidable, for how can one verify determinism without predictability?

Chaos theory is "acausal" in the true sense of the word. Before 1960, if we saw an apple flying sideways, we would ask for the cause of the motion. Nowadays, we merely say it is undergoing chaotic motion. On the other hand, chaos theory is not acausal in the sense of synchronicity; it does not allow an event at one place to influence a situation somewhere else instantaneously.

Much of the above seems self-evident today, but there are scientists who remain reluctant to accept that chaos theory tells us anything new. In fact, in your desperate wanderings, you stumbled across a dispute between a modern scientist and Laplace when the latter asked, "Do the modern-day followers of Planck truly claim for all to hear that particle trajectories are a figment of the great Newton's imagination?"

"Well, the point is that in chaotic systems one cannot predict the trajectory of particles for any length of time, so why talk about them?"

Laplace answers, "But this *is* amusing. If I were but slightly unsure of the speed of Mars in its orbit around the Sun (an uncertainty which of course precise astronomical observations could eliminate), then at a later time I would not know exactly where the planet was. Indeed, after so many orbits, the uncertainty in position would have grown so large that I could say no more than that Mars was somewhere in orbit around the Sun."

"But in an ordinary classical system this may be after billions of orbits. In a chaotic system after one or two collisions you are sunk."

"I see no fundamental distinction. Newton's second law regulates the universe. If one cannot follow the particles in a chaotic system, then surely it is merely because one lacks sufficient information."

"In an ordinary system, the uncertainty in position, as you call it, grows slowly, in direct proportion to time. In a chaotic system, the uncertainty grows exponentially with time."

"But this is merely a matter of degree, not of kind. Whether the uncertainty in behavior grows linearly in time or exponentially, the result can be computed. Surely an intelligence vast enough to submit all these data to analysis would be able to predict the position to arbitrary accuracy."

"No. To predict the position of a particle in a chaotic system requires an infinite amount of information. No computer can calculate it."

"Here you are mistaken. If you require that the position of Mars be computed with a certain accuracy, then this computer as you call it, given enough time, shall predict the position of Mars to the requested accuracy. In the case of chaotic motion—how I dislike that term!—the computer would merely require more time for the analysis, but although the obstacles in the second case are formidable, they are nothing more than practical. An intelligence greater than ours would be able to carry out the calculation to arbitrary accuracy."

"This is not merely a matter of practicality. Imagine the vastest, fastest computer you will; it would take far longer than the age of the universe to make such a calculation."

"Again, such are obstacles only to the feeble intelligence of the human mind and humiliating only to our self-love. But are we restricted to imagining only computers built by humans? In the eyes of the Creator, if one needs make reference to an unnecessary hypothesis, there could be uncertainty in neither future nor past. His intelligence would encompass all trajectories, regardless of how complicated they appear to our limited perception."

"What if I told you that to make such a calculation for a chaotic system took an infinite amount of time?"

"In that case, I would concede a fundamental distinction between ordinary classical systems and chaotic systems and would be therefore willing to yield the argument. But in this instance, you are obviously mistaken due to your stupidity in matters of natural philosophy. To calculate the position of a finite collection of particles simply cannot take an infinite amount of time. It is obvious to anyone after even a single moment's reflection."

Thus insulted you might sensibly walk away. However, to the best of our knowledge the question of the time required to compute the positions of particles in a chaotic system has yet to be answered.

Other Conceptions of Time Are Contrasted and Compared

When under the sacred fig tree the Buddha saw his past and future incarnations he seemed to transcend our normal notions of time. Is such transcendence possible? Are there more than one notion of time? Are they compatible with known physics? By the second afternoon, with the strength of the scientific factions waning, the philosophers are eager to discuss these questions.

Most philosophers are aware that once Einstein created special relativity in 1905, time ceased to have the absolute meaning that had enshrined it since the time of Newton. Whereas Newton assumed that a functioning clock always ticks at the same rate, regardless of its position or state of motion, Einstein showed that the rate at which a clock ticks depends on the clock's velocity. As a result space and time were put on an equal footing and lost their independence. Moreover, the concept of "now" is ambiguous in relativity because whether two events take place simultaneously also depends on an observer's velocity and, as we've discussed, because you cannot know that an event has taken place "now" until news reaches you later at the speed of light. With general relativity, Einstein showed that the rate of a clock's tick depends on the surrounding gravitational field. Thus time lost its independence from gravity and acceleration. In particle physics, an antiparticle moving forward in time can be viewed as a particle moving backward in time. Because antiparticles have the opposite electric charge from their particle counterparts, time has lost its independence from charge.

Such considerations suggest to Fritjof Capra that modern physics forces us to extend our concept of time in the same way that Eastern mystics transcend the ordinary awareness of time while in meditative states. Daisetz Suzuki puts it this way: "In this spiritual world, there are no time divisions such as the past, present and future; for they have contracted themselves into a single moment of the present. . . . The past and the future are both rolled up in this present moment of illumination. . . ." By the same token, Capra maintains, when mystics transcend time, they transcend the world of cause and effect.

To take up the last point first, Capra asserts that modern physics too goes beyond the limits of cause and effect. If a particle moving forward in time can be interpreted as an antiparticle moving backward in time, then particle interactions cannot be viewed as proceeding either forward or backward in time. There is no past or future, hence no cause and effect.

But does this have anything to do with modern physics? We pointed out yesterday (or was it tomorrow?) that in any deterministic universe the past is determined from the future no less than the future is determined from the past. This is true without invoking antiparticles. And as we hope by now is abundantly clear, Newtonian physics itself is time-symmetric; you cannot tell whether a movie of a collision among particles is being run forward or backward in time. We dismiss movies of many billiard balls coming together to form a perfect triangle, not because of anything in the physics, but because of our experience. To say that there is no time in modern physics is to dismiss the experience of irreversible thermodynamic processes as illusory. To say "the space-time of relativistic physics is a . . . timeless space of a higher dimension. All events in it are interconnected, but the connections are not causal" is neither more nor less correct than saying Newtonian mechanics is not causal.

To say that an antiparticle moving forward in time is the same as a particle moving backward in time does extend time symmetry to a wider symmetry combining time and charge, but it also propagates a common misunderstanding. We do not observe antiparticles traveling backward in time. We observe antiparticles, like any other particles, traveling forward in time. And like an electron, for example, its antiparticle the positron has a positive energy. But if you want to interpret the positron as a particle moving backward in time, you must interpret it as a particle with *negative energy moving backward in time*. What is negative energy? Let us say that if—contrary to observation—negative-energy particles existed, we might direct them into an ice cube. This would be the same as positive energy—heat—spontaneously flowing out of the cube into a warm room. But this is a violation of the second law of thermodynamics, which in one of its earliest formulations says that heat never spontaneously flows from a cold body to a hot one. So we see that bringing antiparticles moving backward in time into the picture does not simplify life.

With regard to the perception of time in meditative states, here we are in greater sympathy with Capra, for there can be no doubt that in such states one's time sense is altered, just as it is in an isolation tank. The objective of yoga, the "stilling of the mind," is accomplished principally through the slowing of respiration, which induces an altered sense of time. Pantanjali's Sutras and other yogic-tantric treatises give details on the "control of the moments and of their continuity" and it is of these "suspended states" that Suzuki speaks above. Yet one does not have to be a mystic or a yogi to some-times experience such states. Lovers often speak of a "timeless moment," and most of us occasionally become so engrossed in work or the creative process that we look up to find hours have passed when we thought it was minutes.

One must accept such experiences as real. We are then faced with a situation similar to that of thermodynamics. We can choose Boltzmann's approach and try to explain these states in terms of something more fundamental, perhaps physical. We can choose the Planckian approach and say that there is no evidence that the psychological sense of time can be boiled down to atomic oscillations. Or we may choose a more mystical path and assert that all reality is consciousness, that physical time does not exist outside our perception.

With the present state of ignorance, we do not claim to know which of these approaches is correct or how, if at all, they are related to each other, but if altered states are real, they are worthy of description and investigation. To the reductionist it is entirely reasonable to think that time perception can be linked to physical processes. We measure time through repetitive motion, be it the swinging of a pendulum, the pulse of the heart or images that flash through the mind. It is not then surprising that when our internal clocks begin to tick at a different rate, our time sense is altered. Much work has been done along these lines by experimental psychologists. To test whether temperature alters time perception, subjects have been placed in dangerously hot rooms, in thermal suits and heated helmets; they have been immersed on bicycles in tanks of cold water and have had electric currents passed through their bodies. Other researchers have found that people's estimate of the length of a sound will change if the sound is preceded by a series of clicks or flashes.

Such experiments do not pinpoint the nature of the human internal clock, if one exists. This allows room for a Planckian interpretation: the mind is a complex phenomenon, the most complex in the universe. If one cannot reduce the behavior of a gas to Newtonian mechanics, it makes little sense to try to reduce awareness of time, and consciousness in general, to a collection of particles and fields. Indeed, if chaos is operative in the mind (presumably not *too* much chaos can be operative in the mind) it may simply be impossible to approach the mind in a Boltzmannite fashion. A more appropriate model may be to treat it in the large, analogously to Prigogine's approach to thermodynamics; a statistical treatment of states of the mind would hardly resemble a treatment based on equations for the fundamental electro-chemical entities of the brain. There have also been attempts to model the mind as a hologram—an entity that stores information everywhere.

For the mystic, time perception depends on the state of consciousness of the individual. In the Vedic tradition there are three states of awareness: the waking state, in which senses are turned outward and one is aware only of the external world; the dreaming state, in which the senses are turned inward; and the state of dreamless sleep, in which there is no awareness of any world. When awake, we live in normal clock time, "profane time," where events follow one another chronologically and our experience of the world is fragmented. While dreaming, events may take place without chronology and time may expand or contract so that years may pass in seconds. And at moments of illumination we experience "sacred time," in which the flow of time seems to stop and in which we feel connected with the entire universe:

At the time of self-realization
At the moment of insight
Arises the conviction truly
That you are in essence the World

Many of us have had such experiences when, after long study, we suddenly understand a difficult problem in mathematics, or when embarking on a novel the structure suddenly materializes and the story, like a landscape, unfolds before you. At such moments the ego dissolves and there truly arises the conviction that you are in essence the World.

In the Upanishads, moments of illumination are associated with waking from the state of dreamless sleep into a fourth, superconscious state—*turiya*. Here the "I" dissolves into union with the *atman*, the universal Self, the Brahman. The famous motto *Tat tvam asi*, "Thou art That," describes the union with That: that which cannot be described, absolute Reality. It is in this state that the distinction between the observer and the observed vanishes and the limits of space, time, and causality are transcended.

The view that profane time is something to be transcended is fundamental to Mahayana Buddhism, which stresses the unreality of the present instant as it is transformed from past into future. "Existence and non-existence are not different appurtenances of a thing, they are the thing itself." "The nature of anything is its own momentary stasis and destruction." The temporal world is unreal. In order to escape the illusion of time, one must, like the Buddha, attain enlightenment and stand outside the flow of events.

It would be tempting to conclude from this that to the Eastern philosopher and modern physicist alike, real time is sacred time, when the universe is seen as a pattern of connected events without cause and without effect. The matter is more complicated. Influenced by the Indians, some early Taoists and Neo-Confucianists conceived time to be cyclical, but by and large to the Chinese time was rigidly chronological; family trees were meticulously kept and recent ancestors were very much part of the family. Although some Taoists also spoke of "escaping the present"—for to the eternal Tao there is no past, present or future—overall the Chinese seem to have been obsessed with time-keeping and time-keeping devices; the first known mechanical water clock dates from tenth-century China. Chinese time, by and large, was profane time, much like mundane Judaeo-Christian time.[†]

In Indian tradition, on the other hand, if sacred time governed moments of enlightenment, profane time governed the universe. As did other cultures, India developed a notion of cyclical cosmic time. Each cycle consists of four ages, each degraded from the last. The Fourth Age, which comes first in the Indian reckoning, is the "golden age," a beatific epoch of prosperity and justice, in which *dharma,* or duty, law, is respected. In the Third Age only three-quarters of the dharma is observed; humans now know suffering and death. The Second Age follows, in which only half the dharma exists on Earth, and evil and suffering increase; the human lifespan grows shorter. We, of course, live in the final, Evil Age (the *Kali Yuga*), where wealth becomes the sole criterion of virtue, sex replaces love and calculators replace minds. At the end of the four ages, which last a total of 12,000 divine years or 4,320,000 human years, there is a dissolution, a general cleansing. This is in fact where the famous line from the Bhagavad Gita comes in, "I am Time the destroyer of all; I have come to consume the world"— which Oppenheimer misquoted at the test of the first atomic bomb. After the great dissolution, the cosmic cycle begins anew.

To use the language of science, the doctrine of cosmic cycles corresponds to irreversible time. Humans, animals, the entire world, the entire universe are subject to the passage of irreversible time. The four ages are only the beginning of irreversible time. After one thousand of such cycles, there is a Great Dissolution and fourteen great cycles make up one day in the life of Brahma, the Creator, and Brahma himself lives only one hundred years, to be followed by another Brahma.

But the Buddha knows not only the past but the future and can return to his previous existences. The enlightened one is able to escape the shackles of irreversible time. When a Buddhist monk or a yogi reaches nirvana, time becomes reversible. In the Indian epic the *Ramayana,* it is written, in free translation:

One who transcends the limitations knows time as his chariot.
But one who is limited knows cause-effect relations.

Again, enlightenment seems to be associated with an escape from the profane, irreversible time to the sacred, reversible time. The division is meant to remind you of the division between thermodynamics and field theories. But are we to conclude that the Boltzmannian approach is higher than the Planckian? That with perfect knowledge, one sees no direction of time; that only when we are limited by ignorance, time appears irreversible?

No.

If direct experience is meaningful, then we should no more dismiss chaos theory and thermodynamics from modern physics than we should dismiss the rise and fall of the seasons and the waxing and waning of life as illusory. The profane is as necessary as the sacred. Moments of insight come only after much hard work and struggle brings a person to the point at which rational thought yields to intuition. The flash of illumination takes place in a timeless moment, but the glow of illumination, shedding light on both past experiences and future, is irreversible.

One must live in both profane and sacred time.

In Which the Fifth Debates Are Summarized

Some time ago you lost track of time. Knowing neither day nor hour, but overwhelmed by exhaustion, you decided that future discourse would only prove fatal and you moved in the direction of the main Academy gates. However, somewhere along the way you encountered another visitor, evidently in the same predicament as yourself, and you fell into discussion. Before you knew it you were both back at the palazzo reviewing the ongoing arguments and discussing the nature of time, wondering whether the millennium computer crash marks the end or the beginning of the Evil Age of Man.

The great paradox of the Fifth Debates was that daily experience indicates that time goes forward but the most basic theories of physics display no arrow of time. The irreversibility of virtually all real-world systems is embodied in the second law of thermodynamics, which states that entropy in an isolated system never decreases. But Newtonian mechanics, electrodynamics, relativity and quantum mechanics are all time-*reversible* theories and admit no quantity that increases in only one time direction. Thus, physics is divided into two totally incompatible branches. One contains reversible laws of nature; the other contains irreversible laws of nature.

Boltzmann attempted to reconcile the two by deriving the second law of thermodynamics from Newtonian physics. As useful as his prescription turned out to be, however, to this day questions remain about its basic assumptions. Many of the arguments centered around these premises. Recently Ilya Prigogine and collaborators have attempted to reconcile thermodynamics with Newtonian physics by assuming thermodynamics is primary, but this approach also met with strong opposition. Inherent in Prigogine's program was another important factor in the Fifth Debates—chaos theory. The lack of predictability in chaotic systems undermines the Laplacian notion of determinism and seems to rule out any possibility of proving such determinism. Despite Laplace's protests chance is, evidently, a fundamental feature of nature.

During the proceedings, six arrows of time other than the entropy arrow came to light. At present there is little, if any, understanding of how these arrows are connected. Finally, the authors turned philosophical and pointed out that if one wants to connect Eastern mysticism and physics, one should be careful to distinguish among philosophical systems and acknowledge that the irreversible figures alongside the reversible.

At understanding this you went out to the ocean shore to greet the next cycle of time.

WHY IS THERE LEFT AND RIGHT?
Chirality, Complexity and Emergent Properties

Anyone who, upon looking down at his bare feet, doesn't laugh,
has either no sense of symmetry or no sense of humour.

—*Descartes*

Right-Handed Tantras:
Posture,
Control of Breath,
Stilling of the Mind,
Allowing thoughts to cease
Be relaxed and aware.

Left-Handed Tantras:
Fish,
Meat,
Parched grain,
Liquor,
Sexual coupling

The Question

In a previous cycle of time, it is almost certain that you were incarnated as a warrior, defending a castle against barbarian invaders. As the enemy broke through the portals, you found yourself backing up the great spiral staircase, which the previous warlord had constructed in the form of a left-handed helix. Not without reason. As a defender backs up the stairs, it is crucial that the sword arm can swing freely and not be blocked by the central pillar. Unfortunately, you realize as you back up the spiral, such a construction assumes the sword arm is the right one, and you are left-handed. You remind yourself that next time you buy a castle it should have right-handed turrets.[†] All goes dark.

In your next incarnation you find yourself in the Allen Memorial Art Museum at Oberlin College, gazing at "Jacob Blessing the Sons of Joseph," by the Dutch master Adriaen van der Werff. In dramatic lighting, Jacob lies on his deathbed, blessing Joseph's sons Manasseh and Ephraim, his left hand resting on Manasseh's head, his

Figure 6.1
Chiral mismatch

right arm extended to Ephraim. For some moments you and a stranger stand side-by-side admiring the painting, though neither of you is sure what karmic acts forced you to become art critics in this lifetime.

Strolling out of the museum, you appear in the present cycle of time, at none other than the Academy art museum (and you thought you had seen the last of the Academy) with the same stranger, whom you now recognize as your newfound friend from yesterday's debates. Before you hangs the same painting by van der Werff. However, something strikes the two of you as odd. The painting is reversed, as if someone put a slide into a projector backward, but apart from the interchange of left and right, the Academy's version appears identical to the one at Oberlin. Faced with this absurdity, the two of you launch into an argument over which is the original. You suggest the lighting looks subtly better from one side; your friend disagrees, replying that the composition is better the other way. After some minutes of this, your companion concludes eruditely, "The Academy's must be the original. Here we see Jacob is blessing Manasseh with his right hand, which is correct, because it is evident that Manasseh is the elder, and one always blesses the eldest child with the right hand."

Just to be certain, the two of you check the Bible and discover, to your amazement, that when Jacob blessed the children, he laid his right hand on the head of Ephraim, the younger child, and when Joseph saw this he was displeased and bade Jacob put his right hand upon the head of Manasseh. And his father refused, saying that the younger brother shall be greater than the elder.

You report the implications of the discovery to the museum curator—the Academy's version is a fake. At first she refuses to believe you but when you set before her the Biblical revelation, she has no choice but to accept that the Academy has been duped. She orders the forgery burned.

Problem resolved, you exit the museum to take a brief stroll on the beach. In the morning light you begin collecting seashells, only to notice that almost all of them are in the form of right-handed spirals: turned clockwise they advance the way an ordinary right-hand screw advances. Scratching your heads, you wonder why this should be so, but on your way to the palazzo for breakfast, you stop by the Academy arboretum to be captured by the scent of honeysuckle. While inhaling its fragrance you perceive that the honeysuckle winds to the left, like the castle turret of a previous incarnation; in fact all the honeysuckle plants in the arboretum wind in the same direction.

Puzzlement increases as you pass by the physics corpus and overhear an academician telling a student that left-handed electrons far outnumber right-handed electrons in nature. What is a left-handed electron? you ask your friend, who shrugs, replying that it doesn't seem to make sense.

By the time you reach the palazzo as the sun rises on these Sixth Debates, it appears that everyone at the Academy has overnight begun discussing the problem of left and right. Is there a fundamental difference between the two? Why are some things always left-handed and other things always right-handed? Is the preponderance of left-handedness electrons connected with the fact that artists tend to portray the left cheek in portraits?

What happened to quantum mechanics? you ask the authors.

Well, we never promised.

A Contemplation of the World's Handedness

Probably the most obvious asymmetry in the world around us is the asymmetry between left and right. As you stroll around the Academy on your morning intelligence-gathering mission, you notice, as many have before you, that human beings exhibit bilateral symmetry. Appearances can be deceptive. On the inside, the symmetry disappears: the heart is on the left while the liver is on the right. One of the Academy Greek scholars tosses in your direction the word *chiral*, for hand. When certain simple objects—such as spheres—are reflected in a mirror, the image is indistinguishable from the original object. These objects exhibit no left- or right-handedness and are said to be *achiral*, or exhibit chiral symmetry. Objects that differ from their mirror images are said to be chiral, or chirally asymmetric.

The vast majority of objects, functions and processes are chirally asymmetric. Right-handers, of course, far outnumber left-handers, by about twelve to one, and this is true all over the world. Most seashells are right-handed, or dextral, meaning they spiral like a right-handed screw, and this is true on both sides of the Equator. In right-

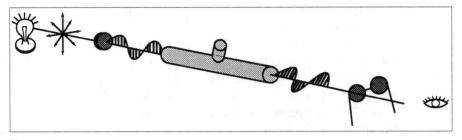

Figure 6.2
Handedness of crystals is typically determined by this simple apparatus. Light from a lamp is first polarized by passing it through a polarizing filter. The light is then passed through a tube containing a solution of dissolved crystals. If the solution is optically active, it will rotate the plane of polarization clockwise or counterclockwise. Therefore, in order to observe the light, a second polarizing filter (which can be a pair of Polaroid sunglasses) will have to be rotated by the same amount. (For more on polarization, see the Commentaries to the Sixth Debates.)

handed species, left-handed animals appear only as rare mutations—fewer than one percent—and as few as one in a million are sinistral, left-handed. In India, left-handed conch shells are considered so rare as to be sacred. There are, certainly, left-handed species, such as the lightning whelk of the Atlantic coast, and of course slithy toves are also left-handed. But only rare species exhibit equal numbers of left- and right-handed individuals; the Cuban tree snail is one of these.

As you discovered in the arboretum, chirality is not restricted to animals. Honeysuckle winds to the left, but bindweed winds to the right. Academicians have also discovered that the bacterium *Bacillus subrillis* usually forms a right-handed helix, but as the temperature increases, it changes over to a left-handed helix.

"Hmm," you say, doubtfully.

Chirality is no less a fundamental feature of the microscopic world. Most molecules are not identical to their mirror images and hence display chiral asymmetry. At the chemistry corpus, Louis Pasteur demonstrates the discovery that made him famous. He directs you to a microscope, under which you see tiny crystals of paratartaric, or racemic, acid, which form on wine casks during the fermentation of grapes. Pasteur tells us that in 1848, "I noticed that the confused mass of crystals of paratartrate could be divided into two groups according to the orientation of their facets of asymmetry." In other words, the crystals came in two types, each a mirror image of the other. Pasteur separated the two types of crystals, dissolved them in water, and passed a beam of polarized light through the solutions. "Great was my happiness on the day when I discovered that the solution containing the one crystal rotated light to the left as much as the other solution rotated it to the right."

Before your eyes, he passes a beam of polarized light through the two solutions and, using your sunglasses as a detector on the other side, you find that to observe the beam in the one case requires you to rotate the sunglasses clockwise and in the other case, counterclockwise. And so crystals can be designated right- or left-handed by the direction they rotate polarized light.[†]

Pasteur is not done. In 1857 he found that mold had contaminated a petri dish containing an optically inactive solution (meaning it did not rotate light). Rather than throwing away the contaminated dish, Pasteur checked the optical activity. As he shows you now, the dish with the mold rotated light! Microorganisms had changed an optically inactive solution to an optically active one. "In experimental science," Pasteur says, "chance favors only the prepared mind."

"So," your companion concludes, "the solution before the mold had equal numbers of right- and left-handed molecules, but afterward the solution contained more of one than the other."

"This was my conjecture, exactly," replies Pasteur. "Indeed, no other chemical characteristic is as distinctive of living organisms as optical activity."

For that reason in 1874 Pasteur appeared before the Gallic branch of the Academy (the French Academy of Sciences) and declared: "Life as manifested to us is a function of the universe and of the consequence of this fact. *L'univers est dissymétrique*," a motto we have quoted as an epigraph to the Third Debates.

Pasteur's conviction that the chemistry of life displays handedness turned out to be true to an extent that even he couldn't have imagined. Like crystals, molecules themselves can be chiral. Chemists refer to mirror-image forms as L and D enantiomers. The L and D, standing for levorotary and dextrorotary, are relics of Pasteur's discovery that some molecules rotate polarized light to the left, while their mirror images rotate it to the right.

Chiral molecules play a particularly crucial role in the chemistry of life. Proteins are long chains of amino acids. Although several hundred amino acids exist in nature, all proteins are made from the same twenty. Of these, all but one (glycine) are chiral. Furthermore, except for a few lower organisms that use D-amino acids in some specialized molecules, like cell walls and antibiotics, proteins in living things are always composed of L-amino acids.

"This is curious," you interject.

Indeed.

There is a tactical advantage to coordinating the handedness of a castle turret with the handedness of the defender. Similarly, there is a chemical advantage to sticking to one handedness or the other. The function of the proteins known as enzymes is to catalyze biomolecular reactions, including the synthesis of other proteins. An enzyme's ability to do this depends crucially on its three-dimensional structure, in particular on its ability to fit hand-in-glove with the molecule it is assembling or cleaving (the substrate). A left-handed enzyme cannot shake hands very well with a right-handed protein. The enzyme's structure depends in turn on the amino-acid sequence, which would be different if one switched to D-amino acids. In fact, synthetic chains of amino acids, which are composed equally of L and D enantiomers, cannot form the structure known as the alpha helix, which is present in most enzymes and necessary for efficient catalytic activity.

Just as there are only L-amino acids, RNA and DNA have only one handedness due to the fact that there are only D-sugars in the four bases that make up nucleic acids. As a result of the D-sugars, DNA and RNA form right-handed helixes. To replicate, RNA and DNA rely on proteins composed of L-amino acids, and so the relative chirality of proteins and sugars is intimately connected. Without chirality, life as we know it would not exist.

"Hmm," you say, a point on the tip of your tongue.

Because life's key molecules are chiral, human chemistry turns out to be highly sensitive to chiral differences. Probably the most famous, if horrible, example concerns the drug thalidomide. One enantiomer cures morning sickness, the other causes birth defects. Since the thalidomide tragedy in 1963, when thousands of infants were born without arms and legs, pharmaceutical companies have paid close attention to separating enantiomers. The pleasing orange scent now wafting toward us is limone, used in perfumes. Its enantiomer smells like turpentine. Aspartame—

"But," you interrupt, finally catching the point you wanted to make. "Is there any reason biochemistry couldn't work if the handedness of all the molecules involved were reversed? I mean, what if all amino acids were D instead of L and all RNA and DNA had L-sugars instead of D?"

Ah, the $64,000 question. The preference for L-amino acids and D-sugars is in fact extremely peculiar for the following reason. Chemistry is governed by the electromagnetic interaction between atoms and molecules, and electromagnetism does not distinguish between left and right. Electromagnetism is said to be *parity-conserving*. Thus, if one reflects any chemical process in a mirror, the mirror-image process should take place. It does. Any chemical reaction—such as molecular synthesis—proceeds with equal probability that L or D enantiomers are involved. When one synthesizes amino acids and sugar compounds in the lab, unless extreme care is taken, one always gets equal amounts of left- and right-handed molecules.

So your question has a definite answer: yes, biochemistry should work if the handedness of all the molecules involved were reversed.

"How strange!" you exclaim. "Then why is life based exclusively on L-amino acids?"

Ah. This is the $128,000 question, and on hearing it the academicians let their speculations run wild.

But before abandoning the shores of the known and certain for the seas of the uncertain and doubtful, let us continue the downward journey. Nature exhibits chirality even on the most fundamental level. Before 1957 no one would have believed this. Back at the physics corpus, an academician puts to you a question that was a favorite on oral exams to graduating aspirants: "Given only a radio, how would you describe the difference between left and right to an alien on another planet?"

"Well, I would say that left is to the . . ." and here you stumble, realizing that "left" is an arbitrary designation that depends on the direction you are facing.

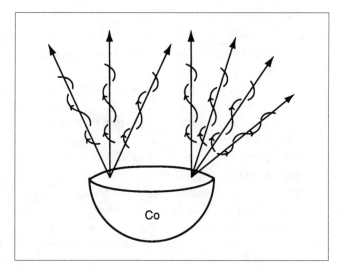

Figure 6.3
In Madame Wu's experiment, a sample of radioactive cobalt was allowed to decay. Measurements showed that far more left-handed electrons (electrons that advance like left-handed screws) were emitted than right-handed electrons (electrons that advance like ordinary right-handed screws).

"Well, then," says your friend, "I would say that if a screw advances when you turn it clockwise . . ." and then your partner also stumbles, realizing that "clockwise" is also an arbitrary designation.

Actually, before 1957, the correct answer was, "Stupid question. There is no fundamental distinction between left and right and no way to explain it to an alien." So it appeared. Not only is electromagnetism parity-conserving but so are gravity and the strong nuclear force. For this reason, physicists were certain that nature did not discriminate between left and right. However, in 1956 Tsao D. Lee and Chen N. Yang showed that no evidence existed for parity conservation in the weak force, and suggested possible experiments to test it. As we discussed in the Third Debates, the weak force is about 100 billion times weaker than the strong force and governs the radioactive decay of, for example, the neutron, which decays into a proton plus an electron plus an antineutrino. In early 1957 Chien-Shiung Wu and her colleagues at the National Bureau of Standards and Columbia University tested Lee and Yang's conjecture and verified that, to everyone's absolute surprise, in the radioactive decay of cobalt far more left-handed electrons were emitted than right-handed ones (see Figure 6.3).

Madame Wu's experiment not only caused a sensation among the Academy's physicists, but gave a way to explain to an alien the distinction between left and right: Radio the alien with directions for setting up Wu's experiment. The majority of electrons emitted will be designated left-handed.

Nature does not distinguish left and right only in electrons. In 1957 Sudarshan and Marshak pointed out that left-handed electrons were coupled to another particle, the left-handed neutrino,* and so in Wu's experiment the emission of left-handed electrons was accompanied by the left-handed neutrinos. By contrast, right-handed electrons were not coupled to anything, a reflection of the fact that right-handed neutrinos do not exist in the universe. Particles known as muons (the muon is identical to the elec-

---

* Technically, a right-handed antineutrino.

tron except that its mass is about 200 times higher) exhibit a similar handedness. The theory of Sudarshan and Marshak, in accounting for the handedness of electrons and muons, was the first theory of the weak force and demonstrated that the chiral aspect of the field was important in weak interactions.

At hearing all this, the antisymmetrist forces from the Third Debates raise a cheer. Another fundamental asymmetry in the universe. It appears so. Since the theory of Sudarshan and Marshak, the parity-violating weak force has been unified with the electromagnetic force into the electroweak force (by Glashow, Salam, and Weinberg) and the electroweak force has in turn been incorporated into the standard model of particle physics, which also came up for discussion in the Third Debates.

In the standard model, two particles, the W and Z bosons, are responsible for transmitting the electroweak force. The W and Z particles can be thought of as possessing charge, analogous to electric charge. Left- and right-handed electrons have Z charges of opposite signs and approximately equal magnitudes. By contrast, the weak W force is nonzero for left-handed electrons, but zero for right-handed electrons. Right-handed electrons simply do not feel the W force.* It is the W force that governs the radioactive decay that figured in Madame Wu's experiment and resulted in mostly left-handed electrons.

However, it would be wrong to say the standard model "explains" parity violation. The W and Z charges are considered to be a fundamental property of nature, just as mass is. There is no deeper understanding of why the weak force distinguishes left from right. It just does.

Grumbling from the symmetrist quarters.

Since atoms contain electrons, one result of the weak force is that atoms and molecules themselves must be chiral, even if their preference for left or right is infinitesimal. Nevertheless, we see that chirality exists at all levels, from the level of fundamental particles, all the way up to the macroscopic world of plants and animals.

And to the world of human creation. After you complete your reconnaissance mission of the natural world, you again find yourselves at the Academy museum, where art historians and aestheticians are debating the significance of a strange fact: most of the portraits hanging from the walls exhibit the left cheek. At first both you and your friend, displaying a skepticism that was perhaps absent when you first entered the Academy, are so reluctant to believe it that you traverse the museum from top to bottom, counting each of the thousands of portraits of the Academy's unrivaled collection. To your consternation and amazement, it does indeed appear that left cheeks outnumber right cheeks, 60 to 40 percent. "How peculiar!" you exclaim. "How extraordinary!" your friend says, simultaneously. What's so special about the left cheek? Is it aesthetically more pleasing than the right cheek?[†]

---

* The Z and W forces are often referred to as the "weak neutral current" and "weak charged current," respectively.

This is exactly what some of the academicians are arguing, though they seem to be having a hard time proving it. The discussion widens to the question Is there any intrinsic difference between a painting and its mirror image?

"Obviously," interjects Vincent van Gogh. "When my painting 'The Potato Eaters' was inverted during printing, I complained to Theo that it resulted in five left-handed people. And everyone knows that left-handed engravers cross hatch in the opposite directions from right-handed engravers."

But in a painting devoid of such obvious gaffs would it be possible to distinguish the original from its mirror image? You are reminded that this morning only a biblical clue allowed you to detect which blessing of Jacob was the original. Nevertheless, one faction, led by art historian Heinrich Wölfflin, argues that subtle differences do distinguish the original work from its reflection. He and his followers maintain that a painting is more than an individual work; it embodies a shared collection of symbols and cultural cues that are to be read in a certain way. A painting is entered at the left foreground, the eye travels up and to the right, and therefore objects acquire certain prominence depending on their position along this "glance curve."

But now the perceptual psychologists enter the dispute. They lead you to a secret attic room in the art museum, where subjects strapped in chairs are shown pairs of mirror-image slides and asked to identify the original. The investigators find that subjects can identify the original painting only if they are well acquainted with it. Otherwise the results are statistically random. Furthermore, it seems, a majority of subjects prefer the right cheek. And thus you learn the truth of the biblical precept "Whosoever shall intend to paint thee on the right cheek, turn the left cheek to him."

"But what I want to know," your friend asks, "is how the turned cheek is connected to left-handed electrons."

How Did It Happen?

As the sun rose toward the zenith, you could not but notice how congenial the discussion was this morning, compared to the near riot of two days past, hardly a debate at all, and the thought crossed your mind that martial law may have its advantages. Tourists are few in number, the central plaza largely deserted, the Knowledge Marker stable, leaving the small band of happy debaters to rove freely between the biology and chemistry corpuses. It also vaguely occurred to you that after your prolonged absence from the Real World you have probably been reported to the missing persons bureau, if not forgotten altogether, and almost certainly your date from whenever it was (you have a hard time remembering) has by now found a new significant other.

No matter. The intertwined questions of whether chiral asymmetry on different scales is connected and how we ended up with the chiral world we got strike you as eminently worth pursuing. (You did pose them yourself.) Biologists in particular are eager to discuss the matter and agitate to discuss the last question first.

The authors, on the other hand, in control of the situation, decide to take up the first question first: can chirality on one level induce chirality on another level? The answer appears to be yes, at least to some extent. As we mentioned, the existence of the weak force means that atoms must be chiral. Just as honeysuckle winds in one direction, so do electrons of an atom; the weak Z force causes them to form right-handed helices and we can say that atomic electrons and consequently the atoms are right-handed.* Despite the weakness of the weak force (the standard model predicts that polarized light passed through an atomic gas should be rotated 10^{-5} degrees, the angle subtended by your hand [left or right] at a distance of approximately 1,000 kilometers), the effect has been measured; and so here is an instance where chirality at the level of elementary particles has caused chirality at the level of atoms.

During the exposition you have been undoubtedly wondering about the next connection—between atoms and molecules. No one has proven that the chirality of atoms induces chirality in molecules, but possible connections have emerged. Consider two molecules of opposite chirality, such as L- and D-amino acids. Because the Z force attracts right-handed electrons to the nucleus but repels left-handed electrons, it causes one enantiomer to be at a slightly different energy level than another. Now, according to basic principles of statistical mechanics, the enantiomer with the lowest energy level should be found in greater abundance. Stephen Mason and George Trantor of Kings College, London made such calculations in the 1980s and found that the life-building L-amino acids are the ones with the lowest energy.

"This means they should outnumber the D-amino acids!" you exclaim.

Correct, but they do so by only one part in 10^{17}, an infinitesimal amount, which explains why they are found in the laboratory in essentially equal numbers; neither has the energy difference in enantiomers predicted by Mason and Trantor actually been observed.

Still, you cannot but wonder, perhaps this is a clue that the weak force has something to do with the total dominance of L-amino acids in biology.

"But does the weak force prevent you from turning the other cheek?" your friend ponders, just as you are about to run off to perform the experiment.

We don't know; however, progress has been made in isolating the gene responsible for chiral asymmetry in vertebrate embryos. As you noticed when you tried to explain left and right to an alien, the concepts "left" and "right" have meaning only when defined relative to an axis with a direction. Facing north, left is on one side of the north-south axis, facing south it is on the other side. A chick embryo for the few days of its development is little more than a flat sheet with a groove running from one end to the other. At one end of the groove, which is called the primitive streak, sits a structure known as Hensen's node. The node can be thought of as defining "north." It turns out that after about forty-eight hours some genes are expressed on the right side of the primitive streak and some on the left, which eventually leads to the heart's tubes looping toward the right.

In 1995 Cliff Tabin's group at Harvard Medical School apparently isolated the relevant genes. The gene with the mysterious name *Sonic hedgehog* is restricted to the left side of Hensen's node, whereas the gene known as *activin receptor IIa* is expressed on the right, as shown in Figure 6.4. (The gene is thought to be a receptor for activin, a hormone known to play a role in tissue differentiation.) By placing acrylic beads soaked with activin on the left side, the researchers could suppress *Sonic hedgehog* on the left and cause the appearance

* The W force between the nucleus and the electrons is negligible.

Figure 6.4
At about forty-eight
hours a chick embryo is
little more than a
groove, called the
primitive streak, with a
structure called
Henson's node at one
end. However, this is
enough to define a di-
rection. In normal
chick embryos, the
Sonic hedgehog gene is
restricted to the left of
the primitive streak,
and the gene *activin
receptor IIa* is restricted
to the right. This sug-
gests that these genes
are crucial in deter-
mining internal left-
right asymmetry in
chickens, but no one
knows how. (The noto-
chord develops into the
spinal column.)

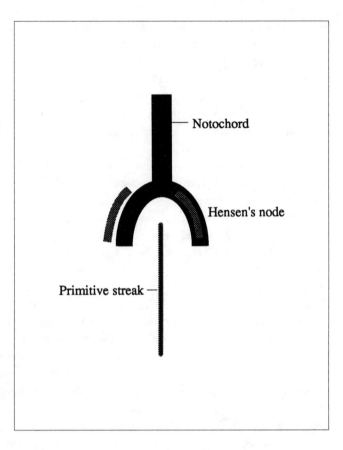

of *activin receptor IIa*. Embryos were thus no longer asymmetric with respect to the hedgehog gene and, when allowed to develop, the direction of heart looping in the embryos became random. The researchers expect that placing beads soaked with *Sonic hedgehog* on the
right will reverse the usual asymmetry of the embryos and provide good evidence that
Sonic hedgehog and *cAct-RIIa* play a key role in the development of vertebrate handedness.

Thus we have the origin of Archilochus' famous

Aphorism:
**The fox knows many things, but the sonic hedgehog knows the difference between
left and right.**

"Let's not get carried away," objects one academician. "To date there has been *no*
evidence that there is a similar expression of these genes in mammals or zebrafish.
What's more, when *activin receptor IIa* is taken out of mice they develop normal asymmetry. In any case, none of this explains how *cAct-RIIa* or *Sonic hedgehog* itself distinguishes chirality."

Ah well.

Not to be deterred, biologists have developed theoretical mechanisms in which molecular handedness would be passed up or "converted" to the cellular and multicellular level
and, finally, to the embryo as a whole. Nevertheless, to the best of our knowledge none of

these models has been confirmed, and given that the offspring of dextral snails can be sinistral, it is difficult to see how microscopic chirality determines macroscopic chirality.

"But if it did," your companion asks, "would that mean that the weak force is responsible for handedness all the way up to the macroscopic level?"

Let us turn our attention to the second question: how did chirality emerge? We have already affirmed that there should be no chemical preference for one enantiomer over the other and so it is hard to understand how we ended up with the chiral world we got. Biochemistry's indifference to left and right is demonstrated by the famous Miller-type experiments, in which the primordial atmosphere of the Earth is (supposedly) simulated. Such experiments produce *racemic* mixtures—mixtures with equal amounts of L- and D-amino acids. Moreover, a number of biomolecules (for example the protease from the HIV virus) have been synthesized with reversed chirality and they have functioned perfectly well (on substrates with reversed chirality).

"This proves that the handedness of life was caused by a random event," insists a tall academician as the speculation takes off.

"No, it proves that the earliest life forms were imported from outer space," answers another academician, raising her finger.

She refers to the fact that amino acids have been found in carbonaceous chondritic meteorites, the oldest meteorites, which are characterized by carbon-rich inclusions called chondrules. In particular, in 1970 investigators discovered that the famous Murchison meteorite contained some of the amino acids necessary for terrestrial biochemistry.

"The meteorite is incontrovertible evidence that some cosmochemical reaction that took place in the solar system or in extrasolar space produced molecules necessary for life," continues the same academician.

"Yes," objects a third, "but the Murchison meteorite contained equal numbers of L- and D-amino acids. It proves nothing about the origin of handedness."

"You should keep up with the literature," retorts the second. "In 1997 Cronin and Pizzarello reexamined the meteorite and found that it contained about 7 percent more L-amino acids than D. Left outnumbers right in space, just as on Earth."

"But on Earth it is 100 percent," your partner points out, "isn't it?"

"Because the meteorite is 4.5 billion years old," she goes on, ignoring the question, "the new analysis proves that handedness arose before the origin of life."

"But," you say, attempting to interrupt, "even if this is true, how does this explain how extraterrestrial handedness arose?"

It doesn't. The meteoritic evidence may indicate there was a universal mechanism for the origin of handedness that began operating before the origin of life on Earth; otherwise it merely pushes the question to the great beyond.

"This is not very certain," you say.

True, but as the discussion moves toward the experimental chemistry building, you discover that there is no lack of ideas among academicians for initiating chirality.

"I believe the origin of handedness was an accident," insists the same tall academician who tried to make the point earlier. "Kondepudi has discovered that if an achiral sodium chlorate is allowed to slowly crystallize, over a period of days from a solution, then virtually all crystals will be of one handedness, but which handedness cannot be predicted." Dilip Kondepudi, on sabbatical at the Academy from Wake Forest, demonstrates the simple and startling experiment. He first dissolves sodium chlorate crystals—a close relative of ordinary table salt—in beakers of water. Next, he allows the crystals to precipitate out.

"This is how we used to make rock candy as children," you observe.

Precisely, except no strings are attached. Kondepudi then directs you to measure the handedness of the resulting crystals, in exactly the way Pasteur did, by examining whether they rotate polarized light to the left or the right. After several hours at the lab bench you and your friend are forced to conclude that there seem to be equal numbers of levorotary and dextrorotary sodium chlorate crystals.

"What's so interesting about that?" you exclaim. "It's exactly what you'd expect."

Yes, but now Kondepudi has you measure the handedness from another group of beakers in which the solution as been stirred for several days as the sodium chlorate gradually crystallizes. After several more hours the two of you tabulate the results and you are shocked: each of the beakers contains nearly 100 percent left-handed crystals or 100 percent right-handed crystals.

"How did that happen?" you exclaim.

To be honest, nobody knows. However, Kondepudi's experiment is a good example of what scientists call *spontaneous symmetry breaking*. He began with a solution that made no distinction between left and right—it was chirally symmetric—and yet in the process of slow crystallization one handedness came to dominate the other. Presumably the first crystal had a fifty-fifty chance to be L or D, but once that occurred, the crystal stimulated production of others of the same handedness and suppressed production of crystals of opposite handedness. The process does appear random; in about 50 percent of the beakers L-crystals dominated and in 50 percent D-crystals dominated.

Spontaneous symmetry breaking, in which a system goes from a symmetric to an asymmetric state, is a common process in nature, and has been known for a long time. The earliest known example is the Tower of Babel from the Book of Genesis. Before the tower, mankind was in a symmetric state, all speaking the same language, but afterward it was in a state of broken symmetry, all races speaking their own language. Symmetry breaking is often accompanied by a *phase transition*, in which a system changes from one form to another. The precipitation of solid sodium chlorate crystals from a liquid solution, or the rock candy, are good examples, although the most famous phase transition is probably that endured by the frog prince in the Grimm Brothers' fairy tale (where phase transitions are commonplace).

Symmetry breaking and phase transitions have become central in all branches of modern physics. You will recall that the symmetrist forces in the Third Debates argued that all asymmetry is merely broken symmetry; symmetry is primary, asymmetry secondary. The concepts of symmetry breaking and phase transitions reemerge yet again in the Eighth and Ninth debates when the academicians take up theories of everything and cosmology.

"So, as I maintained," reiterates the same tall academician, "the origin of handedness was an accident. Originally, there were equal numbers of L- and D-amino acids, but some random event caused the dominance of one over the other."

"Umm, wait a minute," objects your friend. "This doesn't make sense. Granted, symmetry breaking may explain why there is an excess of L- or D-amino acids in a beaker, but if the handedness is random from one beaker to another, then how does it explain why L dominates in all beakers—everywhere?"

"Good point," you say.

We agree. Symmetry breaking is a random process and cannot in and of itself account for why L-amino acids dominate over D- everywhere. One therefore searches for a universal mechanism that can account for the excess of L-amino acids in biological systems.

"The weak nuclear force," you friend says. "It must be."

Other academicians have had different thoughts. It turns out that one enantiomer of an optically active molecule tends to absorb circularly polarized light[†] more than the other. Hence the suggestion, dating back over one hundred years, is that if a racemic mixture of L- and D-enantiomers were exposed to circularly polarized light, one might be preferentially destroyed, enriching the other.

"Unfortunately," says André Cotton, one of the early investigators, "I found no such effect. C'est la vie."

But in 1929 W. Kuhn and associates did find a slight preferential decomposition of one enantiomer of some organic compounds by circularly polarized ultraviolet light. Since then initiation of enantiomeric differences by circularly polarized light has been a popular scenario. Academician Stephen Mason envisions a pool of racemic amino acids on an eastward-facing slope that catches sunlight only at dawn. At dawn, sunlight is "right circularly polarized," meaning that D-amino acids might absorb it preferentially and be decomposed, leaving an excess of L-amino acids.

"Umm," you say, "what if the pool is facing west?"

"This is the problem," Mason concedes. At twilight sunlight is polarized in the opposite direction, and so one would expect the other enantiomer to be preferentially decomposed.

"In any case," adds your friend, "there must be pools facing in all directions around the world. The average effect should be zero. There must be another explanation."

Others have thought so. In 1957, when parity violation was discovered, Frederic Vester and Tito Ulbricht at Yale proposed that the left-handed electrons preferentially emitted in radioactive decay might preferentially produce left circularly polarized light on interaction with matter, which would in turn preferentially decompose one enantiomer, presumably D, leaving an excess of L. Unfortunately, no effect was observed; calculations by Roger Hegstrom at Wake Forest indicate the effect should be only about one part in 10^{11} and experimental limits put it at less than one part in 10^9.

"All this seems very uncertain," you say, "not to mention doubtful."

"It *must* be the weak nuclear force," your friend says.

All right. It may be the weak nuclear force. You have already seen in the chemistry lab how spontaneous symmetry breaking can result in the dominance of one enantiomer of sodium chlorate over another. Spontaneous symmetry breaking occurs only under certain circumstances. It cannot take place in systems closed to the inflow of energy. Such systems tend to proceed toward thermodynamic equilibrium, a state in which the concentration of the molecule depends only on the molecule's energy and entropy. Because L- and D-enantiomers have essentially the same energy, their concentrations will be found in equal abundance. Thus symmetry breaking generally occurs in far-from-equilibrium systems, systems that are open to the inflow of matter and energy.

In 1953 Sir Frederick Charles Frank of the University of Bristol devised a model to show how spontaneous symmetry breaking might proceed in a chemical system. His model required two molecular species that were each capable of replication. In other words, via ordinary chemical reactions, each species produced more of itself, a process called autocatalysis. Since autocatalysis of one species depleted the reactants necessary to produce the other species, the two species competed with each other and the presence of one diminished the population growth rate of the other. The replication rates of the two species were identical, as were their effects on each other. Nevertheless, the situation is unstable: any random event that results in a slight excess of one species leads to a runaway in which that species becomes totally dominant.

Now, the weak nuclear force is so weak that one would expect L-amino acids to dominate D-amino acids by only one part in 10^{17}, an unimaginably small difference. Nevertheless, in 1985, Kondepudi and George Nelson showed that the weak nuclear force might actually be strong enough to push a system, such as Frank's, in one direction. Imagine a pool of water in which two enantiomers compete with one another. There will be many random fluctuations that tend first to favor one enantiomer, then the other. These fluctuations are vastly larger than an effect of the weak force but because they are random, they tend to cancel out. And so, given enough time, the weak force can have an effect.

"Can it produce all L-amino acids?" your friend asks.

Unbelievably, yes, under certain circumstances. Take a pool about the size of a typical lake, one that is far from equilibrium and hence open to the flow of reactants. Give it about 100,000 years and the weak force can provide enough of a bias to influence the direction of symmetry breaking; one enantiomer will always come to dominate the other.

"I knew it," says your friend.

"I find it hard to believe," you say.

"Humph," adds a skeptical academician. "How are they going to prove it? We can't run experiments for 100,000 years."

That is true and the effect has yet to be demonstrated in any real chemical system. However, Frank Moss of the University of Missouri and Peter McClintock of the University of Lancaster have simulated the system with electronic circuitry and found that the predicted effect takes place. That is, a systematic effect as small as one part in 10^{17} can decide the outcome of the situation.

"I still don't believe it," replies the academician.

Well, as Planck said, no theory is accepted until the last of its opponents dies off.

But the weak nuclear force far from exhausts the proposals to explain the origin of chirality. Literally dozens of mechanisms have been suggested. Any terrestrial mechanism, however, suffers from the same problem that invoking pools of water facing the morning sun did: it is essentially a random process that averaged over the face of the Earth should not favor either enantiomer. In light of the Murchison meteorite, you might suggest searching for extraterrestrial mechanisms. This too has been done, but the proposals (such as invoking polarized radiation from neutron stars to induce handedness in interstellar molecular clouds) suffer from the same problem of being randomly oriented across the galaxy.

There remains the question of whether biological handedness came into existence before or after the origin of life. Given that achiral molecules do not seem to replicate well or engage in efficient catalytic activity, it is hard to see how life could have arisen before the establishment of molecular handedness, a conclusion apparently supported by the excess of L-amino acids in the Murchison meteorite. However, we have reached the point where the uncertain merges into the entirely speculative and, as the entourage moves to the palazzo for dinner, we modestly step aside.

A Practical Exercise—Speculation Rampant

Over a rich bed of philosophy garnished with metaphysics, however, the academicians eagerly continued their trek toward the unknown. What follows is an edited version of some questions that arose during the final discussion of the 1991 Ciba Foundation Symposium on Biological Asymmetry and Handedness. It will help to realize that *Drosophila,* the famous fruit fly, displays no internal distinction between left and right—it is bilaterally symmetric. "Anteroposterior" refers to the axis that runs from the front to the rear of an organism, while "dorsoventral" refers to the axis that runs from the stomach to the back.

WOOD: The question is whether there is anything special about the left-right axis. Is it fundamentally different from the anteroposterior or the dorsoventral axis?

WOLPERT: Yes, because it has meaning only with respect to those other axes. [Recall the difficulty in explaining left and right to an alien.] We are back to where we started.

WOOD: I mean is it different in the sense of having a fundamental asymmetry such that bilateral symmetry must be imposed on it?

YOUR FRIEND: I think he means, which came first, bilateral symmetry or left-right asymmetry?

YOU: They sound like the physicists, arguing over which was prior, symmetry or asymmetry.

THE AUTHORS: Ah, the eternal questions.

WOLPERT: I personally believe that there is a symmetrical organism on which the left-right axis is imposed, and there is embryological evidence to suppose that. The evolutionary evidence may disagree, but I doubt it.

MCMANUS: You have to account for why *Drosophila* is not asymmetrical.

WOOD: You could imagine superimposing perfect bilateral symmetry on a basically asymmetrical developmental process.

WOLPERT: The primitive condition is symmetry. Before there were left- and right-sided organisms, there were symmetrical organisms with a front and a back, primitive multicellular animals, bilaterians or coelenterates or whatever. [Bilaterians: any bilateral organism; coelenterates: jellyfish, sea anemones, etc.]

As the participants recline with port and cigars:

CORBALLIS: I wanted to speculate about human laterality and why it might be important in human evolution. . . . You don't see handedness in the primates—there is no population bias as there is in humans. This is something that seems to distinguish humans from other species.

YOU: Gorillas aren't left- or right-handed?

YOUR FRIEND: Neither are monkeys.

CORBALLIS: Language itself also distinguishes humans from other species. . . . So human handedness and cerebral dominance for languages are, as it were, the stuff of humanity. . . . An important question about human laterality is why is it directional? Why is it the right hand and the left brain? There are good reasons to suppose that we started to communicate and compute messages by using manual gestures rather than the voice. That idea has been around for centuries. It is supported by the observation that sign language even now is as natural as vocal language. Deaf children who learn sign language learn it very easily; they go through the same critical periods, the same stages, they even babble in sign. So as language evolved there was a gradual transition from manual gesture to communication by voice, and it would have been an advantage to have the control of the voice in the same hemisphere as control of the hands.

YOU: Why?

YOUR FRIEND: I don't know. Is this connected to the weak force?

CORBALLIS: So right-handedness may be a relic of a past that was involved in the development of language. And it may have been more important then than now to have these control systems in the same hemisphere.

WOLPERT: That explains lateralization but does not explain why lateralization is handed, not random.

YOU: He means why right-hand, left-brain is dominant in humans?

YOUR FRIEND: Yes, we are back to the beginning.

BORG: In sign language, is one hand used a lot more than the other? If I were using sign language, could someone follow me in a mirror or would they get confused?

YOU: I would.

PETERS: The right hand is more active. In formal sign language, there is a relation between the left hemisphere and signing. The literature reports that aphasia [loss of ability to use or understand words] in signing is common after left- but not after right-hemisphere damage.

JEFFERIES: Has anybody looked at the neural structure of the brains of people who have never learned to speak, such as deaf mutes?

GALABURDA: There is amazingly little known about that. I have tried to get brain specimens from Ursula Bellugi. She has followed a huge population of congenitally
deaf people who have only ever spoken with sign language, but has been unable to
provide me with any brain specimens. . . .

The debate adjourns for dessert.

Chirality Paves the Way for Complexity and Emergent Properties

As the sun set on the Sixth Debates and you found yourself in the Academy ceme-
tery plundering the graves of deceased academicians for suitable brains, it struck you
that today's discussions differed somewhat from the other debates you have been subjected to on these hallowed grounds. Although the weak force figured prominently,
there seemed to be less attention paid to fundamental principles and more reliance on
experimental results, not to mention outright speculation. Not today did you suffer
lengthy discourses on action principles, time-reversible laws and the CPT theorem.
Rather, as the small band of debaters gyrated from life science corpus to necropolis, the
participants waxed poetically over beakers of crystals, chicken embryos, meteorites and
amino-acid abundances. No one mentioned "the laws of biology."

For good reason. If the elementary systems physicists deal with are "credit approved" in the sense that scientists attack them confident that fundamental laws and
relatively simple mathematics will suffice for their analysis, then the complex systems
of biology and chemistry are "cash only"; they require mathematics so involved that it
is abandoned as useless and, consequently, practitioners of chemistry and biology are
much more accustomed to trial and error.

Such humility in the face of nature's complexity made you wonder whether the
reasonably harmonious and quiescent day was connected with the personalities of experimentalists.

In retrospect, the first signs of trouble appeared at the words "complex system,"
which induced a phase transition in the debate. Visitors, who disdained such an obscure
topic as chirality, now began to arrive in droves. Particularly evident were those of the
left, on whose syllabi for "Advanced Holistic and Field Studies" the topics of "complexity," "chaos," and "emergent phenomena" were required experiences. Hurriedly you put
aside brain collection and perceived, as the reductionists' blood froze, that tonight symmetry could be broken.

But probing both visitors and academicians with the questions, What is a complex system? What is an emergent phenomenon? you found no consensus. Perhaps
there shouldn't be. If a system is too complex to be described by simple mathematics,
then its behavior is almost certainly too complex to be described in a few words. But
one thing can be said with certainty: complex systems, whatever they might be, are the
rule in nature, not the exception. Comparatively few systems manifest the predictable
behavior of the systems traditionally analyzed by physicists. When you encountered
the authors in your survey, they suggested that complex systems might be better

named "ordinary systems" and the typical systems of interest to physicists should be relabeled "simple."

The authors also found it is easier to produce examples than definitions. Certainly societies must be considered complex, the behavior of the mob of antireductionist New Agers that just entered the Academy gates armed with aura-imaging equipment should be considered complex, and even most ordinary chemical systems. But the reductionist lurking within many of us would still like to know, Is there a single property that characterizes complex systems?

The plaza has been lately equipped with audience response meters, and a quick Reality Referendum shows that many people believe that a complex system is one consisting of a large numbers of particles, like a box of gas or the brain or a lynch mob. This is not an unreasonable thought. In the Fifth Debates we referred to thermodynamic systems as systems containing many particles. Thermodynamic systems are indeed complex—or can be—and perhaps for this reason academicians typically associate complexity with thermodynamic systems. However, it is not true that a system needs to contain many particles to exhibit complex behavior. Three particles—the Sun, Jupiter and a comet—are enough to produce a chaotic system whose motion is unpredictable.

Here, surprising no one, Laplace rises with an objection, arguing as he did yesterday that given enough time and a powerful enough computer, he can forecast the behavior of any system to arbitrary accuracy.

Wearied of Laplacian unregeneracy, mathematicians have introduced the concept of *algorithmic complexity* to determine whether a system should be termed complex or not. If, given any computer Laplace chooses, a simulation of the system requires more time to run than the system itself takes to evolve naturally, then the system should be called incomputable, transcomputable—or complex.

"This sounds reasonable," you nod.

Although Laplace is undergoing a fit of apoplexy, it does to us as well. Given that no hard-and-fast definition of complexity exists, we might take this as a convenient operational definition—a complex system is one whose behavior is too hard to compute in less time than the system computes itself.

In a small corner of the plaza symmetry broke as Laplace continued to harangue the modern mathematicians that the concept of complexity is nothing fundamental, and the schismatic mathematicians continued to ignore him. Most of the crowd did too. They were not interested in definitions; the people wanted emergence.

Behold.

As the complexity of a system increases, properties emerge that were not exhibited by the system's individual components. Such "emergents" are characteristic of the system as a whole and change as the complexity increases from one level to another. Several emergent phenomena are visible to you as the evening sets in: on the verge of a new millennium, the Unarians, guided by the Archangel Uriel, attempt to convert the plaza into a landing zone for the Space Brothers; the number of kiosks purveying Grand Unified Theories on the plaza has also markedly increased since the afternoon.

Emergence did not originate with the millennium. In the 1920s the zoologist C. Lloyd Morgan championed emergentism in the context of evolution, but since then it has been applied to other areas of biology, chemistry, physics and sociology. As with complexity itself, a rigorous definition of emergent properties does not—and perhaps cannot—exist, but this challenge has stimulated rather than curtailed debate. Schrödinger himself considered emergent properties in his seminal work, *What Is Life?*

> What I wish to make clear . . . is, in short, that from all we have learned about the structure of living matter, we must be prepared to find it working in a manner that cannot be reduced to the ordinary laws of physics. And that not on the ground that there is any "new force" or what not, directing the behaviour of the single atoms within a living organism, but because the construction is different from anything we have yet tested in the physical laboratory.

Schrödinger then takes the example of an engineer who specializes in steam engines and who comes across an electric motor. The engineer will be familiar with the copper, the iron and so on, but "the difference in construction is enough to prepare him for an entirely different way of functioning. He will not suspect that an electric motor is driven by a ghost because it is set spinning by the turn of a switch, without boiler or steam."

Unfortunately, Schrödinger does not make entirely clear what he considers the new principles of life to be. On the one hand, he emphasizes the importance of the second law of thermodynamics and points out that life represents a local reduction in entropy—order from disorder. On the other hand, like Boltzmann, he evidently accepts that the second law is a direct result of microscopic dynamical laws and, moreover, that life itself is "a clock-work," in which case both the second law and life are reducible to "ordinary laws of physics."

Schrödinger's lack of clarity illuminates the problems inherent in discussing emergence. Roald Hoffmann, addressing the audience amid the signs and portents, is frank about the difficulties:

> Emergent properties—a buzz word by itself. Yes, we just live with them, though there is this terrible reductionist vein accepted by chemists. But synthesis is so essential a part of chemistry of our time (whereas analysis was dominant in the last century) that it's hard to think in reductionist terms. And every new molecule has some properties that are different.
>
> To me emergent properties are simply the new concepts/questions/properties (but really questions by an experimentalist) that arise as complexity increases.
>
> Be warned, my antireductionist perspective is not popular or typical in chemistry.

The sounds of symmetry cracking ricochet around the plaza. As he predicted, the right wing is not pleased with his remarks and it appears that even the chemistry faction will fragment into small domains. The mathematicians, dissatisfied with his loose talk, have by now split off from the experimentalists completely and attempt to define emergent phenomena in much the same way as they defined complexity itself, via the concept of computability. An emergent phenomenon is one that cannot be simulated or predicted in less time than it takes for the phenomenon to simulate itself.

"Nature is the best computer," remarks an academician on the left.

Despite the fracturing that has begun, a large area of symmetry remains among the participants. On a practical, operational level, few scientists object to the concept of emergent phenomena. In everyday life we talk about emergent properties as soon as we have a collection of things. We don't call a hair or two a mustache, although in China seven longish hairs on either side makes you a wise man. Generally the concept of a beard requires many hairs. A forest is a collection of trees; one or two Indians don't make a tribe. A more abstract emergent might be the S & P index, a futures index. Brokers are allowed to speculate on the S & P index. But you might ask what exactly is being traded. It isn't an individual stock, nor is it even the average of a number of stocks; it is the future behavior of an average. Trading on the S & P index is a very evanescent concept.

But the physicists gathered beneath the turbulent skies (unprecedented; the Academy's Elysian Weather Purification System must be malfunctioning) have already objected to the common use of "emergent phenomenon"; something, they say, is lacking. True—what is lacking is the concept of interaction. The properties of a beard are, by and large, little different than the sum of the properties of the individual hairs. But in virtually all physical systems, the components interact, giving rise to a collective behavior that is far more complicated than the behavior of the components acting individually.

And now the New Age, which has for several days been agitating for quantum mechanics, unfurls its slogans: "The Whole Is Greater Than the Sum of the Parts," "The Universe Is Not a Loose Collection of Objects, Forces, and Energy." One speaker, Erwin Lazlo, declaims, "Reductionism has been proven wrong in the new physics, where even the properties of ordinary atoms were shown to be irreducible merely to the properties of their parts."

The authors, surveying the scene through their field glasses, feel obliged to respond. Not so implicit in all this is the mistaken perception that only in quantum mechanics or field theory is the "whole greater than the sum of its parts." Yet when two classical light or sound waves combine, their intensity is not twice but four times the intensity of the individual waves. What better confirmation of "the whole is greater than the sum of its parts" is there? A confirmation that was recognized many decades before the "new" physics. Individual molecules of sodium chlorate are achiral, manifesting no distinction between left and right. But, as an experiment demonstrated earlier in the day, sodium chlorate crystals—composed of many molecules—are either levo- or dextrorotary. Here is a well-defined emergent property. A Newtonian system consisting of three gravitating particles can behave in a completely different manner from a system of two gravitating particles. A few moments ago we mentioned that a system of three particles can be chaotic—and this is due entirely to their interaction energy. Togetherness has always been important, the new physics notwithstanding.

Even in mathematics, properties of a system are not always reducible to basic elements. Where is the twist in a Möbius strip? Roger Penrose asks. At what place does the impossibility of one of Escher's impossible figures reside? Reductionism in mathematics, of course, met its ultimate limit early in the century with Gödel's theorem, which established the impossibility of placing mathematics on an axiomatic footing.

Despite our response, Steven Weinberg would chide us for not getting tougher with the likes of the New Age. "At the nuttiest extreme," he says of the antireductionists, "are those with holistics in their heads, those whose reactions to reductionism take the form of psychic energies, life forces that cannot be described in terms of the ordinary laws of inanimate nature. I would not try to answer these critics with a pep talk about the beauties of modern science. The reductionist worldview *is* chilling and impersonal. It has to be accepted as it is, not because we like it, but because that is the way the world works."

Although Academy physicists appreciate Weinberg's straight talk, he is correct; the New Age does not appreciate it. Fissures began to appear on the central plaza between residents and tourists.

By then it was apparent to all that the debate had shifted slightly from emergence to reductionism, but the shift is only one of emphasis, for the two concepts are mirror images of one another. Two questions had emerged from the remarks of Hoffmann and Weinberg: Is reductionism true? Is reductionism practical? In other words, do the laws that govern emergent phenomena rest on deeper laws? If they do not, then one requires *new laws of nature* at the higher levels, a possibility that pure reductionists reject out of hand. But then, if the emergent laws are derivable from deeper principles, is this fact of any practical value?

New fault lines began to appear on the plaza. As if by divine intervention they separated the biologists, who deal with complex systems on a day-to-day basis, from the physicists, who claim to deal with nature on its most fundamental level.

Ernst Mayr, a leading evolutionary biologist, declares, "Reductionism is at best a vacuous, but more often a thoroughly misleading and futile approach." He points out that it is hopeless to reduce concepts such as territoriality, display and so on to chemistry and physics. One might add, as François Jacob did in the Second Debates, that it seems futile to "describe a feeling, a decision . . . a guilty conscience in terms of chemistry and physics."

Weinberg answers that Mayr, a friend, "did not understand the distinction I was making, between reductionism as a prescription for progress in science, which is not my view, and reductionism as a statement of the order of nature, which I think is simply true."

So the physicist has carefully distinguished between reductionism as a fact and reductionism as a tool. All science is to some extent reductionist; that hardly can be denied. From Darwin to Mendel to Crick and Watson, biologists have followed physicists along their reductionist course until today some geneticists would assert that the entire field is little more than a study of the role of DNA. Others, like Mayr, argue that a knowledge of DNA is of no use to population biologists and ecologists.

Weinberg points out that his views on reductionism are no stronger than those of the geneticists. "In precisely the same way, whether or not the *discoveries* of elementary particle physics are useful to all other scientists, the *principles* of elementary particle physics are fundamental to all nature."

The antireductionist mob present on the plaza boos Weinberg soundly for his politically incorrect views. Nevertheless, Weinberg's position cannot be dismissed out of hand. If one abandoned the concept of laws, of rules, then no science whatsoever would be possible. Is the universe describable? To the extent that science answers this question in the affirmative it is because science recognizes some order in the world. And if physics has been the most successful of sciences, within its domain, it is because physicists have, in the Western tradition, dissected, analyzed and reduced.

But if reductionism has been successful, it should also be recognized conceptually as a one-way street. As the fissuring continued, Jean-Pierre Changeux repeated his argument from the Second Debates that, although Mendel deduced laws of heredity from the traits of peas, it would be absurd to think that one could begin with the laws of probability and deduce the existence of chromosomes or DNA. Roald Hoffmann's remarks on synthesis versus analysis a little while ago are also a reflection of the one-way street. Reductionists invariably tend to think in terms of analysis, "deconstruction," but synthesis is much more difficult. The eminent physicist Philip Anderson (who at least at the time admitted, "We must all start with reductionism, which I fully accept") put it like this some years ago:

> The main fallacy in this kind of thinking is that the reductionist hypothesis does not by any means imply a "constructionist" one: the ability to reduce everything to simple fundamental laws does not imply the ability to start from those laws and reconstruct the universe. In fact, the more the elementary particle physicists tell us about the nature of the fundamental laws, the less relevance they seem to have to the very real problems of the rest of science, much less to those of society.

More recently biologist Robert Wesson has declared, "It may be confidently asserted that properties of compounds depend on forces and bonds as described by quantum mechanics, but in practice chemists try to predict the characteristics of a new compound by analogy with similar known compounds." As if to echo Wesson, Mario Bunge maintains, "We still do not have an adequate understanding of a single organic molecule."

However, as much as the audience registers high approval of such positions, it *is* possible to make *some* deductions from basic principles—that is the mark of a successful theory, in physics at least. The reason Wesson can say that "it may be confidently asserted that properties of compounds depend on forces and bonds as described by quantum mechanics" is because given quantum mechanics, one can deduce much of basic chemistry, in particular crystal structure. To a great extent one can sympathize with Weinberg's reaction to the extremists. Antireductionists do tend to bend their interpretations of physics in predictable directions. Wesson, who appears to be a Planckian, on the one hand asserts that the properties of compounds depend on quantum mechanics, but on the other hand states, "The famous second law of thermodynamics is not derivable from knowledge of particles; it rather contradicts their nature." Then he goes on to say, ". . . some theorists see organisms as systems of dissipation of energy into entropy," which is equating apples and oranges. Energy is energy; entropy is entropy.

Another crack of symmetry breaking resounds throughout the plaza as the physicists, once and for all, split off from the rest of humanity. The exodus of the physics quarter is not so much because Wesson is wrong (much of what he says is right—more or less) but because it blurs the fine distinctions on which science rests. Whether or not the universe is describable, scientific theory does have its own meaning, which is not entirely a matter of referendum.

But in mentioning the second law Wesson has encircled the precise arena where emergentism confronts reductionism head-on: thermodynamics. If organisms and organic molecules are too difficult to describe in terms of basic physics, thermodynamic systems are extremely well described in terms of physics; thermodynamics is an entire branch of physics unto itself. Its properties—temperature, pressure, entropy—emerge from the collective behavior of particles and, in fact, one cannot define temperature, pressure and entropy for single entities. Precisely because thermodynamics exhibits emergent phenomena, which may or may not be derivable from more "fundamental" quantities, thermodynamics has become the front line between reductionism and emergentism. The Boltzmannites and Planckians, who have not ceased their struggles, are once again visible and neither side shows any sign of yielding.

The positions have been plainly drawn. Weinberg minces no words when he says, "Thermodynamics has been explained in terms of particles and forces."

Prigogine, it goes without saying, disagrees. He repeats his declaration from the other day: "Thanks to Poincaré's work, we know that the fundamental description of the universe is based on unstable systems."

Here is a complete disconnect. If the unstable systems lie at the basis of our description of the universe, then the principles of elementary particle physics cannot. Prigogine's contention, that entropy is a fundamental quantity, that thermodynamics is not reducible to particle physics, requires new laws of nature at the thermodynamic level.

Had not the leftists been impeded by the domain wall separating them from the rationalists, there is no doubt the New Age would have lynched its foes on the spot. Yet, despite mortal danger, Weinberg remains unapologetic to the end: "The reason," he says, " [particle physicists] give the impression that we think that elementary particle physics is more fundamental than other branches of physics is because it is." As the crowd attempts to breach the domain wall, he reflects, "We may formulate what we learn about mobs in the form of laws (such as the old saw that revolutions always eat their own children), but, if we ask for an explanation of why such laws hold, we would not be very happy to be told that the laws are fundamental, without explanation in terms of anything else. Rather, we would seek a reductionist explanation precisely in terms of the psychology of individual humans."

Here we must disagree. Weinberg is ignoring the interaction term. Friedrich Schiller, from the sidelines, said it best: "Anyone taken as an individual is tolerably sensible and reasonable—as a member of a crowd he at once becomes a blockhead." As to whether the Planckians or the Boltzmannites are correct, we can only reiterate what we have said in the Fifth Debates. Prigogine may be wrong, but Weinberg's contention that thermodynamics has been reduced to particle physics is in all probability an unverifiable statement. If thermodynamics reveals true emergents, which Weinberg accepts, then their behavior must be incomputable. He has made a leap of faith.

Vice President Al Gore, coming on to the remains of the day, also has feelings on the matter: "When I say that our current, chaotic political culture reminds me of Ilya Prigogine—that because our system has more and more energy coming in, it will eventually reorganize itself into a complicated and unpredictable new system … nobody has a clue what I'm talking about."

Gore may be correct, for by now symmetry breaking is complete and even the physicists have ceased speaking to one another.

As the authors gazed over the now silent plaza, in which the various factions were frozen in their respective domains, they decide to reexamine the popular belief that thermodynamic systems require large numbers of particles.

One can create a statistical model that exhibits entropy with as few as four particles. Similarly, one can define a temperature and entropy for a single particle bouncing around in an irregular container. After a short while the position of the particle will no longer be predictable and all we can say about it is that it is somewhere in the box. Entropy has increased. This is not strictly speaking a single-particle system; the "irregular container" consists of many particles, but the dynamical part of the system is one particle. What is important is that the system exhibit complex behavior, regardless of numbers.

If we changed the irregular container to a polygonal container in which the sides of the polygon were extremely short, one could even derive an exact solution for the position of the particle at any time. But it would be useless because any slight deviation of the particle from its initial position would result in a totally different trajectory, just as in a chaotic system. Thus the essential difference between mechanical and thermodynamic systems lies not in the number of particles, or even in the fact that we can exactly solve the former but not the latter. Rather the difference lies in the fact that in thermodynamic systems the quantities of interest—such as entropy and temperature—require that the particles behave in essentially a random, or at least unstable, way. In such cases the fundamental quantities are not the exact position or velocities of the particles, but rather their distributions—the region in which they may be found, assuming we do not know the exact initial conditions.

Laplace, had he not been frozen in his private domain, would object again, maintaining there is no distinction whatsoever, since an intelligence vast enough, etc. etc.

We would reply that, fundamental issues aside, thermodynamics is not a luxury. Chemical reactions are governed by the laws of thermodynamics, as are large-scale motions of air, which have consequences as hurricanes and tornadoes. Planck discovered his blackbody theory through thermodynamics, and the Bekenstein-Hawking discovery that black holes have entropy (a subject that will appear in the Eighth Debates) would have been impossible without thermodynamics.

As the authors say their permitted say, Freeman Dyson, appearing late on the field of battle, also takes stock of the situation: "The best way to understand science is to understand the individual human beings who practice it. Science is an art form and not a philosophical method. The great advances in science have come from new tools rather than from new doctrines. If we try to squeeze science into a single philosophical viewpoint such as reductionism, we are like Procrustes chopping off the feet of his guests when they do not fit on to his bed. Science flourishes best when it uses freely all the tools at hand, unconstrained by preconceived notions of what science ought to be. Every time we introduce a new tool, it always leads to new and unexpected discoveries, because Nature's imagination is richer than ours."

In Which the Sixth Debates Are Summarized

And so the various disciplines were scattered upon the face of all the Earth. As you stepped among them and watched the whirlwinds touch down and sweep away the tracts, cleansing the plaza for tomorrow, you and your friend discussed what, if anything, had been concluded today.

The main message of the debates was that nature distinguishes handedness at all levels. Only the simplest objects, for instance spheres, are identical to their mirror images; most objects are not. Your left hand cannot be superimposed on your right hand; a right-hand screw is fundamentally different from a left-hand screw. Objects that display handedness are termed *chiral*. From animal species and human organisms, all the way down to the level of elementary particles, objects are overwhelmingly chiral.

The chirality of biomolecules, in particular, is crucial for life. Although laboratory experiments tend to produce racemic mixtures—mixtures containing equal amounts of left- and right-handed molecules (termed L and D for levorotary and dextrorotary) — essentially all amino acids found in life on Earth are L, and all the sugars found in RNA and DNA are D. While efficient catalytic activity requires biomolecules to be all of one handedness, there is currently no understanding about how one type came to predominate over the other, and speculation became rampant. The difficulty in coming up with a mechanism to explain chirality led at least one academician to propose that the handedness of terrestrial biomolecules was an accident. As Kondepudi demonstrated, the process of *spontaneous symmetry breaking* during crystallization can produce sodium chlorate crystals of entirely one handedness, although which chirality results in each trial is a random event.

Chirality exists in subatomic processes as well. All interactions involving the electromagnetic, gravitational and strong nuclear forces are indifferent to left and right, or, as scientists say, parity conserving. But the weak nuclear force distinguishes left and right, or does not conserve parity. The parity-violating weak force appears to be a fundamental asymmetry of nature.

As twilight set in, participants realized that the argument over chirality really concerned complex systems and the debate became more general. What is a complex system? What is an emergent property? Everyone had his own ideas on the subject and no consensus was reached. Roughly speaking, an emergent phenomenon, or "emergent," is a property that arises as the complexity of a system increases and is not observed in the behavior of the system's individual components.

Most academicians and visitors were willing to accept the existence of emergents. The real question was whether their properties could be reduced to more basic principles. And there disagreement became fierce. On the ultraviolet end of the spectrum were the New Agers, who said, "Never!" On the infrared end of the spectrum was Steven Weinberg, who declared, "Always!" at least in principle. The authors argued that attempts to reduce thermodynamics to particle physics have thus far failed, Weinberg's statements to the contrary, but that reductionism was largely responsible for the success of Western science. The rest of the company was arrayed between the extremes, but the debate became so unstable that symmetry spontaneously fractured, leaving the world divided into the noninteracting disciplines we find today.

Despite your increasing level of fitness over the past few days, like Illarion after his twenty-five-year argument with Merobaudes, you found yourself wearied beyond words by the evening's debate and again resolved to quit the Academy immediately. In any case, surveying the fractured landscape around you, it seemed inevitable that future debates would be canceled once and for all. But as you chanced across the authors by a campfire, they assured you that thermodynamics would restore symmetry by the morning. Indeed, as the whirlwinds continued to generate entropy, the authors divided the yarrow stalks and the hexagram to emerge was *chin*, the sun rising over the Earth, progress. That meant tomorrow—quantum mechanics.

You walked out to the beach to collect a few seashells and get some sleep.

IS THE UNIVERSE WEIRD?
Quantum Mechanics

The subtlety of nature is far beyond that of
sense or understanding.
—*Francis Bacon*

Behold the light emitted from the Sun,
What more familiar, and what more unknown?
While by its spreading Radiance it reveals
All Nature's Face, it still itself conceals.
—*R. Blackmore, 1712*

The Question

Many tales are told of the Theban Sphinx. One, of highly dubious origin, relates that the Sphinx was a fabulous monster with the voice and face of a beautiful woman, the wings and claws of a great bird, and the body of a lion, who lay in wait for victims along the narrow road to Thebes. As a traveler approached, the Sphinx blocked the path and put forth a riddle, "What walks on four feet in the morning, two in the afternoon, and three in evening?" and when the traveler failed to answer correctly, she devoured him. Only Oedipus, according to this corrupt version of the story, saw the solution: "Man, who as a babe crawls on all fours, who as an adult walks on two feet and who in old age requires a cane for support." At which the Sphinx, despondent, hurled herself off the cliff and perished.

In another apocryphal version, Francis Bacon claimed that the legend of the Sphinx was an allusion to science. Science, being the wonder of the ignorant and un-skillful, may be not absurdly called a monster. The many shapes of the Sphinx allude to the variety of matters with which science deals. The beauty and elegance of science are reflected in the face of a woman; the Sphinx's wings represent scientific discoveries, which spread and fly abroad in an instant; and the arguments of science penetrate and hold fast the mind, like claws, sharp and hooked.

According to yet a third version, the Sphinx was a close relative of the Rabbi of Minsk. It is told how Rabinovich and Abramovich went to the Rabbi to settle a dispute over a particular tract of land. Rabinovich said, "O Wise One, you must tell Abramovich to return this parcel of land. It has belonged to my family for generations and you cannot allow this thief to steal it." The Rabbi nodded solemnly and said, "Yes, yes, you are right." But then Abramovich turns to the Rabbi and says, "O Wise One, my family has used this land as a wheat field for generations. If not for us the land would have gone to ruin." Hearing this, the Rabbi nods gravely and says, "Yes, yes, you are right." Afterward, a third party, Yankelevich, who has been listening to the conversation, turns to the Rabbi and says, "This all seems very complicated. Everyone says you are very wise, but I saw you agree to two totally contradictory claims." The Rabbi nods solemnly and says, "Yes, yes, you are right."

But the true story of the Sphinx is this: Last night, you fell into a dreamless sleep and awoke into a state of awareness truly heightened by the discovery of a lion sleeping between you and your friend. Startled and fearful, the two of you took to your heels, only to find as you looked more calmly over your shoulders that it was not a lion you saw, but a great bird. Shaking your heads in disbelief, you closed your eyes and opened them only to see that the bird had been replaced by a beautiful woman. At that, the three of you fell into conversation and you asked, "How is it possible for you to sometimes be a woman, sometimes a bird and sometimes a lion?"

"Ah," she replied, smiling, "quantum mechanics."

Fin de Siècle Puzzles

The Sphinx was not the only unusual aspect of the morning. Yesterday's debate began in an entirely congenial fashion and most of the others had gotten under way as academicians divided themselves into two or three opposing factions. On this morning of the Seventh Debates, however, you detect an altogether different atmosphere. Busloads of tourists, especially from Southern California, have already arrived, each carrying a copy of Jung or Deepak Chopra. (A special bus for spiritual advisers has been chartered.) The Knowledge Marker is also behaving strangely. At one moment it is registering a high value for Academy ideas; at the next moment it registers zero; at other moments it seems to vanish altogether. The academicians themselves have not formed large alliances, but seem to cluster into small groups of hardly more than one person apiece. Some academicians have come armed with dueling pistols and accompanied by seconds. All in all, the picture has not left you at ease. Once again you turn to the Sphinx and ask, "What is going on?"

"Ah," she replies, this time in the form of a great bird, "quantum mechanics."

The Sphinx is not incorrect. Quantum phenomena have left people ill at ease since the late 1800s when Heinrich Hertz in Germany and Aleksandr Stoletov in Russia independently discovered the photoelectric effect. At first glance, the photoeffect seems rather simple: a beam of light was directed onto a piece of metal, typically a block of zinc, and the metal was found to emit an electric current, or in modern language, a stream of electrons. On the other hand, in terms of the understanding of the day, the photoeffect was impossible to explain.

For example, if the light was too red—if it was of too low a frequency—then regardless of how bright the source, the zinc would emit no electrons whatsoever. Above the threshold frequency (typically in the ultraviolet), when electrons *were* emitted, their numbers increased with the light's brightness, but did *not* change with the frequency. Yet the *energy* of the electrons depended on the frequency and not on the brightness.

How are we to understand all this? "Not from Maxwell's theory," replies the Sphinx. Indeed, it is all exactly contrary to Maxwell's theory of light. According to Maxwell, a light wave's energy is proportional to its brightness—or as scientists say, its *intensity.* By increasing the beam's intensity, one should be hitting the zinc with arbitrarily large amounts of energy. *Something* should happen. Below the threshold frequency nothing did. For the same reason, once the electrons are ejected, increasing the light intensity should increase the electron energy. Again nothing. Contrariwise, according to Maxwell, the light energy does *not* depend on frequency. So how can changing its frequency affect the energy of the electrons?

These problems, and others, suddenly vanish if light energy came prepackaged in small units. Instead of regarding the intensity of a light beam as related to the amplitude of a wave, as Maxwell did, we regard it as proportional to the *number* of packets. In which case, increasing the light intensity merely increases the number of packets, and since each packet kicks out a single electron, the number of ejected electrons goes up as well. If a packet's energy is smaller than the minimum energy required to knock an electron free of the zinc, nothing happens. Furthermore, again contrary to Maxwell, we associate the light frequency with the energy of the packet; the higher the frequency, the higher the packet energy. Thus, raising the frequency produces higher-energy electrons, as observed.

In 1900 Max Planck introduced the quantum of energy, "quantum" from the Latin for "how great" or "how many." The idea that Planck was forced to accept much against his will was that in nature energy comes in discrete values, the smallest value being one quantum. Einstein took over the concept and in 1905 postulated that light also comes in quanta; in doing so he explained the photoelectric effect. The small energy packets are of course the light quanta, which since the 1920s have been called photons. The quantum hypothesis was also used by Bohr to derive the hydrogen spectrum, which arose in the First Debates. Today, the idea that natural phenomena are often quantized is such a fundamental part of our worldview that the word "quantum" no longer surprises anyone.

"It should," says the Sphinx.

Isaac Newton, among others, championed the idea that light—like everything else in the Newtonian cosmos—was composed of small particles, "corpuscles," he called them. Nevertheless, the corpuscular theory had trouble explaining certain optical phenomena (in particular refraction) and so other philosophers, including Descartes and Christiaan Huygens, opposed it with their own theory, that light was composed of "vibrations"— waves. The contest between the rival systems went on for over a century (and you have been tired after less than a week) with little in the way of deciding the issue, until around

1800 the doctor and linguist Thomas Young—later famous for deciphering the Demotic script on the Rosetta stone—performed a crucial experiment.[†] Young passed a beam of light through two pinholes and noticed that on a distant screen the light formed an *interference pattern.*

You have undoubtedly seen interference patterns at the bottom of a swimming pool. Ripples of water crossing the pool are waves, with crests and troughs. When a crest meets a crest, the height, or amplitude, of the waves add to produce a new wave whose amplitude is the sum of the individual heights. We say the waves are in phase. When a crest meets a trough of equal amplitude, the waves cancel out and the water at that point remains undisturbed. Whether you see bright and dark regions on the pool floor depends on whether the waves are in or out of phase. The phenomenon of wave addition, with phases taken into account, is termed *interference* and is characteristic of all waves; it is probably their most important characteristic.

And so when Young observed an interference pattern on the distance screen, he concluded that light behaved like a wave. In fact, assuming the dark spots were places where crests fell on troughs, he could use simple geometry to calculate the wavelength of the light—the distance between two successive troughs or crests. Young's experiment actually works better with thin slits than pinholes, and nowadays with a laser as the light source, it takes students five minutes to discover that the wavelength of red light is a bit less than one millionth of a meter.

Interference phenomena are almost impossible to explain by particles, yet a few opponents of waves had yet to die off. The end came in 1850 when Jean-Bernard-Léon Foucault showed that light in water travels more slowly than light in air (whereas according to the corpuscular theory the attractive force of the water pulling down on the light must cause the corpuscles to travel faster). After Foucault, the corpuscular theory of light was abandoned once and for all.

Until the photoelectric effect.

Then light suddenly became a particle again.

"How can light sometimes behave like a wave and sometimes like a particle?" your friend asks as the sun begins its morning ascent.

"Ah," the Sphinx replies and growls, "quantum mechanics."

In Which the Quantum Nature of Light Is Revealed (Again)

When all was said and done, the visitors to the Academy remained unimpressed by the photoelectric effect and the exposition elicited much grumbling. Deepak Chopra had already given numerous lectures about quantum healing, so what could be so mysterious about photons knocking a few electrons out of a piece of zinc? Around the central plaza the general sentiment was to skip the preliminaries and get on with the show. Nevertheless, the authors responded that in the spirit of know-nothingness, we need to reexamine fundamental concepts. This is especially true of quantum mechanics, which is both the most certain and least certain of the sciences. No prediction of quantum mechanics is known to be wrong and quantum mechanics has become the indispensable tool of modern physics. The debate is over what the basic concepts of quantum mechanics mean. For seventy nonstop years the disagreement over these issues has had academicians at each other's throats (blood has been drawn in the palazzo this morning).

In this regard one can do no better than recollect how the great scientists of the early twentieth century greeted Einstein's explanation of the photoelectric effect. It is 1911, six years after Einstein has introduced the light quantum. You are there. Among the scientific community there is still only one word for the idea that light can behave like both a particle and a wave:

Preposterous.

Einstein himself warns, "I insist on the provisional character of this concept, which does not seem reconcilable with the experimentally verified consequences of the wave theory." Robert Millikan is so opposed to the idea of quanta that he spends years measuring the photoeffect only to prove Einstein right and himself wrong. In 1916, when the results are in, Millikan declares that despite the fact that Einstein's explanation is apparently perfect, it is "so untenable that Einstein himself, I believe, no longer holds it." Einstein has not retracted, but in 1951 writes, "All these fifty years of pondering have not brought me any closer to answering the question, What are light quanta?" Planck and others, in proposing Einstein's membership to the Prussian branch of the Academy, give him the highest praise, concluding, "His hypothesis of light-quanta cannot really be held too much against him."

Light quanta, photons, were and remain a Sphinx. Anyone who does not find them puzzling has lost the ability to be puzzled. Not wishing to admit to such a fundamental loss, a few visitors have agreed to sit patiently through the next exposition. The Sphinx-like nature of photons is best illustrated by a modern version of Young's experiment. Once again we consider a screen with two slits in it, which project an interference pattern against a distant wall. For today Academy technicians have set up a laser whose intensity can be adjusted, and on the far side of the slits they have mounted a video camera which can record the resulting interference pattern and broadcast it to the crowd gathered on the central plaza.

The techs first shine the laser through the slits to get the same pattern of bright and dark spots that Young observed hundreds of years ago, the "two-slit pattern." On the other hand, when the techs cover up one slit or the other, the big closed-circuit TV displays a much different pattern—a "one-slit" pattern, which is shown in Figure 7.1b. They also cover one slit and then the other to produce two one-slit patterns side-by-side, but these do not resemble the two-slit pattern. The experiment indicates that to produce a two-slit pattern the light wave must really pass through *both* slits. Some of the crowd wonders what happened to Sunday afternoon football.

Now comes the crucial moment. The technicians lower the intensity of light until only one photon passes the slits at a time. What does the crowd observe?

As each photon strikes the video detector, it produces a single point of light on the big screen above the plaza. Yet as hundreds and thousands of photons hit the detector, an ordinary two-slit interference pattern gradually comes into existence. The process might be called the pointillist school of quantum mechanics. People close to the TV see

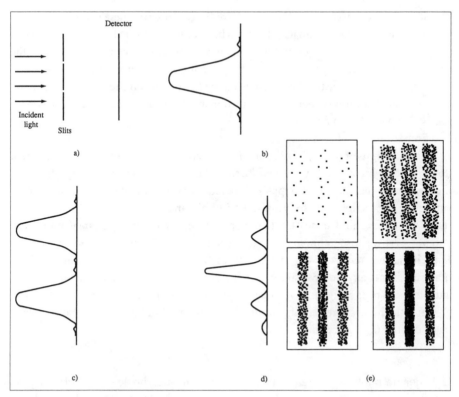

Figure 7.1
The two-slit experiment. In (a), collimated light is passed through a screen containing two slits. If only one slit is open, the intensity of light on a distant detector follows a "one-slit" pattern, as in (b). The peak represents the spot of highest intensity, which then falls off. Repeating the experiment with one slit closed, then the other, produces two single-slit patterns, as in (c). If both slits are open, however, one gets the very different two-slit pattern (d), showing that to produce it, the light must pass through both slits. All this is true even when the intensity is reduced so much that only one photon passes through the slits at a time. As photons pass through, a pattern is gradually built up. If one slit is open, a one-slit pattern gradually emerges. If the slits are alternately opened and closed, then two single-slit patterns emerge. Only if both slits are open as each photon passes does a two-slit pattern result, proving that each photon must have passed through both slits. The pattern, as one would actually see it building up on a detector, is shown in (e). The dark, central strip corresponds to the central peak in (d), while the left and right strips correspond to two of the side lobes.

nothing but a collection of dots. Those farther back are able to perceive Seurat's *Sunday Afternoon on the Island of La Grande Jatte.*

"You are witnessing the greatest dilemma of the twentieth century!" one academician cries, dying of ecstasy on the spot. Most of the crowd merely shrugs as the body is carried off, but for the first time a murmur of astonishment rises from several quarters.

"This is exceedingly peculiar," your friend says, taking the words right out of your mouth. "Surely if photons each produce exactly one spot of light on the screen, they are behaving like particles."

"And in that case," you add, "they must each go through one and one slit only. A particle cannot go through two slits. How are we getting a two-slit pattern?"

This time the technicians randomly block one slit or the other as photons pass. As the pattern builds up on the TV screen, it soon becomes apparent that the result is two one-slit patterns, not one two-slit pattern. They repeat the experiment several times until you are convinced that to get a two-slit pattern, both slits must remain open as *each* photon passes.

"A photon," says academician Dirac, appearing briefly, "can interfere only with itself." To paraphrase, each photon behaves like a wave, an extended object that passes through both slits; it interferes with itself and is detected at one point like a particle.

Unconvinced, your friend says, "I bet we can decide which slit the photon passed through. Why don't we mount the video detector on tiny springs? Then, if a photon came through the left slit, it will kick the detector a little bit to the right, and if it came through the right slit, it will kick the detector a little bit to the left. We will have then succeeded in observing both the wave nature of the photon—the interference pattern—and the particle nature—the slit from which it emerged."

Academy technicians, being the finest in the world, ready the new experimental setup within seconds. As your friend predicted, each photon gives the detector a little kick and so we are able to determine from which slit it appeared. As the tally mounts, it appears that equal numbers go through both slits, but on a random basis. Unfortunately, as the bright spots accumulate on the screen, no interference pattern is visible; somewhere along the line it has been washed out. The crowd moans, convinced there has been a technical difficulty.

"You cannot measure all aspects of the photon at once," says the Sphinx.

She speaks wisely. The interference pattern is a record of the photons' position on the screen. The kick given to the detector is a measurement of the photon's impulse, or momentum. If the detector is mounted on springs, however, it invariably jiggles like a badly held camera and the interference pattern is washed out. Try as you might, it is impossible to measure position and momentum with perfect accuracy simultaneously. This is an example of the Heisenberg uncertainty principle.

"Ah," someone in the crowd exclaims, "I knew it."

As you look back on events, this moment stands out as a crucial one. Recognizing familiar territory, the visitors have suddenly displayed heightened interest in the discussion. In fact, one of the Californians nearby remarks without warning, "Hey dude, this is quantum consciousness. You decide to measure the photon's wave or particle nature. Your choice has made reality manifest. The observer is part of the system. Thou art That."

At which a nearby physicist pulls out his revolver and shoots the tourist. "Yeah dude, see if you can alter that reality." The Academy security guards applaud. One less interloper.

You turn to the Sphinx and ask if it is really true that the observer can determine the outcome of an experiment. The Sphinx replies, "It is true that you can observe only one aspect of the photon at a time."

The cryptic response has left you dissatisfied, but the Sphinx proposes another version of the experiment, popularized by John Wheeler: the "delayed-choice experiment."

The setup is virtually the same as the last, with two small modifications. First, the technicians have put an electronic locking lever on the video detector, so one can lock it in place to record the interference pattern, or unlock it, allowing the detector to ride on springs and measure the photon's kick. The technicians have also placed the video camera at the far end of the Academy, so that after passing through the slits the photons must travel across thrice times nine kingdoms to reach the detector. This gives you ample time *after* the photons have passed through the slits to lock or unlock the detector and decide what to measure.

Surely, you say, the photons will have already gone through one slit or the other, so the momentum can be measured, or they will have already gone through both, so that an interference pattern results. One can't influence their behavior after the fact. This violates causality (in which you have come to believe, sometimes).

Nevertheless, when the experiment is carried out, it gives exactly the same results as before: you wait for the photons to pass, quickly unlock the detector, and find you are able to determine that the photon emerged from one slit, like a particle. But you don't get an interference pattern. Locking the detector gives an interference pattern and you must conclude the photons went through both slits, like a wave.[†]

This is unbelievable! you gasp. How can you determine the nature of the photons *after* they have already passed through the slits?

"Time is nonlocal!" cries one of the Californians.

You look to the Sphinx for explanation, but she is eating the tourist. When she returns she says, "Watch your language."

In Which the First of the Paradoxes Are Interpreted

In retrospect, lunch was marked by a certain schizophrenia. As the Sphinx dined, you and your friend engaged in the animated, puzzled conversation that always results when people are confronted for the first time with the paradoxes of quantum mechanics. At the same time a sullen "show me" attitude manifested itself among the spectators, a poll of whom showed that 64 percent had their own solutions to the paradoxes, 15 percent thought that "quantum mechanics" referred to their neighborhood auto repairmen, and the remainder had come to the Academy to receive their Unified Field of Invincibility certificate, notarized.

A number of philosophers, in fact, preempted the authors by announcing to the assemblage that the two-slit experiment marks the end of causality as we know it. Whether a photon went through the right slit or the left one was a purely random process. The laws of quantum mechanics do not tell us through which slit each photon will pass; they merely tell us the odds. The deterministic universe of Newton is finally, irrevocably overthrown. Once and for all, too. At its most fundamental level, nature is not deterministic, but probabilistic. The philosophers build a huge pyre and begin burning copies of Newton's *Principia*.

Jung adds, "A great change in our standpoint is setting in. Causality is statistical and must allow for exceptions!"

(Stormy Applause.)

It is true that quantum mechanics builds probability into the theory at the ground floor. As mentioned in the Fifth Debates, quantum mechanics does not answer certain simple questions like, "Is the photon here or is it there?" It merely gives you the probability of finding it in one place or the other.

"But how is this different from ordinary probabilities?" you ask. Excellent question. When wind whistles by telephone wires, they sometimes sing. The singing is accompanied by the release of small wind vortices into the air, much like those that accompany the Sphinx when she takes off. The production of vortices is totally random; one cannot predict where or when one will form on the telephone line. The only difference is that in the case of the singing telephone wires we have the suspicion that if we knew everything about the system, we could make predictions about the air's behavior. In view of chaos theory, this may be more an act of faith than scientific reality. Nevertheless, in quantum mechanics, there simply is no more fundamental explanation. Probability does seem to be a fundamental feature of nature.

Many scientists find this hard to swallow and for that reason they have developed "hidden variable theories." As with the singing telephone lines, there really should be something else behind quantum phenomena that behaves deterministically, and once these "hidden variables" are known we will be able to predict with certainty the behavior of quantum systems. Plans are made to debate the issue later in the day.

In spite of all this, it is possible to make too much of quantum probability. Many aspects of quantum mechanics are as deterministic as Newtonian mechanics. For example, when you measured the interference pattern on the screen, you were forced to conclude that the photon went through both slits. This is not a question of probability; the photon went through two slits.

What is probabilistic in quantum mechanics and what isn't?

Another excellent question.

In sneak previews of today's debates we briefly discussed the wave function and the Schrödinger equation. They are not too mysterious. Early in the twentieth century Louis de Broglie proposed that if photons sometimes behave like particles, it would not be unreasonable to suppose the reverse—particles could behave like waves. Subsequent to de Broglie's proposal, researchers confirmed that particles do manifest wavelike properties; the two-slit experiment could just as easily be carried out with electrons instead of photons. Newton's second law governs the motion of classical particles. In quantum mechanics we replace the particle with a wave equivalent known as the wave function. Schrödinger's equation is the Newton's second law of quantum mechanics; it governs the evolution of the wave function in any physical situation.

Like Newton's second law, Schrödinger's equation is deterministic and time-reversible. This is why quantum mechanics, neither more nor less than Newtonian mechanics, reveals no arrow of time. Unlike Newton's second law, Schrödinger's equation does not tell you exactly where you will find the particle at any time in the future. Rather, the wave function is assumed to encode all possible outcomes of a measurement and give the probability that the particle will be found at a certain position.* Schrödinger's equation thus determines how probability varies from one place (and time) to the next. In the two-slit experiment, Schrödinger's equation gives the probability that a photon will be recorded at a certain place on the video detector, but it does not tell you anything more. Specifically, it does not tell you how the extended wave ends up being detected at a single point.

The infamous "collapse of the wave function" has reappeared. After an observation, the photon is detected at a definite location. Potentiality, or probability, has become actuality. Once detected by the video camera, a photon never dissolves backward into a wave function; hence Penrose's quantum arrow of time from the Fifth Debates.

But you see something is very strange about the legendary collapse. The Schrödinger equation, because of its deterministic nature, cannot explain it. As far as the Schrödinger equation goes, it will evolve the wave function forever as if nothing ever happened. Niels Bohr, shortly after quantum mechanics was invented, declared that collapse was due to the act of observation but otherwise lay outside the realm of quantum mechanics.

You wrinkle your nose. This seems more like a swindle than an explanation. Others feel the same way. Suddenly three hundred people are clamoring to put forward their solution to one of the greatest intellectual conundrums of the century. With great speed the Academy security guards erect a potential barrier to the crowd, but a few participants succeed in tunneling through anyway. There is no choice but to turn off the microphone for the moment and, like Bohr, refuse to deal with the issue.

Returning to probability. The fact that the wave function gives the probability of the outcome of a measurement rather than the position of a particle has led to some novel statements about quantum phenomena. Amit Goswami, stopping at the Academy on a promotional tour of his book *The Self-Aware Universe*, says about wave functions: "Initially people thought, 'Oh, they are just like regular waves. But very soon it was found that, no, they were not waves in space and time. They cannot be called waves in space and time at all—they have properties which do not jibe with ordinary waves. So they began to be recognized as waves of potential.'"

"Is he talking about a thirteen-dimensional universe, outside of space and time?" you ask, puzzled. Hey, we're all bozos on this bus. Certainly wave functions oscillate in space and time just like ordinary waves; all of their properties are wavelike. They merely do not represent sound waves or light waves; they represent probability waves (which is mysterious enough).

But the New Age is not alone in misconstruing the role of probability in quantum mechanics. When academicians talk about the probabilistic nature of quantum mechanics, they forget they are speaking only about elementary phenomena. You may have heard that the decay of a radioactive atom is a probabilistic event; one cannot predict exactly when it will occur. True—if you are talking only about a single atom, or a sample

---

* Precisely the square of the amplitude of the wave function at any position gives the probability that the particle will be found there.

in which the decay products are immediately removed from the system. But if the products recombine or if other interactions with the environment are present, then the sample no longer decays at the rate characteristic of random processes.* The rate depends on *everything* and for complicated systems it is no longer probabilistic. The moral is, you can't decide on whether a process is probabilistic until the entire process is complete. You could not decide whether a photon passed randomly through one slit or the other until you decided whether to measure the interference pattern on the detector.

The two-slit experiment illuminates the problem with probabilities in yet another way: the intensity of the beam is given by the square of the wave function, which in turn gives the probability of finding a photon at a particular location. So *intensity is a measure of probability.* However, we can put a thin glass plate behind one of the two slits. This does nothing to alter the intensity of the light at that position and so should not affect probabilities. But it *does* alter the phase of the different portions of the beam passing through the slits. Where a crest previously fell on a crest and the intensities added, now a crest may fall on a trough and the intensities subtract. A previously bright spot on the detector may now be a dark spot. We have changed the result of the experiment *without* changing the photon's probability of passing through a given slit. As before, one does not know the result until the fat photon has sung.

Throughout the exposition the crowd has been growing restless. This all seems very esoteric. "What's the point?" shouts one of the visitors, who wears a "Synchronicity Is All There Is" T-shirt, above the din of the lunchtime crowd.

Well, this is a little bit like smoking marijuana without inhaling (or similar experiences). If one is going to make statements like, "Natural laws are merely statistical truths and therefore must necessarily allow for exceptions," perhaps it would not be a bad idea to learn a little about the statistical nature of quantum mechanics. We have seen that there are statistical aspects to the theory but, in the first place, this is not the whole story, and in the second place, it is not the natural laws themselves that are statistical, rather their predictions. Although quantum mechanics allows a certain latitude in the results (photon passes randomly through right or left slit), the latitude itself is quite definite (photon lands on screen and does not vanish into never-never-land).

These distinctions will become clearer as we briefly examine the

Heisenberg Uncertainty Principle

The crowd roars.

Many things have been said about the Heisenberg uncertainty principle. An otherwise reputable physicist once remarked, "The Heisenberg uncertainty principle proves that anything is possible." Heisenberg himself said, "The more precisely we determine the position, the more imprecise is the determination of the velocity in this instant, and vice versa." In other words, the impossibility of measuring both the position and the velocity of a particle simultaneously is a result of the measurement process itself; the experimenter disturbs the system by the very act of measurement.

---

* Meaning that the number of atoms decreases exponentially with time.

"Therefore the experimenter is a participator!" cries someone. "A participator in the creation of reality!"

What is true is that the Heisenberg uncertainty principle is one of the most important statements arising from physics. With Gödel's theorem and the strictures of chaos theory, it puts an absolute limit to knowledge nature permits us to have about the universe. But let us not go transcendental.

In some respects the uncertainty principle doesn't even have anything to do with quantum mechanics. If your friend were to ask, "Where is New York City?" you might answer, "Northeast of New Jersey," or perhaps, "Southwest of Connecticut." But it would make no sense to say, "New York City is located 150.23 kilometers northeast of Philadelphia or 322.36 kilometers southwest of Boston," for the simple reason that you can pinpoint a city block to .23 or .36 kilometers but not New York City itself.

Similarly, as the discussion adjourns to the fishpond, we find that it doesn't make sense to ask certain questions about waves. The Sphinx dives into the center of the pond, splashing about for her daily bath, in the process creating ripples that spread outward to the edge of the pond. Watching the expanding waves, you take it into your head to ask, "Where is the wave?" and Werner Heisenberg, standing nearby, hands you a few sticks and pushes you and your friend into the water. "Autumn follows summer," he says. After pondering the Zen master's instructions the two of you attempt to stake out the position of the wave, but by this time it has spread over the entire surface of the water, and you realize the task is going to be difficult.

Determined to locate the wave, you ask the Sphinx to quiet down. The great bird obliges, then begins a sharp splashing with her wing, which sends short pulses in your direction. Here at least is a possibility of a measurement and you and your partner begin planting stakes in the mud. You find it is not difficult to say, "Here is a peak," when a pulse passes by a stick. On the other hand, by definition a wave is a repetitive phenomenon; if only one pulse passes by your eyes, you can't measure a wavelength (the distance between two successive peaks); you can't even say what you see is a wave.

Failing to find a wave, you ask the Sphinx to begin splashing rhythmically. Again she obliges, sending out a complicated series of ripples in your direction. Now you have a genuine wave and decide to measure its wavelength in terms of stick lengths as it goes by. But to be sure you have correctly measured a full wavelength, you must observe a number of peaks (to be certain you map out the true shape of the wave and have not measured the distance between intermediate humps of the complex waveform). However, in doing so you have measured an extended object and have no right to say exactly where it is located. You are back to the wave being spread across the pond (see Figure 7.2).

We see that determining the location of a wave and its exact wavelength are two mutually exclusive demands. This is the uncertainty principle. Notice that there is no statement about quantum mechanics. It works for ordinary waves as well as quantum waves. Quantum mechanics comes in via de Broglie's discovery that particles have wave properties. In particular, the particle's velocity (momentum) is connected to its wavelength. If we plug in de Broglie's relationship to the classical uncertainty principle, we get the Heisenberg uncertainty principle.

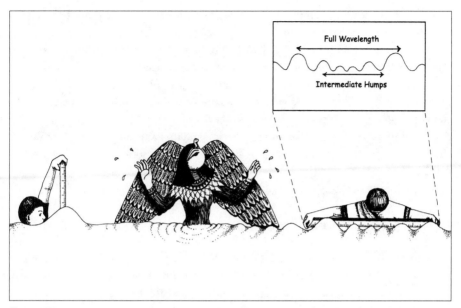

Figure 7.2
A pulse is easily located, but its wavelength can't be determined. A complicated wavetrain has a wavelength that can be measured, but it is an extended object and its position can't be localized.

Niels Bohr is perturbed. "I have always maintained that it is the very act of measurement, so to speak, which disturbs the system and leads to the uncertainty relationship. To measure the position of an electron requires a photon to hit the electron and bounce into the eye. But when the photon hits the electron, it imparts a momentum to it, so that the momentum of the electron is changed and becomes uncertain, as Heisenberg has said in his original essays."

Heisenberg nods assent.

Surely this misses the point. There is no mention of measurement in Heisenberg's actual derivation. The uncertainty relations are a purely mathematical statement about waves, not of measurements. It is true that any measurement of a system must be compatible with the uncertainty principle, but measurements do not produce the uncertainty principle; otherwise we would have to ask why every conceivable measurement technique results in the same mathematical statement. The Heisenberg uncertainty principle is a law of nature, not of measurement. Which leads to the maxim:

Heed the calculation, not the interpretation.

"What's more," adds Heisenberg's rival, Erwin Schrödinger, walking up to water's edge, "Heisenberg's is not the best uncertainty principle. I have a better one."

Schrödinger is correct; however, we refer you to the Commentaries on the Seventh Debates for a few more words on this subject.

Regardless of such refinements, the most important message of the uncertainty principles is that it is impossible to measure all aspects of a system simultaneously. Such an impossibility leads directly to the concept of *superposition*. If one cannot observe all aspects of the Sphinx simultaneously, one is not allowed to conclude that the Sphinx is one-third woman, one-third bird, and one-third lion. That would be a different animal. One can only conclude that she is somehow all of them together, but nature permits us to perceive only one aspect of her at a given time. The "collapse of the wave function" dilemma can then be rephrased as, Why is a Sphinx, which according to quantum mechanics exists as a superposition of states, inevitably observed in one state only? The photon must also be regarded as a superposition and the same question applies: Why is a photon sometimes observed to behave like a particle and sometimes like a wave, but never both?

And thus all of those gathered at the Academy are abruptly confronted with

The Measurement Problem

In retrospect, it seems the audience was well behaved until late afternoon, when the debate on the measurement problem began. The physicists began reviewing their notes, responding irascibly to anyone who disturbed their concentration, for they knew what was at stake in the trial to come. The visitors, too, understood that the measurement problem touched not only quantum mechanics but the deepest issues of transcendence, consciousness, diet, potentiality . . . of Reality itself. Trading of Academy ideas was suspended on the world market until the outcome of the debate could be observed.

Like many debates in quantum mechanics, this one can be traced to Niels Bohr and his Copenhagen school. For Bohr, the external world, the world we live in, is classical (i.e., non-quantum-mechanical). Although photons and electrons in the quantum realm exist as superpositions, in the classical realm photons and sphinxes are observed in unambiguous states. The transition from the quantum world to the classical world is the act of observation.

"No photon exists," Bohr explains, "until a detector fires, only a developing potentiality. Particlelike and wavelike behavior are properties we ascribe to light. Without us, light has no properties, no existence. There is no independent reality for phenomena nor agencies of observation."

But as you noticed earlier, the "Copenhagen interpretation" does not explain how the observation causes the collapse; this is simply outside the theory. "What a swindle," you curse under your breath, again dissatisfied.

Many in the crowd agree. Now that the potential barrier has been removed from the discussion, the spectators are agitating to be heard. "Consciousness! Consciousness!" the crowd begins to chant. "We want consciousness!"

There are hints of it in Bohr's statement. Somehow, the registering of an event in the consciousness of the observer turns potentiality into actuality. But Bohr denies this. A grain of silver bromide on the film would suffice. "The irreversible act of amplification" is what is important. Of consciousness, he says, "I am absolutely prepared to talk about the spiritual life of an electronic computer; to state that it is reflecting or in a bad mood. . . . The question whether the machine *really* feels or ponders, or whether it merely looks as though it did, is of course absolutely meaningless."

In the mounting noise, such fine distinctions go unheard. Arthur Eddington remarks, "What we are observing is not nature herself, but nature exposed to our type of question. Space exists only in relation to our particularizing consciousness. . . . Science is concerned with the rational correlation of experience rather than a discovery of fragments of absolute truth about an external world."

The eminent chemist Cyril Hinshelwood chimes in to suggest that a more appropriate name for elementary particles might be "manifestations."

Consciousness or observership? The distinction may be a fine one, but as the parade of speakers continues into early evening, it becomes clear that most physicists accept the idea that an experimenter participates in the creation of quantum reality. Eugene Wigner, with his usual modesty, says, "It appears that our theory denies the existence of absolute reality—a denial which is unacceptable to many. . . . I do not know how one could define operationally the reality of anything."

"Time is nonlocal!" someone shouts, risking death by Sphinx, referring to the delayed-choice experiment in which *you* evidently decided which property the photon would assume *after* it passed through the slits.

Wheeler replies, "The past has no existence except as it is recorded in the present." Indeed, Wheeler has summed up the entire Copenhagen interpretation with his slogan, "No phenomenon is a phenomenon until it is an observed phenomenon."

At which someone in the audience asks, "*No* phenomenon?"

And Wheeler amends the motto to "No elementary phenomenon is a phenomenon until it is an observed phenomenon." Wheeler does not wish to claim that the entire classical universe is produced as the result of observations. Or does he? He has remarked, "The universe is a self-excited circuit. As it expands, cools and develops, it gives rise to observer-participancy. Observer-participancy in turn gives what we call "tangible reality" to the universe." He draws a famous picture that illustrates his idea (see Figure 7.3) and coins the term "participatory anthropic principle" to express the possibility that observership is necessary to bring the universe into existence. Ignoring demands from some quarters of the audience, the academicians again defer discussion of the anthropic principle to a future debate.

Although Wheeler, like Bohr, is careful not to use the word "consciousness" in his speculations, others have gone the distance. Physicists John Barrow and Frank Tipler now take the floor. "We ourselves can bring into existence only very small-scale properties like the spin of the electron. Might it require intelligent beings 'more conscious' than ourselves to bring into existence the electrons and other particles?"

The idea that the collapse takes place upon entry of the measurement signal into the consciousness of an observer actually seems to have originated with Wigner, who went so far as to propose a model of how this might happen and even suggested searching for unusual effects of consciousness on matter.

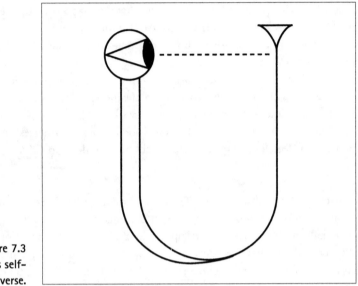

Figure 7.3
John Wheeler's self-
excited universe.

"B . . . but wait a moment," you ask courageously, "it seems to me you've been assuming the observer is part of the classical world." Absolutely true. "But isn't the observer a quantum system? If natural laws are quantum-mechanical, then shouldn't everything behave according to quantum mechanics?"

Many have asked the same question; it is the basic problem with the Copenhagen interpretation. Who decides when a system is quantum and when it isn't? Is a measurement accomplished when the signal enters the mind of Wigner, who is observing the system, or Wigner's friend, who is observing Wigner observe the system? And so on. Hawking suggests that the collapse is caused by the way we ask questions about observations, a remark even more cryptic than Eddington's.

With academicians so eager to consider the role of consciousness in creating quantum reality, one cannot blame the self-awareness movement for doing the same. Indeed, the past hour has witnessed a swift increase in the number of observer-participants crowding the plaza. Vendors are selling miniature sphinxes and cats, as well as guides to the various interpretations (with capsule biographies of leading proponents). By this time the academicians are outnumbered but fail to realize it. One of the most vocal participants on the left is Amit Goswami, who spoke briefly before:

"In quantum physics, objects are not seen as definite things. The material world of quantum mechanics is just possibility. It is consciousness, through the conversion of possibility into actuality, that creates what we see manifest. In other words, consciousness creates the manifest world." And a little later: "I have added [to the concept of delayed choice] the concept of 'self-reference.' Actually the concept of delayed choice is very old. It is due to a very famous physicist named John Wheeler, but Wheeler did not see the entire thing correctly, in my view. He left out self-reference."

We leave you and your friend to debate whether "the material world of quantum mechanics is just possibility." As for the rest, it is so late in the day. . . . If by "the universe is a self-excited circuit" Wheeler did not mean a self-referential universe, then the English language has lost any meaning it once had. And consciousness creating the manifest world? Well, long before Goswami there were the Upanishads:

> That which makes the eye see but cannot be
> Seen by the eye, that is the Brahman indeed.
> This Brahman is not someone other than you.

On the other hand, perhaps Goswami has something. To turbulent applause, Deepak Chopra assures the audience, "Your body is not separate from the body of the universe, because at quantum mechanical levels there are no well-defined edges. You are like a wiggle, a wave, a fluctuation, a convolution, a whirlpool, a localized disturbance in the larger quantum field. The larger quantum field—the universe—is your extended body." There is a certain poetry here. The applause redoubles.

Bryce DeWitt, generally acknowledged to be the father of quantum gravity, will have none of this. Consciousness? Bah, humbug. "The problem is not that we take our theories too seriously," he has been known to remark, "but that we don't take them seriously enough." If the Schrödinger equation does not predict a collapse of the wave function, then the matter is quite simple: the wave function doesn't collapse.

The universe splits.

DeWitt was not the originator of the many-worlds interpretation of quantum mechanics—that distinction belongs to Hugh Everett III, who published the controversial idea in 1957—but DeWitt remains its foremost champion.

The wave function contains all possible outcomes of a measurement. The usual assumption is that at the wave function's collapse one possibility is singled out and recorded as a measurement. But the theory doesn't say this; that is interpretation. According to the many-worlds view, every possible observation is realized in some possible universe. As DeWitt once said, "This universe is constantly splitting into a stupendous number of branches, all resulting from the measurement-like interactions between its myriad components. Moreover, every quantum transition taking place on every star, in every galaxy, in every remote corner of the universe is splitting our local world on earth into myriads of copies of itself."

DeWitt vividly recalls the shock he experienced on his first realization that there were 10^{100+} copies of himself constantly splitting into yet further copies, but he has steadfastly maintained that, as bizarre as it sounds, the many-worlds hypothesis is the only consistent interpretation of quantum mechanics and the only one that does not require additional metaphysical assumptions.

Yet other many-worlders do not believe in splitting universes. Murray Gell-Mann feels that the appellation "many-worlds" is unfortunate and that one should not speak of the alternate universes as "all equally real." Instead one should imagine all possible histories of this universe, each history with a different probability. Only one history is realized. (By history Gell-Mann evidently does not mean a past chronology of events, of which there is only one. We take up his cryptic statement again in Part II of the Ninth Debates.) So, you wonder, highly confused, does that mean that there are interpretations of interpretations of quantum mechanics?

Yes.

But whether one views the alternate universes as real or as mere possibilities, we may ask, does the many-worlds view deal with a quantum-mechanical issue? You see, the Schrödinger equation has much in common with insurance companies. Using actuarial tables, an insurer can calculate the odds of your death this year and adjust the premiums accordingly. A year from now, a many-worlder must interpret this to mean that in so many universes you died and in so many others you didn't. (Small consolation for those copies of you who kicked the bucket and the heirs of those of you who didn't.)

Every time you are faced with a probability distribution (the odds of dying) and a sample (the number of people the company has insured), a similar situation arises. But this is the nature of statistics, not of quantum mechanics. When you flip a coin, do you say the universe splits? When the muffler on your car breaks, do you say the universe splits? No you don't. A many-worlder would answer that a coin flip is a classical phenomenon whereas the many-worlds interpretation deals with quantum phenomena. But it doesn't. The many-worlds interpretation deals only with the probability that a certain event takes place, *not* with the quantum wave function itself.[†] Thus it is not answering a quantum-mechanical question and there is nothing in either classical or quantum physics that says a noisy exhaust system results in one universe where the Midas muffler company goes bankrupt and another universe where it doesn't.

In retrospect, no one had estimated the hostility between the proponents of the various interpretations. By sundown several duels had already taken place, though their outcome was indeterminate. On the grounds that consciousness figured in both Copenhagen and California, the New Agers called for an alliance with the Copenhagen school, to be called Nouvelle Copenhagen, but Bohr and his adjutants angrily rejected the suggestion. Hidden variabilists petitioned to be heard but their numbers fell short of the legal minimum. A flurry of excitement ensued, and the expectation of renewed trading on the World Idea Exchange, when the Copenhageners claimed the Sphinx was about to sign an exclusive endorsement, but she refused.

"Really," she said, "what do you take me for?"

Just then, as twilight descended, a new phalanx arrived on the scene with unfurled banners proclaiming *Decoherence!*

For the New Age has passed. No one in physics circles talks too much these days about consciousness forcing the collapse of the wave function. The magic word is "decoherence," a term coined by Wojciech Zurek at Los Alamos National Laboratory. For the first time, perhaps, physicists have provided a mechanism that may help us understand the collapse of the wave function.

Let us return to the two-slit experiment and Dirac's cryptic remark that "a photon can interfere only with itself."

"Yes," you agree, the authors didn't explain that as well as they should have.

Patience. In order for light to interfere in the two-slit experiment it must be *coherent*. That is, the light waves emerging from the two slits should be in phase. Only in this case (or in the case when they are exactly out of phase) can all the crests or troughs fall together on the detector to produce bright areas and dark. Nowadays, coherence is automatically achieved with the laser, which naturally produces coherent light.* If one instead illuminates the slits with two ordinary incandescent light bulbs, the phases are random and the light is *incoherent*, Crests fall on troughs indiscriminantly at any point on the detector and the interference pattern is washed out. You merely get a uniform illumination across the detector element, just as if you directed a table lamp onto the wall.

Now we can understand Dirac's statement. If a single photon passes through both slits, the phases of its different parts are naturally correlated, and the light interferes with itself. Photons from different sources passing through the slits have random phases and cannot interfere.†

At this juncture, you and your friend are undoubtedly wondering how a coherent beam *decoheres* to become an incoherent source. A natural question. Almost anything will do it. A leaky faucet dripping water in front of one slit would disturb the phases and decohere the light. It is much more difficult to produce a coherent beam than an incoherent beam; virtually any factor in the environment will cause a light beam to decohere.

Now we almost have it. Incoherent light actually behaves like a system of classical particles—billiard balls. Because incoherent light cannot interfere with itself, it does not manifest wave properties. There are no dark interference fringes where light plus light produces the absence of light. Light produces more light, just as if the photons were Newtonian corpuscles—the more of them you have, the brighter it gets. At the opposite extreme—coherence—photons display both their wave and particle nature with all the peculiar quantum-mechanical properties demonstrated by the two-slit experiment.

And so here we are: the process of decoherence changes a system from a quantum-mechanical system into a classical system—the collapse of the wave function. In this picture the quantum system gradually begins to behave like a classical system, but not in one fell swoop. Rather, for a macroscopic system, the collapse might take a nanosecond. Any environmental disturbance is expected to contribute toward decoherence be it a speck of dust or microbes applauding the photons as they pass. In many ways decoherence is not unlike waking from a dream. Sometimes when you are half awake you are aware of both the dream world and the outside world. If you wake up slowly enough you can retain a memory of your dream, but if a noise disturbs you and you wake up with a start, the dreams vanish forever.

---

* Young produced a coherent beam by passing the light from his source through a collimator—a small hole in a screen—before sending it through the pinholes. Light coming from a small region of a source tends to be roughly in phase.

Orthodox Copenhageners are dismayed. Their interpretation of quantum mechanics assumes the measurement to be a single event (what exactly that event is has been the subject of the afternoon's debate). According to Bohr's school, decoherence should set in instantaneously. And whenever it does, the world splits.

Some physicists have drawn attention to the fact that the decoherence program does not really give the outcome of the measurement, but only changes a quantum superposition to a set of classical probabilities. That is, if incoherent photons act like classical billiard balls, then each has a fifty-fifty chance of going through the right or left slit, but that exhausts the possibilities; interference is ruled out. When physicists say the environment selects the outcome, they really mean only that, classically speaking, the photon must go through one slit of the other. Decoherence has reduced quantum mechanics to a coin toss; as in the old days, if you knew everything about the coin and its environment, you could predict whether it would land heads, tails, or on edge.

Only toy calculations have been done along these lines, and perfect knowledge of the environment is certainly impossible. Yet decoherence is more than just an interpretation of quantum mechanics. Finally we have a physical mechanism that holds the promise of describing the collapse. What is perhaps most striking about decoherence is absence: absence of consciousness. To observe the Sphinx requires neither Bohr nor Wigner nor Wigner's friend, nor you nor your friend nor observers nor participators.

The exposition of decoherence has left the audience stunned. No consciousness, no mysticism, only reductionism. West Coast observer-participants reel at the setback. Some of them board the buses in disgust and prepare to exit the Academy, but apparently the universe splits. Departure prevented in one universe, they retrieve their cell phones and desperately begin calling the 900 Quantum Hotline for emergency spiritual advice. Foiled by a busy signal (in at least one universe), the stranded visitors begin agitating for consciousness to be reinstated into the interpretation program.

We cannot provide solace in any universe; however, we can ask another question: is the measurement problem a quantum problem?

A profound silence descends upon the central plaza. What can the authors possibly mean by this? What have academicians been arguing about all day?

We ask again: is the collapse problem a problem with quantum mechanics or a problem due to the fact that our detectors give only yes or no answers? Be it a Geiger counter or an element in a video camera or a grain of silver bromide on a film emulsion, our detectors fire or they do not, either the wave function collapses or it does not.

So? the audience responds, righteously. Whether a detector fires is a random event, given by the quantum mechanical probability that the photon or the Sphinx will be detected in this state or that one.

Absolutely true, but suppose you had a detector that measured the wave function itself—

Here we are cut off by objections from all sides. Every textbook writer knows that only the square of the wave function, which gives the probability of a measurement's outcome, has physical meaning.* The wave function itself—well, that is a convenient artifice, but like quarks, it cannot be directly observed. The West Coast crowd resumes its mass exodus. And so the first to depart do not hear that

It is possible to measure the quantum-mechanical wave function. In 1988 Yakir Aharonov and colleagues proposed a type of experiment in which a detector couples gently to the system, so gently that it does not cause the wave function to collapse. Since then, R. Hulet and colleagues of Rice University have actually carried out "weak measurements" on an optical system. It turns out that little information is revealed by each weak measurement but by repeated samplings one can construct a profile of the wave function.

Because weak measurements reveal a continuous, noncollapsing wave function, they suggest that the measurement problem is not exactly quantum-mechanical in origin but results instead from the introduction of yes or no detectors. As the many-worlds interpretation would have it, weak measurements show that the Schrödinger equation merely continues to evolve the wave function deterministically; unlike the many-worlds interpretation, a weak measurement reveals how all potentialities gradually evolve in this universe.

If we could see the wave function in this manner, our perception of the universe would be very different than it is now. The outcome of an experiment would be somewhat like walking on the beach with waves lapping at our feet. Sometimes the waves reach our feet, sometimes they don't; at other times we are covered up to our ankles in water. We would see a continuous evolution rather than the discontinuous evolution portrayed by current detectors.

This is not idle speculation. Someday such detectors will be built and our view of the universe may radically change. Such a view of the quantum world may approach the experience of the world found in meditative states or enlightenment.

Is Quantum Mechanics Nonlocal?

When the last photons had settled on the Academy for the night, the picture appeared somewhat different. After decoherence had spread and the universe had split several more times, the correlation of forces was ripe for an onset of synchronicity. Suddenly and without warning, from one end of the Academy to the other, visitors simultaneously shouted, "Explain the EPR paradox or die!"

No aspect of quantum mechanics has caused more confusion, speculation and downright misrepresentation than the famous 1935 thought experiment of Albert Einstein, Boris Podolsky and Nathan Rosen. And no wonder. The EPR experiment lies at the heart of the attempts to connect quantum mechanics with Eastern mysticism. It is because of the EPR experiment that Fritjof Capra is able to say that "the universe is fundamentally interconnected, interdependent and inseparable." It is to EPR that Amit Goswami refers when he claims that physics has "conclusively established the veracity of the spiritual notions, and particularly the notion of transcendence." It is to EPR that

* See footnote on page 163.

Figure 7.4
In viewing a quantum sphinx, each observer may choose woman, bird or lion, and each observer sees front or back at random. But if the two experimenters both choose the same animal, one must observe a front and the other a back.

the Jungians turn when searching for a mechanism to explain synchronicity. EPR is the holistic universe.

"That may be," your friend remarks, wishing fervently for some sleep, "but the universe will be here tomorrow."

Are you so sure? you reply. It may split again soon.

Too true. Best listen up now to the immortal thought experiment that has the Academy on the verge of civil war. Best forget everything you've ever heard about it.

We are fortunate that the Sphinx has agreed to participate in our explanation of the EPR experiment. As we know, it is possible to view only one aspect of the Sphinx at a time. Suppose at the sound of a gong she allows us to choose which aspect. If you wish to see bird, woman or lion, you merely select the lenses on a special pair of Sphinx viewers (available at the Academy gift shop, special discount offer today only). However, each time you flip the viewers to observe one aspect or another, you find you see only an animal's front or back. You have no choice in this matter and, indeed, whether you get a front or back appears to be totally random.

Let's now suppose your friend, who is standing at the far end of the Academy golf course, is wearing an identical pair of Sphinx viewers and so, like yourself, is able to choose which aspect of the Sphinx to observe at each sound of the gong. Your friend's selection is independent of your own; while you are choosing bird, your friend may be choosing lion and so forth. We can make this more dramatic. Because the two of you are standing at opposite ends of the golf course, you are out of shouting range and, lacking cell phones, cannot communicate your selections to one another. Therefore you *cannot know* what your friend is viewing and your choices *must* be independent of one another's.

What happens when you and your friend make measurements on the Sphinx? Every time you choose an aspect you see a front or a back, so you write down a series of observations like **(woman, front)**, **(bird, back)**, **(lion, back)**, **(bird, front)**. Your friend is doing the same. The pairs appear random, but when the two of you get together afterward and compare notes you find that every time you happened to choose the same aspect—say lion—one of you got front and the other got back. When you question the Sphinx about why this happens, she replies, "A lion must remain whole."

This seems reasonable, you nod. A lion must always have a front and a back (as must a bird and a woman). But when you stop to think about it, the situation is extremely peculiar. Each time you get a measurement like **(lion, back),** you realize that if at the same gong your friend decides to measure "lion," then the result must be **(lion, front)**. You seem to be able to predict the outcome of your friend's experiment without any communication between the two of you.

You ask whether this is any different than having a black ball and a white ball in an urn. If you choose the white ball and your partner walks off with the urn, you know with perfect certainty that the black ball is rattling around inside. No magic here.

But remember, the Sphinx is a superposition; according to quantum mechanics she *does not exist* as a lion or bird or woman until you select the lenses on the viewer. Thus, you are determining what aspect of the Sphinx to observe. Your partner is making the same decision at the same instant, so it is impossible for the two of you to inform each other of your choices. Yet if your partner is measuring the same aspect of the Sphinx, you know before comparing notes what the answer should be. This seems like a violation of causality. No signal can travel faster than light. Or, if it is not a violation of causality, then it violates the fundamental quantum assumption that the Sphinx is in no definite state before you observe her. Paradox.

For those who have gone through the EPR experiment before, it may be helpful to relate this illustration to the usual version of the experiment due to David Bohm. As we have mentioned in previous debates, an atom, like a top, may have spin, or angular momentum. Bohm assumes that an atom with zero spin—no angular momentum—disintegrates into two smaller particles, which fly off in opposite directions. By conservation of angular momentum, the spins of the daughter particles must sum to zero. This is true in both classical physics and in quantum physics. The Sphinx is our analogy to Bohm's "spin zero" atom.

Now, you may choose to measure how much spin lies in any of three directions (say the x-direction, y-direction or z-direction). If you find that daughter particle A has "up" spin along one axis, then daughter particle B must have "down" spin along the same axis in order to ensure that the spins sum to zero.

In our illustration, the three forms of the Sphinx correspond to the three directions you are permitted to choose. Her front and back correspond to up and down spins. Now, classically, particles have a definite spin up or a definite spin down, a situation corresponding to the case of the white and black balls in an urn. But because quantum mechanical particles exist only in a superposition before detection, we must regard each daughter particle as having a combination of up and down spins. A detector, though, measures only spin up *or* spin down.

Again we face the paradox: if you measure A's spin to be up in a certain direction, then you know immediately and with complete certainty that a measurement of B's spin along the same axis will be down, even if particle B is by now light-years away. The conclusion seems either to contradict the fundamental quantum postulate that B has no definite state before a measurement, or it suggests that somehow a signal travels from A to B at faster-than-light velocities.

Such was the paradox that EPR set before the physics community and which has refused to die. There is, actually, a third resolution: no signals are transmitted between A and B, but A and B should not be regarded as separate entities. In which case quantum mechanics is "nonlocal"; it operates, somehow, across vast differences in space and, if you remember the delayed-choice experiment, across vast distances in time. All this is extremely counterintuitive.

EPR couched their conclusions in somewhat vague language, but it is clear that Einstein could not stand the third resolution, that quantum mechanics was nonlocal. What EPR could say, however, is that given the impossibility of superluminal signaling, and *if* the two particles are totally independent, then quantum mechanics must be incomplete. Quantum mechanics denies physical reality to particle B prior to detection even though its spin state *can* be determined by a measurement on A.

Might there not be a deeper theory that gives physical reality to the Sphinx's spin state before measurement?

Hidden variables.

(Organ music, please.)

During the previous exposition, you seem to have been wondering why we unduly complicated matters by introducing three directions of measurement, or three aspects of the Sphinx. Wouldn't talking about observations of one aspect of the Sphinx have been sufficiently confusing? You would observe only the bird, finding that you got front or back about 50 percent of the time. And each time you would be able to predict what your partner measured. You get a front, you know your friend got a back. It's that simple.

Listen attentively. In the case you mention, the measurements on the Sphinx really are no different from the balls in an urn. You take a white ball out, you immediately know that the other is black. No one talks about faster-than-light signals. The balls in an urn represent a *classical* theory—the balls have a definite spin state (color) and there is a definite probability, a fifty-fifty chance, of drawing white or black. So regardless of whether the Sphinx is a quantum sphinx or a classical sphinx, if you measure only one of her aspects, you will get 50 percent backs and 50 percent fronts. Observing only one aspect of the Sphinx is insufficient to distinguish the predictions of quantum mechanics from those of a classical probability theory.

Now, hidden variables.

(More organ music, please.)

Hidden-variable theories were developed because many scientists, including Einstein, found the fundamentally statistical nature of quantum mechanics unpalatable; there should be something else lurking in the background that behaved deterministically. A particle really should have a definite spin state before measurement; it should not be in a superposition of states; somehow all the aspects of the Sphinx really do exist before you observe one.

In this sense, the classical balls in an urn constitute a hidden-variable theory. Two balls exist in the urn with an equal chance of either one of them being pulled out. The balls are not in a superposition of states. When a Bohm atom decays, particle A is emitted with spin up and particle B is emitted with spin down. Measurements of one hundred such disintegrations would show that in about fifty cases particle A is emitted with spin up and particle B is emitted with spin down—and in about fifty cases the reverse. *Hidden-variable theories thus assume that particles are emitted in definite spin states and that they are emitted in these states with definite probabilities.* A hidden-variable theory of Sphinxes would say that she is never in a superposition. Instead, each time the gong sounds, a hidden variable (perhaps the position of Venus) leads her to decide to face you or your partner. In contrast to quantum mechanics, hidden-variable theories (at least of this type) are referred to as "local" because once particles A and B are emitted they cease interacting with each other.

We have made these fine distinctions for two reasons. One, the usual expositions that talk about measuring spins along one axis only (such as that in the *Tao of Physics*) are not really talking about a quantum-mechanical problem. The second reason will become clear all too soon.

The upshot is: to distinguish quantum mechanics from hidden variables, one must make successive measurements in various directions. When one does this one finds that the quantum-mechanical probability of measuring the Sphinx facing forward in one aspect (**bird, front**) then forward in another aspect (**lion, front**) *differs* from the classical probability for these two successive measurements.

Mathematically it is easy to calculate both the probabilities for various combinations. To introduce one bit of shorthand, call **P(bird-front; woman-front)** the *classical* probability that you observe (**bird-front**) while your partner observes (**woman-front**).

Then it is a simple matter to show that this probability is always less than the sum of two other probabilities involving fronts. Specifically, **P(bird-front; woman-front) is always less than P(bird-front; lion-front) + P(lion-front; woman-front).**

This is one of the famous Bell inequalities, discovered by John Bell in 1964. To put it in words, Bell inequalities give relationships among the *classical* probabilities that particles A and B will be found in certain spin states. The important point is that the quantum-mechanical probabilities for measuring the spin states *violate* the Bell inequalities. The amazing thing about Bell's inequalities is that they are quite general. It does not matter exactly what the various **P**s actually are; as long there is *some* definite probability for particles A and B to be emitted with their spins in certain directions, the inequalities hold.

Because classical probabilities are associated with hidden-variable theories, Bell inequalities therefore give an experimental test to distinguish hidden-variable theories from quantum mechanics. You and your friend merely make various combinations of measurements on the aspects of the Sphinx, then compare notes. If the probabilities you measure satisfy Bell's inequalities, then she is a hidden-variable Sphinx, if the probabilities you measure break Bell's inequalities, she is a quantum-mechanical Sphinx.

Alternatively, in the laboratory one might measure various combinations of spin states of particles emitted in an atomic decay. This is the basis of the celebrated experiments performed by Alain Aspect and collaborators in 1981. Their experiments conclusively showed that the predictions of quantum mechanics were correct: one can predict the state of particle B even when it is separated from A by such a distance that no communication between them is possible.

Now we are hit by the full force of the dilemma. Quantum mechanics is correct; particles have no definite states before measurements. Faster-than-light signaling is impossible. Yet the outcome of measurements on B is determined before any signal reaches it. Are measurements on particle A influencing particle B? Does a "spooky action at a distance" inhabit quantum mechanics? Is quantum mechanics nonlocal?

"Yes!" cry a multitude of voices from the central plaza.

Well . . .

Certainly philosophers and physicists have not ceased debating the issue since Bell published his paper in 1964, and Monday afternoons at the Academy dueling grounds are traditionally reserved for disputes among interpreters of quantum mechanics (lists are posted twelve hours in advance). "The conclusions [of Bell's theorem] are philosophically startling," two philosopher-physicists exclaim. "Either one must totally abandon the realistic philosophy of most working scientists, or dramatically revise our concept of spacetime."

But tonight the conflict is between New and Old and the rules of engagement have been suspended. "The universe is fundamentally interconnected, interdependent, and inseparable" reiterates Fritjof Capra. Physicist Victor Mansfield agrees: "The quantum vision removes the independently existing dancers and populates the world with possibilities, tendencies or propensities for particular dances. Here the melodies (waves) and dancers (particles) are fully united until we call one of them forth or bring one of them into existence by our measurement."

"The wholeness suggested by synchronicity lies in meaningful connections of events isolated in time and space," say Combs and Holland. "In spite of their separation, they seem to be linked together. Something like this is known in quantum mechanics and perhaps not surprisingly it is referred to there as synchronicity."

Two well-known quantum mechanicians keel over, stone dead.

Not to be outdone, Amit Goswami mounts the platform. "1982 [sic] was the year that Alain Aspect and his collaborators performed the great experiment that conclusively established the veracity of the spiritual notions and particularly the notion of transcendence. Should I go into a little bit of detail about Aspect's experiment?"

"Please do," the interviewer answers.

"The photons in Aspect's experiment influence one another, at a distance, without exchanging signals, because they are doing it faster than the speed of light. And therefore it follows that the influence could not have traveled through space. Instead, the influence must have belonged to a domain of reality that we must recognize as the transcendent domain of reality."

"Would all physicists agree with that interpretation of the experiment?" the interviewer asks.

"Well, physicists must agree with this interpretation of the experiment."

Not wishing to see a repeat of yesterday's symmetry breaking, physicists ready their weapons as the Academy security guards stand by, idle.

Again we impose martial law. What are we to make of all this?

You shake your head. Who knows?

Your friend asks, What does "influencing one another without exchanging signals" mean? and we reply, Who knows?

To a large extent we can agree with Capra and Mansfield that the universe is interconnected in a way that defies classical ways of thinking. At precisely this juncture we run into the greatness of quantum mechanics; it brings us face to face with the inadequacy of ordinary language to describe quantum reality. But when people say "quantum mechanics is nonlocal," what do they mean?

If they mean merely that the system is extended, nondecomposable, then this is true. The flaw in Einstein's reasoning was that he, like most of us, could not help but think of the two widely separated particles as two distinct systems. Not the case. One must always remember that the two particles were both emitted from the same atom and thus have properties that are correlated since birth. They do not constitute two distinct systems; they constitute one extended system. There are not two wave functions for two particles, there is one wave function for two particles. Granted it remains difficult for classical minds to comprehend how the two particles are connected at such great distances, but this is a limitation of our thinking, not of reality. In this sense quantum mechanics is the least understood of all theories.

If one means by "nonlocal" that signals must be exchanged between the two parts of the system at superluminal velocities, then this is simply false. Particles A and B exchange no signals between them. Remember that each experimenter merely records a random sequence of measurements. It is only after comparing notes one realizes that these measurements are correlated in a surprising way. In this sense the word "predict" to describe the outcome of the experiment is misleading; "retrodiction" comes closer to the actual situation.

"The universe is twenty-three-dimensional!" cries an ecstatic voice from the plaza. "Distant events are connected through a new kind of spacetime."

One might invoke a twenty-three-dimensional universe (string theorists would prefer ten) connected with wormholes in some complicated, incomprehensible fashion to make more palatable the outcome of Aspect's experiments. And the universe may indeed be constructed along these lines, but quantum mechanics gives the correct results in four dimensions.

When one terms quantum mechanics nonlocal in the sense of signals being propagated acausally, one is abusing language. Recall the map projections from the Fourth Debates. Take two nearby points on a globe, Calcutta and Bangkok. Flatten the globe into a wall map and the two points are suddenly found on the opposite ends of the Earth. When people talk about nonlocality in quantum mechanics, they usually mean that two widely separated points are causally connected even though they shouldn't be. They are using a bad map to describe quantum-mechanical reality. They are talking in a classical language when only a quantum-mechanical language will do.

It would also not hurt to bear in mind that not all aspects of quantum mechanics are nonlocal. The Schrödinger equation is perfectly local and when an atom suddenly absorbs a photon that is 5,000 times its own size, no one talks about signals being exchanged. On the other hand, this example can be used to point out a weakness in the field picture, so integral a part of physicists' mindsets and so beloved of the New Age. If we believe that photons are represented by fields that propagate at the speed of light, then the time it takes the photon to be absorbed by the atom should be equal to the distance traveled divided by the speed of light. That the process takes place instantaneously shows that the picture of propagating fields is not always useful.[†]

"Do you mean fields, fields of all things, aren't real?" a representative from California gasps, incredulous. "It goes beyond comprehension."

To return to the previous subject. In the Sixth Debates we discussed some holistic, classical phenomena. It is important to bear this in mind; nonlocal classical phenomena do exist. If you are driving on a highway, a common experience is for traffic to slow down and get bottled up for no apparent reason. Only after half an hour do you find that an accident has occurred on the opposite side of the highway and drivers are merely slowing down to have a good look. Similarly, if one sends a stream of waves toward a potential barrier, for instance a stream of water waves toward a bridge pylon, one observes that the waves sense the pylon long before actually reaching it and begin to change their shapes far upstream. The phenomenon is known as prescattering. As Shakespeare put it, "Thou call'dst me dog before thou hadst a cause, but, since I am a dog, beware my fangs."

Now a ripple of consternation prescatters through the New Left crowd but neither are academicians smiling, for they also overlooked prescattering and hence lost a publication opportunity. Without Publicity There Is No Prosperity.

In Which the Meaning of Nonlocal Is Obscured

The difficulties brought down on us by loose talk can be illustrated by the following story: Three friends went fishing, having agreed to share the catch equally. After a successful day, they put the catch on the boat deck to be divided up later, had a big party, and went to sleep. In the middle of the night the first fisherman wakes up, feeling ill, and decides to go home with his share. But he sees that the catch can't be divided evenly into three because there is one fish too many. So he throws the extra fish into the lake, takes his share, and goes home. A little later the second fisherman wakes up, also feeling sick, and decides to go home with his share. Not realizing that the first fisherman has already left, he also sees that the catch can't be divided equally because of an extra fish. He throws the extra fish into the water and departs. Finally the third fisherman wakes up, ill as well, and, unaware that his friends have gone, counts the fish, finds one extra, throws it away and takes his share.

The question is, What is the smallest number of fish that the fishermen could have caught during the day? The problem is not difficult and if you work it out you find that twenty-five fish is the minimum. No smaller number works.

"I disagree," Dirac responds, "they could have caught minus two fish."

In other words, let us ask, Is it possible somehow to regard the quantum-mechanical particles as being emitted, not as superpositions of spin states, but with definite spins? Can we assign definite probabilities that particles are emitted with their spin up or down in the various directions? If so, we could transform quantum mechanics into a hidden-variable theory, a nonlocal theory into a local theory.

This seems impossible, you reply. The entire debate on nonlocality, the "spooky action at a distance," is due to quantum superposition: particles do not have definite spins before detection. You have said this yourself.

Nevertheless, several investigators, among them ourselves, have shown that such a transformation is possible. Exactly as hidden-variable theories do, quantum mechanics assigns probabilities that particles are emitted with spin up or down in various directions. The only difference from the hidden-variable case is that . . . the quantum-mechanical probabilities are sometimes negative.

"I have a headache," you say.

"What in God's name is a negative probability?" your friend asks, alarmed.

The palazzo erupts into a regular barroom brawl as academicians begin hurling eggs in the authors' direction. At the mention of negative probabilities a side debate begins, a continuation of the debates on Mathematics and Reality. We do not tune in.

Despite the fact that negative probabilities originated decades ago with Dirac, it remains indisputably difficult to understand what could be meant by a statement like, "The probability of rain in today's forecast is negative twenty percent." Probabilities, we are taught, are always positive and add up to 100 percent. This feature at least is shared by the quantum probabilities; they do add up to 100 percent, and so are quite well-behaved apart from the fact that you can't imagine what they are.

To give the general picture. Recall that the Bell inequalities dealt with expressions like **P(bird-front; woman-front)**, which was the "joint" probability that you observed the front of the bird at the same time your partner observed the front of the woman. Bearing in mind that aspects of the Sphinx correspond to directions, and front or back corresponds to up or down spins, this particular joint probability could be written more formally as $P(x + ; y +)$, the probability that particle A was emitted with spin up in the x-direction while particle B was emitted with spin up in the y-direction ($+$ means "up"). The earlier Bell inequality could be recast as $P(x +; y +) < P(x +; z +) + P(z+;y +)$ where the symbol $<$ stands for "less than."

We have violated the dictum that every equation cuts the number of readers in half, but strictly speaking this is only an inequality. Anyway, the single thing to note is that each P depends on two directions only, such as x and y, whereas there are three directions overall, x, y, z. In deriving these expressions, one first calculates "three-directional" probabilities, which do depend on all the directions, then from those probabilities extracts the Ps. For each P, one direction always drops out as irrelevant.[†] For hidden-variable theories the three-directional probabilities are always positive, as well-behaved probabilities usually are, and everyone assumed that analogous probabilities in quantum mechanics did not exist.

To one's initial shock and consternation they do exist and can be negative. But we are saved from losing too much sleep over them because the three-directional probabilities are never actually measured in either theory. Only the two-directional Ps actually figure in the measurements and these are positive in both cases. Bell's inequalities remain unaltered and the quantum Ps violate them as usual. From this point of view it is hardly surprising that the hidden-variable theories give a different result from quantum mechanics and fail Aspect's experimental test: they used the wrong set of probabilities for a quantum-mechanical problem.

You are unconvinced, we can tell. From the bonfire where academicians are heaping on copies of the journal in which our paper appeared, we can tell they are unconvinced too.

Yet it would do well to recall from the Second Debates that the Greeks did not know what to make of negative numbers, and it took several centuries for mathematicians to accept imaginary numbers. The very name "imaginary" is a relic of their original status. Before any sense could be made of negative or imaginary numbers a sensible interpretation had to be attached to them. An imaginary number is more like a rotation through 90 degrees than it is like a real number. Recently, complex probabilities have become the subject of investigation among physicists, and negative probabilities do not seem any stranger, although no sensible interpretation has been attached to them. Whatever negative probabilities are, it is clear they do not behave entirely as ordinary probabilities do. Yet in regard to the EPR paradox it was not obvious beforehand that negative probabilities would provide a way out.

The side debate on Mathematics and Reality turns once more to the question of whether physics discovers preexisting mathematics, but again we do not tune in.

We are not necessarily arguing that negative probabilities will turn out to be meaningful or important. Tonight our point is simpler. We can speak in mysterious, imprecise language about "nonlocal" theories versus "local" theories, or we can abolish the distinction between them at the expense of introducing precise but unconventional mathematics.

True beauty, it is said, is always fresh. We have arrived at know-nothingness.

A Practical Exercise—The Fractured Debate

It was a night of splitting universes. In one universe the New Age became older and its adherents departed, overwhelmed by doubt that much of what they thought they knew about modern physics was . . . In another universe they reacted with righteousness and denial that anything said by the authors could be correct, and they fought to the last paradigm. In a third they argued that it was all interpretation and everyone is entitled to his opinion. In a fourth, a doctor pronounced a man dead and had the orderlies remove the body from the emergency room. As the orderlies carried him to the morgue, the man suddenly sat up on the stretcher and said, "But I'm not dead yet." "Who knows better," the orderlies replied, "you or the doctor?"

In one universe, Steven Weinberg repeated that "those who seek extrascientific messages in what they think they understand about modern physics are digging dry wells."

In another universe, Jung replied: "Natural laws are statistical truths, which means that they are completely valid only when we are dealing with macrophysical quantities. In the realm of very small quantities prediction becomes uncertain, if not impossible, because very small quantities no longer behave in accordance with natural law."

In that universe you asked, "Aren't the laws of quantum mechanics natural? Have we not decided that the Schrödinger equation is deterministic?"

Combs and Holland expounded on quantum mechanics: "Unlike the general theory of relativity, quantum theory does not deal with the existence of objects but with actions or events. It is an elaborate mathematical structure within which separate objects have, in fact, no representation. Only events have reality for the theory and these are all intimately connected. For example, if several subatomic particles are involved in an experiment, they cannot be treated as separate entities."

Still puzzled, you said, "Is the wave function an action or an event? Is an electron a noun or a verb? Perhaps I am an adjective."

One version of your friend, nearby, added: "Have we not heard several examples of classical systems that cannot be broken into separate entities?"

"Yes," the authors answered, at the risk of repeating themselves, "an ordinary wave is one example. Another is that the potential energy of any system of particles depends on all of them. Also, in classical electromagnetism, when one considers two moving charged particles, Newton's third law does not hold unless you take into account the participation of the field."

On hearing the word "field," Combs and Holland jumped in again: "Beginning with Einstein the cosmos became an undivided field. Objects such as atoms and stars are viewed as properties of this field. They are seen as local concentrations of it. . . . They have no separate existence in and of themselves . . ."

Bafflement deepening, you asked, "Hmm, does this mean if I hit you over the head with a chair, I am clobbering you with a field? Does the chair have no existence? I clobber, therefore I am not?"

The authors replied, "A highly concentrated field . . . The New Age has forgotten that the word 'model' applies to all models, not just to models it likes best. A scientist constantly jumps between mental constructs, understanding that none is perfect, none describes Brahman, but in a given situation one model may be more appropriate than another. It is quite acceptable to view a chair as an object. In virtually all circumstances astronomers view stars or black holes as objects and as the sources of the gravitational field."

Deepak Chopra said, "The quantum field is just another label for the field of pure consciousness and potentiality. And this quantum field is influenced by intention and desire."

In another universe entirely, Amit Goswami maintained, "What Bell's Theorem and Aspect's experiment imply is not a violation of causality but that simultaneously occurring events in our spacetime world can be related meaningfully to a common cause that resides in a nonlocal realm outside space and time. This common cause is the act of nonlocal collapse by consciousness. . . . The nonlocality of Aspect's experiment fits Jung's description of synchronicity perfectly."

Combs and Holland amplified, "Synchronicity implies a cosmos in which seemingly unrelated events are woven together to form a continuous world fabric. Such a cosmos does not square with the classical, mechanistic physics that views the universe as a loose assemblage of objects, forces and energy."

The authors in that universe merely turned to you and asked, "After today's debates, do you think synchronicity squares with quantum mechanics?"

A New Age advocate remonstrated, "Jung learned his quantum mechanics from the great physicist Wolfgang Pauli himself. Pauli himself believed in synchronicity."

The authors answered, "Jung was Pauli's therapist and through Jung Pauli became quite interested in archetypal images, in particular with regard to Kepler. But in his published lectures and essays, at least, there is no reference to synchronicity."

Erwin Lazlo, in Combs and Holland's universe, objected once more to the reduction of synchronicity to "nothing but coincidences": "Now, nothing-butness has been a bastion of the reductionist's view of reality, where organisms are said to be "nothing but" atoms and molecules and societies and cultures "nothing but" human individuals with particular mind-sets. Fortunately, reductionism has been proven wrong in the new physics, where even the properties of ordinary atoms were shown to be irreducible merely to the properties of their parts."

The Zen master asked, "What is the sound of a broken drum?"

A little earlier, or perhaps later, an interviewer noted the attempts by New Age thinkers to derive free will or volition "from electron indeterminacy" or some such. Ken Wilber, a psychologist, responded: "Yes, it appears the thing to do. . . . Finally, after decades of saying the physical universe is deterministic and therefore human choice is an illusion, you find a little indeterminacy in the physical realm and you go nuts. It's only natural you then try to explain human freedom and even God's freedom as a blowup of the lowest level. You get so excited you forget you have just pulled the reductionist feat of the century."

The authors distilled the essence: "What could be more reductionist than trying to relate synchronicity and consciousness to quantum mechanics?"

In this universe, Steven Weinberg said: "So irrelevant is the philosophy of quantum mechanics to its use, that one begins to suspect that all the deep questions about the meaning of measurement are really empty, forced on us by our language, a language that evolved in a world governed very nearly by classical physics."

Universes Merge?

The universes continued to go their separate ways. In the fifth universe the debate abruptly ended. In the eighth it continued without end. In the sixteenth universe, when the Evil Age had run its course, Shiva danced and the Academy was dissolved by the winds of time, to be restored in the next cycle of existence as a personal growth center. One foot planted in this universe, you asked, Why is it that so many physicists are attracted to Eastern mysticism and, finally, what is the connection with quantum physics?

There can be no doubt that many physicists have been drawn to Eastern philosophy. Schrödinger himself practically adopted Vedanta (the outlook of the Upanishads) as his own. David Bohm, an exceptional physicist and patron saint to the New Age, held a deep interest in Eastern thought, visible through his long association with the spiritual teacher Krishnamurti. Heisenberg in his old age gave a nod to the East, and Bohr himself took over the yin-yang symbol for his coat of arms.

It is not clear that all scientists who have Oriented themselves—and certainly those who have written on physics and Eastern thought—have made an attempt to distinguish one philosophical system from another. Throughout these debates we have tried to speak of Indian philosophy when we mean Indian philosophy, Chinese philosophy when we mean Chinese philosophy, and Zen when we mean Zen. Each of these traditions has distinct beliefs that are often incompatible with one another. One after all would hesitate to lump Presbyterians and Unitarians under the heading "Western religions."

Nevertheless, speaking generally, it is easy to understand the psychological appeal of Eastern religions to theoretical physicists. Physicists are often agnostic or atheistic, and if they are not always rational they always pretend to be. Buddhism, with its lack of a supreme being, and Vedantic philosophy, with its emphasis on Reality, satisfy their deepest spiritual yearnings without overtly compromising them. The Eastern religions' emphasis of unity is also not foreign to the physicist.

But the appeal is more than the merely personal. Schrödinger gave a partial answer to the question you have raised. Science works only because of "the golden rule," the assumption that each investigator is as conscious as the next, that the results discovered by one are as valid (or invalid) as the results discovered by another. "It thinks, therefore it is." This assumption cannot be proven, according to Schrödinger, by scientific means and so science cannot be put on a scientific foundation: "Yet, the most wonderful and most sublime of all teaching, the Brahman doctrine that the all equals the unity of consciousness, culminates in a mystical victory over this dilemma, the words so obscure to the understanding, so close to the intuition: *Tat tvam asi*, That Thou Art. One may see in the Brahman doctrine merely a facile play of ideas. But it would be a vast error to believe that science knows any better or clearer answer concerning these things." Or as is said in the Vedas:

The one highest Godhead
In all beings existent
And when they die yet living
Who sees this is seeing
For the one who has found the highest God everywhere
That man will not harm himself through himself.

Schrödinger's is of course a belief, not a rational scientific conclusion, but it is a belief that has resonated strongly with other scientists. In his philosophical book *Wholeness and the Implicate Order,* David Bohm contrasted the Western and Eastern attitudes to science,* measurement in particular: "In the East, the notion of measure has not played so fundamental a role. Rather, in the prevailing philosophy of the Orient, the immeasurable (i.e., that which cannot be described or understood through any form of reason) is regarded as the primary reality." For Bohm it is significant that the Greek "metron" is etymologically related to the Sanskrit "maya" for illusion; however, this is not really the case. The first syllable of "maya" does indeed mean "measure," but the second syllable designates a bridge to the transcendent realm, to the immeasurable. One should also not forget that the Chinese were obsessed with time measurement and the Indians with exact chemical prescriptions. Nevertheless, it is true enough that Eastern religions regard the immeasurable as the primary reality.

Bohm received his ideas on the undivided, immeasurable Om from Krishnamurti and attempted to give them expression in the concept of implicate and explicate order. If you take a drop of ink and put it into glycerin and stir the glycerin very slowly, the ink will gradually spread over the whole container until the mixture appears uniformly gray. But when you reverse the stirrer—Behold!—the gray mass of ink contracts to the original drop. Another drop would produce the same uniform gray across the container, but when the stirrer was reversed, this second droplet would emerge at its own location. Thus, information about the drop's position was enfolded, or implicated, in the gray mass, which reversing the stirrer explicated or unfolded. Explicate, to reveal.

Bohm felt that the universe was much like the bowl of glycerin. Or a hologram, in which the entire image is recorded at every spot on the photographic plate. The whole is enfolded in each of the parts. He viewed the EPR experiment as an indication that the quantum field is "actually a multidimensional reality which can only be under certain conditions simplified as a three-dimensional reality." Indeed, the entire implicate order had to be "extended into a multidimensional reality. In principle, this reality is one un-

---

* As we have emphasized, there are great differences among schools of Eastern philosophy. It should be pointed out that the Indian Sankhya philosophy, in contrast to traditional Vedanta, is pluralistic and not monistic. It divides the world into *prakriti* (nature) and *purusha* (observer-participant), whereas in Vedanta the ultimate Reality is undivided. Yoga is closely associated with Sankhya, and through meditation the observer-participator can ascend to the state of *kaivalya,* which we discussed in the Third Debates. Buddhism and Jainism stand outside these doctrines.

broken whole, including the entire universe with all its 'fields' and 'particles.' " Consciousness for Bohm was also part of this undivided reality.

But despite the influence of Krishnamurti, Bohm's thought reflects standard Buddhism, not Vedanta. In his drop of ink which can be enfolded and unfolded in the glycerin—or in the hologram from which information can be retrieved at will—is a rejection of irreversible processes, of the sort that Prigogine considers so fundamental. Bohm's view of time, like the Mahayana Buddhists' of the Fifth Debates, is that the temporal order is illusory, something to be transcended. Beyond all lies an undivided, dynamic reality, which Bohm termed "the holomovement," that which is. The Brahman, perhaps.

And so several streams of thought have merged, perhaps inadvertently. If Goswami can—at least vaguely—trace his lineage to Schrödinger and the Upanishads, then Capra, Mansfield, and Combs and Holland are descendants of Bohm and the Buddha.

Fritjof Capra, to be sure, finds a kinship between Bohm's ideas and Geoffrey Chew's bootstrap theory, which was discussed in the Third Debates: "Both," Capra writes, "are based on the same view of the world as a dynamic web of relations. Both attribute a central role to the transformation of order; both use matrices to represent change and transformation, and topology to classify categories of order. Finally, both approaches recognize that consciousness may be an essential aspect of the universe that will have to be included in a future theory of physical phenomena."

Amit Goswami would emphatically agree. But several days ago we saw that Chew's theory was more problematic than the quark model and most physicists have abandoned it long ago. As for matrices, well, name a physical theory that doesn't use them to describe change and transformation; topological classifications are no rarity in physics either.

What of Bohm's view on EPR? Bohm's main contribution was, in fact, to devise the simplified version of the EPR experiment that was universally adopted by physicists and expositors and that enabled John Bell to derive his inequalities. As far as the more speculative aspects of Bohm's thought go, we have already said that one might interpret the EPR experiment as an indication that the universe is many-dimensional ("many" being higher than four) and, as one famous cosmologist exclaimed while pondering the mysteries of the universe at an Oxford pub, "It must be something like that!" Nevertheless, we reiterate that quantum mechanics gives the correct answers to experiments without such extrapolations.

What of EPR and synchronicity? If synchronicity exists, it must surely be a true action-at-a-distance theory because it is hard to see in what way it could be compatible with known physics. Again, we do not refer to synchronicity as "acausal," because it does involve causes, but it does violate relativistic causality. Whether synchronicity exists or not, there seems little reason to connect it with quantum mechanics. As we have shown, "nonlocal" and "local" are words with ambiguous meanings and classical nonlocal systems exist. It makes no more sense to say that synchronicity is connected with quantum phenomena because it is nonlocal than to say it is connected with a train of classical waves prescattering off a bridge pylon. Here is an extended system and if you had never heard of quantum mechanics, would you have used it as a model for synchronicity? Or would you have merely attempted to deal with a nonlocal system?

And what, finally, of quantum mechanics and Eastern mysticism? Is there really a connection? Our position, to the extent that one can express the inexpressible, might be illustrated by considering an ant colony. As Lewis Thomas once put it:

> A solitary ant, afield, cannot be considered to have much of anything on its mind; indeed, with only a few neurons strung together by fibers, he can't be imagined to have a mind at all, much less of a thought. He is more like a ganglion on legs. Four ants together, or ten, encircling a dead moth on a path, begin to look more like an idea. They fumble and shove, gradually move toward the Hill, but as though by blind chance. It is only when you watch the dense mass of thousands of ants, crowded together around the Hill, blackening the ground, that you begin to see the whole beast, and now you observe it thinking, planning, calculating. It is an intelligence, a kind of live computer, with crawling bits for its wits.

One might begin describing the ants as a few independent Newtonian particles, billiard balls with neurons for noggins. But this description soon becomes inadequate. As more ants come together, they seem to be perhaps behaving more like a fluid, many particles jostling their way up the Hill, a hydrodynamic system. And finally, we perceive a real nonlocal, extended system (quantum mechanics is not going to help here), and an adequate description of the colony must go beyond simple physics altogether to include intelligence, if not consciousness.

As our awareness becomes more refined, we discover that physical systems cannot be adequately described by previous models and must be extended. In the First Debates we found this was true even of simple systems, like the hydrogen atom, for which no consistent model exists.

The same is true for personal experience. As our awareness becomes more refined, one has changed and the same experience we have had in the past is now perceived differently, and cannot be described in the same ways. Chemicals affect our experiences. But we may ask, Are there any real things of significance that are being left behind when we consider ourselves to be a collection of neurons governed by electrochemical reactions? Yes, there are other experiences, or we believe there are. The desire to write this book in the simplest terms possible without sacrificing accuracy is not explainable in terms of current science, or anything else that we know of.

The spirit dwells within the flesh, the flesh profits from the spirit. How do you describe this? One thing on which both Zen and Vedanta agree is that it is in those quiescent times, without thoughts, that you are Aware. It is during the quiescent times that you are closest to the Brahman, the universal Self. Descartes had it backward; he should have said: "I am despite the fact that I am thinking." The problem with physicists explaining mystical phenomena, the direct experience of Reality, is that physicists, almost without exception, have not experienced it. They are theoreticians, not experimentalists. They may attempt to couch the experience in terms of quantum mechanics, but Siddhartha did not become Buddha by studying modern physics. When the time was ripe, he went beyond rational thought. He leapt from profane to sacred time.

Just as a physical theory eventually becomes inadequate to describe increasingly complex systems, so must physics be left behind to describe Experience. There is a mathematical sense in which quantum mechanics enables you to see all creation and destruction as mere transformations on the underlying reality. That is, the usual variables of a system—momentum, position and so on—are replaced by operations. The distinction between quantities and operations breaks down in this view. Perhaps this is what Combs and Holland are getting at when they say that in quantum mechanics "only events have reality." But in the same picture you can have things that are distinct and separate and things that are distinct but not separate. The usual language of yes or no is replaced by yes, no or maybe.

Throughout the debates thus far, and throughout the debates to come, we have emphasized and will emphasize the reality of experience. We have also, as theoreticians, attempted to find the best models for describing a given experience. Is it best to depict an ant colony as a collection of individual ants, or in terms of a fluid, or as an extended system? Is it more appropriate to describe the experience of love as involving two separate people, or something more like a wave, in which the intensity of two together is four times as bright as two alone? Is it more appropriate to describe meditative states in terms of Newtonian or quantum physics? There are phenomena for which the quantum model is more appropriate than the Newtonian model. In this sense, continual transformation, creation and annihilation, may be as close as physics can get to a description of Reality. But models come and go. When all is said and done, one must leave behind description for experience.

In Which the Seventh Debates Are Summarized

It was the best of universes and the worst of universes. You were condemned to participate in ever more speculative discussions and you were condemned to sleep and miss the speculation. You left the Academy and you stayed. Your friend entered an ashram and took up the study of quantum theory. Academicians exterminated the New Agers; New Agers immolated academicians with burning moxa. As the two of you—or was it four?—stood on the seashore you saw the entire debates on quantum mechanics unfold before you in all their divine splendor.

One of the main messages of quantum mechanics, as demonstrated by the two-slit experiment, is that it is impossible to measure all aspects of a system simultaneously. This led directly to the concept of *superposition;* a quantum particle exists as a combination of states, but a measurement can reveal only one of them: the particle or wave aspects of a photon, its position or momentum.

The fact that the experimenter decided which aspect of reality to make manifest led to the notion of observer-participant, that reality is not independent of the observer. From there it was not a large leap to the declaration that consciousness is required to bring reality into existence. However, during the course of the debates, the

contemporary idea of *decoherence* was introduced, which goes some way toward eliminating the observer from the picture.

As the day went on, the probabilistic nature of quantum mechanics came to the
fore, but the authors emphasized that, although some aspects of quantum mechanics
are inherently probabilistic, other aspects are perfectly deterministic, such as the
Schrödinger equation. Statements by Jung that the laws of nature are statistical and
hence allow for exceptions were inaccurate. The authors also drew attention to the fact
that only the simplest processes, such as the decay of a single atom, can be regarded as
truly probabilistic. For more complicated systems one could decide whether the process
was probabilistic only after its completion.

Much of the debates, to no one's surprise, centered around the "measurement
problem." The quantum "wave function" gives only the probability that a particle will
be found at a given place at a given time. Furthermore, the deterministic, Schrödinger
equation cannot describe how probabilities are transformed into actualities—the "collapse" of the wave function. But particles are measured at one place or another. How?

Numerous answers to this question were given: the act of observation (somehow),
the consciousness of the observer, the observer's friend. Many-worlders claimed that
the wave function did not collapse at all, but instead the universe split at each quantum
event. The authors, however, argued that the many-worlds interpretation was answering a statistical, not a quantum mechanical, question. They also focused attention on
"weak measurements," measurements that do not actually collapse the wave function.
From the existence of weak measurements, the authors argued that the collapse problem is not a quantum problem but a detector problem.

And then the Sphinx appeared for a demonstration of the famous EPR paradox,
which demonstrates that quantum mechanics is "nonlocal," meaning that its effects, in
weird ways, seem spread across space and time. But here the authors cautioned that the
word "nonlocal" is fraught with misinterpretations and that one must be exceedingly
careful about what one means. Furthermore, the authors demonstrated that the introduction of negative probabilities could abolish the distinction between local and nonlocal theories, showing that calling quantum mechanics nonlocal amounted to an
abuse of language. They also gave examples of nonlocal classical phenomena.

As the universe continued to split, the positions on the left and right moved farther apart and appeared irreconcilable. Some areas of compatability between Eastern
philosophy and modern physics appeared; however, we emphasized that there is no
particular reason to view synchronicity as a quantum phenomenon. We ended by
pointing out that quantum mechanics is a theory, whereas enlightenment is direct experience of reality. Physicists, if they want to experience reality, should stop writing
books and start meditating.

And as your mind quieted on the seashore and the experiment finally concluded,
the Sphinx took flight.

IS THERE AN ANSWER?
Theories of Everything

I am puzzled in a question about Hell;
He says, in Hell there's one material, fire,
And yet it shall not burn all men alike.

—John Webster

In this smallness infinities,

anomalies slough off, the loops vibrate, a keen
undulation, clockwise rippling nothingness
in ten dimensions. Twenty-six the other way.

This fits. But it's not all. The dimensions
must compactify, in a silent crumpling, curling
in of what there's room for, into inwards' innards.

The quantum numbers then come out naturally,
strung out on a loop that is gravity, the source
of all interactions. We are *so* near understanding

everything. I believe, reasons, without words,
classy symmetries. It's a rule. And up scale the sun
shines, frost melts and zing! go the strings of my heart.

—Roald Hoffmann

The Question

It is often told how in A.D. 640, the Arabian conqueror Amru destroyed the Academy Library. Although there is much question as to whether Amru was responsible, existing sources concur that in 640 the library was burned to the ground. Before that fateful year the Academy library had been renowned as, not only the largest in the world, but the largest conceivable library, containing all books that had been written and all books that would be written. Academicians strove for universal knowledge by consuming as many of the volumes as possible, with the predictable result that once having

191

entered the library they were never seen or heard from again. According to the Arabian historian Ibn Khaldun, however, in 640 Amru, who had invaded the Academy, asked the Caliph Omar what was to be done with the mass of books found at the library. The Caliph is supposed to have answered, "Throw them into the water. If they contain anything of truth, we have received from God a better guide. If they contain falsehood, we are well rid of them." But Amru, the accounts tell us, found it difficult to dispose of the nearly infinite number of books revealed therein and resorted to torching the stacks.

In the following centuries the Academy rebuilt the collection to the point where it equaled and even surpassed its former glory, but technology marched on. The Rosicrucians, who surfaced in 1605, claimed that they derived their knowledge from a single volume in which they could read all that ever was or that would be written in other books until the end of time. The Academy made strenuous efforts to acquire the tome (as shelf space in the library had again become short) but apparently the Rosicrucian volume was stored in the tomb of Rosencreutz, guarded by a brass knight who not only slayed any intruder but caused the crypt to submerge itself beneath the earth; consequently the location of the book-of-all-books has been misplaced.

Efforts at compression have not ceased at the Academy, and to no one's surprise the Rosicrucians recently claimed to have put the contents of the fabled book on CD-ROM. But in contrast to the lexicographers and foxes (who know many things), there has always been an opposing school at the Academy, consisting primarily of scientists and hedgehogs (who know one great thing), a school that regards the accrual of vast sums of knowledge as a waste of time. Adherents of the hedgehog school tend to adopt the Vedantic outlook that etymology, grammar, metrics, astronomy and so on are the lower knowledge. It is better to know the one thing from which all other things emerge, the imperishable:

> What cannot be seen, what cannot be grasped,
>   without color, without sight or hearing.
>   without hands or feet;
> What is eternal and all-pervading,
>   extremely minute, present everywhere—
> That is the immutable,
>   which the wise fully perceive,
> As a spider spins out threads, then draws them into herself;
> As plants sprout out from the earth;
> As head and body hair grows from a living man;
>   So from the imperishable all things here spring.

Academy physicists are the most extreme of the hedgehog faction and have long sought to express the imperishable in a single set of equations from which all else springs, the final theory, the Theory of Everything.

And as dawn broke above the Academy on the morning of the Eighth Debates, you found yourself in a universe where both hedgehogs and foxes were asking, Have they done it?

A Recontemplation on the Concept of Unification, and on Supersymmetry

Few tourists were in evidence this morning; they had apparently absconded during the night or ended up in another universe. There remained some followers of Maharishi who expected their Unified Fields of Invincibility to be certified, as well as a number of reporters who with journalistic impatience hoped to be first with an exclusive that the End of Science had arrived. Trading on the World Idea Exchange remained suspended.

As you accompanied the academicians on their calisthenics, you noticed to your surprise and delight that much of the conversation was intelligible to you due to the previous week's boot camp. The hedgehogs were extolling the virtues of symmetry and the progress made toward unification of all the forces of nature. The most vocal of the hedgehogs, the sonic hedgehogs, went so far as to claim that they had achieved this long-sought goal and that all other sciences had become irrelevant. At the opposite extreme, the foxes scoffed at the very notion of a Theory of Everything, and a few of them even attempted to eat the hedgehogs, but were repulsed by ultrasonic barbs.

"But what *is* the Theory of Everything?" both you and your friend want to know.

Well, if you believe that one can unify the four forces of nature into a single "superforce" and if you believe that the four forces are responsible for everything that takes place in the universe, then you will have succeeded in creating a theory of everything.

"A theory or The Theory?"

At the moment no one will deny that there are a number of candidates, so it would be premature to call any one of them The Theory.

"Some theories are more equal than others," replies one of the sonic hedgehogs.

"Ah," responds a fox, "the history of physics has been a process of just peeling off one layer of the onion after another. Each time you have discovered the ultimate particle, you then find another."

"It cannot be denied," sighs a sympathetic hedgehog, "a number of researchers have investigated *preonic* models, in which quarks are composed of more fundamental entities known as preons."

You reel from the shock. "Do you mean, these quarks, which nobody can observe, are already considered to be composed of further unobservable particles?"

Among certain quarters.

"You see!" exclaims the same fox in triumph. "What makes you think you will ever get to the bottom?"

This is the question. Nevertheless, one cannot ignore the trend toward unification in physics, begun with Maxwell's unification of electricity and magnetism into electromagnetism. By now, as discussed in the Third Debates, the electromagnetic and the weak and strong nuclear interactions have all been brought together by shotgun into the standard model of particle physics.* Only gravity remains the "odd man out."

---

*As we go to press, the Superkamiokande detector in Japan has apparently detected a neutrino mass. This discovery will require some modification of the standard model.

However, in the past few years ever more frequent claims have been appearing that even this problem has been solved by string theory, in which case according to the sonic hedgehogs, a theory of everything—or perhaps The Theory of Everything—is in the offing. Sonic hedgehogs are the ultimate reductionists.

"It's all rubbish," replies an old fox.

Let us see.

From the Third Debates you recall the basic facts and procedures of unification. The world seems constructed out of two basic types of particles: *fermions,* such as neutrons and protons, which make up matter, and *bosons,* such as the photon, which are associated with fields. In other words, bosons are responsible for transmitting forces between fermions. Due to basic quantum-mechanical principles, fermions tend to avoid each other as much as possible, while bosons tend to cluster together. The combination of attraction and repulsion gives finite structure to the world, and so any theory of matter must contain both fermions and bosons.

When academicians unified two forces, the technique was to find one symmetry group that included all the symmetries of the bosons for each force taken individually. What this amounts to is allowing fermions that were not allowed to interact before to interact. Since each new interaction requires a new boson, it is not surprising that unification produces new particles. The unification of the electromagnetic and weak force, for example, required two new bosons, the W and the Z, which were experimentally detected in 1983 and 1984. And so the electromagnetic and weak forces were declared one, the electroweak force.

But in the Third Debates, antisymmetrists raised the objection that this was unification without unification. In creating the standard model from electroweak force plus the strong force, the symmetries were merely spliced together. The authors pronounced the words "field tensor" for an object, something like a luggage rack, which contained all the forces in the theory. But like a luggage rack, the field tensor kept the various forces distinct; the electric force was not allowed to mix with the magnetic force and the two of them were segregated from the weak force, and so on.

Even diehard symmetrists were unsatisfied with the jury-rigged situation. For true unification, one would like all the fields to merge within the field tensor—to put all the forces into a single symmetry group. Howard Georgi and Sheldon Glashow of Harvard proposed such a theory in 1974—the first Grand Unified Theory, or GUT, which truly united the electroweak and the strong forces.

Georgi and Glashow's theory had several interesting consequences. Recall that the strong force is a force between quarks, which are rather heavy particles, whereas the electroweak force also involves lighter particles (known as leptons), which include electrons and neutrinos. In combining the strong force with the electroweak force, Georgi and Glashow's GUT allowed quarks to interact with leptons in new ways. In particular it allowed quarks to decay into leptons on a timescale of 10^{30} years. Because ordinary matter—protons and neutrons—is supposed to consist of

Academy Library

Figure 8.1

quarks, one consequence of GUTs was that all matter is unstable and will eventually disappear.*

A great deal of time and money was spent on huge detectors (essentially swimming pools of water surrounded by photodetectors capable of sensing the light flashes emitted by decaying particles) for the purpose of detecting proton decay. Unfortunately, the results of a number of such experiments were negative, which meant either that the theory was incorrect or the decay time of the proton was much longer than anticipated. After Georgi and Glashow's original, many versions of GUTs appeared but no proton decay was ever observed. Largely for this reason, as well as the fact that theoretical fashion marches on, GUTs faded out in the 1980s. It is a bit like the stock market; outsiders wonder why the brokers are buying, buying, and buying, then suddenly selling, selling, selling. Only the brokers themselves are sure they know the true reason.

Nevertheless, the effort to find proton decay was not entirely for naught. The same detectors have been used to detect extrasolar neutrinos, in particular neutrinos produced by supernova 1987A. Although GUTs may have been shunted to the Oblivion File, astrophysical models of supernovas appear to be basically correct.

* Andrei Sakharov predicted the decay of the proton in 1967 in the context of a somewhat simpler scheme.

During the exposition, the debate began to move up the perpetual spiral that wraps around the Great Library, on which scholars ascend to ever greater heights while the summit recedes at an even faster rate fueled by the accelerating growth of knowledge. High altitudes were quickly attained and under the midmorning sun, one question dominated your thoughts: "Where is the lemonade?" You also wondered: "Is this really a theory of everything? I mean, they didn't even include gravity."

"And why 'Grand Unified Theory'?" your friend added, nodding assent. "Isn't it more like a 'Semi Unified Theory' or a 'Quasi Unified Theory'? Shouldn't one reserve the term 'Unified Theory' for one that includes all the forces?"

"How about 'Gibbous Unified Theory?'" mused a nearby astronomer, passing you on the spiral. "After all, they did get three-quarters of the forces."

The authors promised to return to semantics later in the day, if the party survives. Georgi and Glashow's efforts were modest by today's standards. Even without tackling gravity there is another direction one can take along the unification line. For example, although GUTs united the electroweak and the strong forces, they continued to treat matter and forces as separate entities. In the true spirit of symmetry you might try to put fermions and bosons on an equal footing. In that case you would be abolishing the distinction between matter and forces. They are all just manifestations of the same superfield.

Pro-symmetrists, of course, are strongly in favor of such a program.

"But isn't this what the New Agers were saying the other day?" you ask, already feeling the effects of the climb. "Everything is a field?"

"Yes, the rutabaga field," the Zen master responds, banging you over the head with a rutabaga as he continues his endless circulation around the Library, chanting.

The idea to abolish the distinction between matter and fields is actually not new. In the nineteenth century Heinrich Hertz, one of the discoverers of radio waves, speculated that forces and particles were really the same thing. Whereas we usually think of accelerations as being caused by forces, Hertz dispensed with forces in favor of "concealed masses," invisible particles that collided with an object, changing its state of motion. But there were more concrete reasons to regard the difference between mass and force with skepticism. In certain kinds of experiments in which a photon scattered off an electron,* the force transmitted between the two particles is actually carried by an electron, which is a fermion not a boson. So the distinction between matter and forces is not always so hot.

Treating bosons and fermions on an equal footing is not easy because their properties are very different.** Nevertheless, one can find something like a symmetry group that treats fermions and bosons as equals. Physicists call such a symmetry "supersymmetry," and it was introduced by Julius Wess and Bruno Zumino in 1974.

Supersymmetry has been applied to GUTs, in which case one gets, inevitably, supersymmetric GUTs, or SUSY GUTs. In supersymmetric theories, one no longer thinks of fermions or bosons.

"I'm no longer thinking at all," you say, sitting down for a rest.

---

* Known to physicists as Compton and Raman scattering.

** Specifically, the spins of bosons and fermions are quantized in different ways. Bosons must have spin equal to $n\hbar$ where n is an integer and $\hbar$ is Planck's constant divided by 2π; and fermions must have spin equal to $m\hbar/2$, where m is an odd integer.

In any case, instead of independent boson or fermion fields, one now imagines a "superfield." Supersymmetric theories have many attractive features. It turns out that the infinities that plagued quantum field theory in the First Debates are of opposite sign for bosons and for fermions, and so in SUSY GUTs, like mirror-image twins, the infinities tend to cancel out.

"But wait a minute," interrupts your friend, stopping to survey the vast lands spreading forth below. "In all the other unifications you mentioned, you got more particles, since there were more interactions. Does that happen here?"

Yes, it is very much like taking fertility drugs; you tend to get more than you bargained for. And this is the (serious) rub. In supersymmetric theories virtually every ordinary boson and fermion has a supersymmetric partner, termed a "sparticle." Sparticles bring with them a very simple problem: none has ever been detected.

"This is just a matter of time," remonstrates one of the most sonic hedgehogs. "Sooner or later experimentalists will find them."

We reply that the problem goes deeper than a mere lack of experimental detection. The trouble with SUSYs is that no one really has a good way of calculating what the particle masses should be. Naively, as will become clear in a moment, one would expect sparticles to have masses similar to their more familiar partners, like the electron, in which case they would have been found already. The fact that they haven't means that sparticles, if they exist, must be extremely heavy. Vocal proponents of SUSYs, like the sonic hedgehog, tend to claim that the particles always have masses just beyond what can be produced with existing accelerators. In this sense SUSYs are much like theories about aliens in Area 51.

We can say that the heavier the sparticles, the less attractive SUSYs become because the simplifications introduced by supersymmetry no longer apply. For example, if the sparticles are heavy enough, the infinities that supersymmetry alleviated tend not to cancel, as if the sparticles weren't there at all. Most physicists concede that had the Superconducting Supercollider failed to detect sparticles, then SUSYs would have been in serious trouble, but since the accelerator was never built, the answer to this question has been indefinitely postponed.

During the previous discussion, as the hedgehogs and foxes continued to spiral up the Library's great helix, a question gradually made its way from the back to the fore of your mind. Now, so far above the Academy terrain that academicians look like ants, you complained, "I don't understand. Clearly the forces of nature aren't unified. Look around you. There is electromagnetic radiation from the sun—light. There is the strong nuclear force, there is the weak nuclear force. They're all different, right?"

Excellent point. To make it more precise, recall from the Debates on Symmetry that Heisenberg thought the neutron and the proton were two aspects of the same particle because their masses were so nearly equal. If supersymmetry existed, one would expect the electron to have a sparticle partner whose mass is of the same order as the electron mass. The fact that it hasn't been detected is good reason to conclude that supersymmetry does not exist in the realm of the everyday. The same applies to ordinary GUTs;

GUTs had twice as many bosons as the standard model of particle physics, but none seem to exist in our low-energy world. And as we said a moment ago, if sparticle masses are high enough, then they might as well be absent altogether and it doesn't make sense to talk about supersymmetry as relevant to ordinary physics.

What lessons are we to take from this? At the very least we must conclude, to use the language of the symmetrists, that we live in a world of broken symmetry. According to all these theories, the forces were unified only at extremely high energies, energies that would be encountered only at about 10^{-36} seconds after the big bang itself. As the universe cooled, the symmetry was broken. Just as the achiral sodium chlorate solution during the Debates on Chirality spontaneously crystallized into right- and left-handed crystals, so did the superforce break up into the different forces observed today.

At this remark a cell phone rings and an academician reports a resurgence in the debate on symmetry on the ground below. For once again the distinction between states and laws confronts us. The group symmetries employed by the unifiers pertain to the laws at high energies. But these are not the symmetries of the of the world as we observe it. In fact the complicated world as we observe it does not manifest group symmetries. The topic shall return.

The Difficulties of Quantum Gravity Are Pondered

The upward trek continued, aided by the inflation of the Library itself as it was filled with ever more knowledge. By now the entire territory of the Academy was visible and, distantly, parts of the Real World. However, the summit still seemed to be receding ever faster and you remained displeased. "But what about gravity?" you insisted. "You can't have a theory of everything without gravity."

True enough. Gravity has been the most difficult force to incorporate into a unified theory because it is of a fundamentally different nature than the other three forces. Quantum field theories, including GUTs and SUSY GUTs, make a basic assumption that space is always flat. On the other hand, Einstein's theory of gravitation, general relativity, assumes that the presence of gravity curves space. This changes the mathematical structure of the theory completely. Perhaps one can say it simply: quantum field theories assume that gravity doesn't exist, but it does. From the other side, general relativity is a classical theory, meaning it does not take into account quantum effects. However, we know that on subatomic scales quantum mechanics is necessary to describe the behavior of particles and fields. Since the gravitational field in relativity is represented by geometry, ultimately to quantize gravity requires quantizing the geometry of spacetime itself.

"How romantic," you say.

Yes, but one should bear in mind that any unified field theory that includes gravity will be operative only under special circumstances. Gravity is by far the weakest force in nature, approximately twenty-eight orders of magnitude weaker than the weak force. The only time the effects of gravity are noticeable in daily life is when huge numbers of particles are collected together—it takes the entire mass of the Earth to pull you down with a force equal to your weight, which is actually not a large force. One could produce an electrical force of the same size with roughly 10^{-13} kilograms' worth of protons. So in virtually all circumstances quantum field theorists are safe in ignoring the gravitational force on subatomic particles; in other words, they can truly assume space is flat. The only time one would need to take into account gravitational effects on the behavior of elementary particles would be when space is highly curved on atomic scales—either near small black holes or in the first instants after the big bang itself, when the entire mass of the visible universe was compressed into a volume less than the size of an atom.

Nevertheless, the lure of such a unified theory is far more powerful than any remoteness of applicability and much effort has been expended to create one. A number of approaches have been taken, but the success of each of them is debatable. Some ultrasonic hedgehogs, for instance, have attempted to construct supersymmetric theories that include gravity—supergravity, or sugrav. As you might suspect by now, such constructs result in many new sparticles, none of which has been detected. What's more, it has been shown that most of these theories cannot be quantized. We are not speaking of approximations or of difficulty in calculation. Theorists have proven that most versions of supergravity cannot be made mathematically consistent.

Given the difficulty of unifying gravity with the other forces of nature, one might retreat a step from supergravity and attempt the more modest goal of quantizing general relativity itself. Such an approach would be analogous to Dirac's quantization of the electromagnetic field in special relativity, by which he created the first quantum field theory, quantum electrodynamics. In this case one would create a theory of quantum gravity. This "straightforward" approach is in fact by far the older one, pioneered by Bryce DeWitt, John Wheeler and Charles Misner, among others, in the 1950s and '60s. Quantum gravity is also not for amateurs. In ordinary quantum field theories the infinities we have spoken of can at least be removed by the trick of renormalization. In quantum gravity the trick no longer works and the theory to date remains unrenormalizable.

"I tell you there is a more serious problem," interrupts an ultrasonic hedgehog. "String theories have already unified all the forces, including gravity. Ordinary quantum gravity is obsolete, a relic of past generations. It amazes me that anyone would even consider such a stupid idea."

Not all hedgehogs are so ultra, and researchers do continue to explore ordinary quantum gravity. Admittedly, a curious situation has arisen—various factions are working on various theories of everything which probably contradict one another and which may, if the string theorists are correct, have already been relegated to the ash heap of history.

Despite its problems, headway in quantum gravity has been made. The simplest approach, introduced by DeWitt in 1950, is to assume the gravitational field is represented by a classical curvature of space (in other words one does not try to quantize the geometry). Then one treats the remaining forces, like electromagnetism, as quantum fields traveling around on the curved background. This is known as the semiclassical approach, or quantum field theory in curved space.

"Isn't this inconsistent?" interrupts your friend. "How can you quantize one field but not another."

Yes, it is inconsistent. But remember the British Constitution:

She owes her success in practice
to her inconsistencies in principle.

Stephen Hawking abruptly joins in: "A physical theory is just a mathematical model and it is meaningless to ask whether it corresponds to reality."

"Meaningless? If it is meaningless, why bother with experiments?"

"Ask Hawking."

"All that one can ask is that its predictions should be in agreement with observation."

"I don't understand . . . aren't observations a reflection of reality, or aren't they at least their own reality? Does he mean reality or Reality . . . ? Are these debates meaningless . . . ?"

Whatever. Just like the shell model of the First Debates or quantum field theory itself, semiclassical gravity is in fact inconsistent, and arguments persist over even exactly how it is inconsistent.

"This does not seem very certain," you say.

No. Nevertheless, one would hope it would be a reasonable model when space is not so highly curved that one needs to quantize the geometry. Consistent or not, the semiclassical approach has led to quantum gravity's most celebrated discovery: Stephen Hawking's own 1974 prediction that black holes are not actually black but radiate energy, just like ordinary blackbodies. The radiated energy can be regarded as heat and thus Hawking showed that black holes have a temperature associated with them. For a black hole with the mass of the sun, this temperature would be about 1 millionth of a degree above absolute zero, a reflection of the fact that quantum effects are small indeed—black holes are *almost* black.

Having shown that black holes have a temperature, it was a simple matter for Hawking to confirm by the first law of thermodynamics a suggestion previously made by Jacob Beckenstein, that black holes also have an entropy. The connection with thermodynamics was the real significance of Hawking's work: by demonstrating that black holes have a temperature and an entropy, Hawking accomplished one of the first (and certainly the most spectacular) unifications of gravitation, quantum theory and thermodynamics.[†]

But the strange—some would say beautiful—thing about Hawking's calculations was that they gave no inkling about what was going on inside the black hole. What he basically did was to imagine a beam of quanta scattering off the space near a black hole and he discovered that more particles were coming out than were going in. Hawking sensibly interpreted the phenomenon as particle radiation from the hole, but nothing in the calculation explained *how* the black hole was producing particles.

The fact that black holes also had an entropy associated with them was yet more puzzling. If you remember the deck of cards from the Fifth Debates, the Boltzmann interpretation of entropy depended on certain configurations, or states, of the cards being more probable than others. It is easy to imagine a deck of cards or a gas having a number of distinct states (Boltzmann claimed to count them), but even Hawking's calculation, as we just said, gave no way to peer inside a black hole. How then does one count the number of a black hole's internal states? The problem was, to put it succinctly, no theory of the black hole interior existed.

Although everyone accepted Hawking's calculations as too beautiful to be wrong, given the mathematical inconsistencies of the semiclassical approach and the fact that it seemed impossible to know what was going on inside a black hole, no one could come up with any satisfactory picture of black hole radiation or entropy.

Until, perhaps, now.

A Meditation on String Theory, Its Success and Failures

During the exposition on quantum gravity, the debating party spiraled up to yet greater heights, until the Academy itself was hardly visible far below. Even the cell phones ceased to function, being out of range, and as you gazed down at the vast landscape spreading forth beneath you, you were thankful that you did not suffer from vertigo. The fresh air at such dizzying heights was bracing, but you and your friend did begin to wonder whether such quantities of knowledge as seemed contained in the Library truly existed, and also how you were going to get back down.

The upward motion accelerated at the mention of string theory.

"Yes!"

Although string theory burst into prominence in 1984, its roots go back to 1968, before the standard model of particle physics had come into existence. In that year Gabriel Veneziano at CERN decided to guess at a formula that would describe how particles interacting by the strong nuclear force scatter off one another. His formula was disarmingly simple and attracted the attention of other theorists who went on to work out its implications. To everyone's surprise, it turned out that his formula actually represented a new type of physical object, called a *string*.

Everyday strings are composed of neutrons and protons unless they are strings of obligation, in which case they are composed of coercion or guilt. However, the fundamental strings of string theory are the stuff of which everything else (except obligation) is supposed to be composed. Unlike electrons or quarks, which as far as anyone can tell act like point objects, strings act like one-dimensional objects, like . . .

" . . . strings?"

Yes, that is it exactly, except they are very small, about 10^{-33} centimeters long. You might view them as tiny rips in the fabric of space, or as miniature violin strings zooming around in space. The strings can be open, with two free ends, or closed like a rubber band, and as they fly around they oscillate, like violin strings, and as they oscillate they produce music of the spheres. More precisely, just as a vibrating violin string produces many overtones, so do fundamental strings, and these overtones are what we observe as particles: quarks, electrons and so on.

"I was right all along," says Pythagoras, on the Library ramp.

The important point about strings is that they are not points. As we mentioned in the distant Second Debates, all the infinities we have been speaking of arise because field theories describe point particles. The electric or gravitational force between two point particles becomes infinitely large when the distance between the particles shrinks to zero. It was hoped that quantum field theory would eliminate this problem by smearing the particles out into fields, but the difficulty stubbornly remained. Strings, on the other hand, by their very nature have length and cannot be shrunk to zero. Therefore proper string theories are free from the disease of infinities.

"This was one of their greatest attractions," adds a pro-string hedgehog.

True enough.

"Moreover, I can assure you that virtually every string theory predicted the boson associated with the force of gravity—the graviton. This was the first indication that string theories contain gravity."

Except that the gravitational force predicted by early string theories was about 10^{38} times too large. To reduce the force of gravity to the size we observe, one had to twiddle the parameters in the theory so that the string itself had an enormous energy—the kind of energy particles would have only in the first 10^{-43} seconds after the big bang, roughly a trillion times higher than the projected energy attainable by the late lamented Supercollider. Terrestrial experimentalists are not likely to be investigating such energies any time soon. All of this was a bit unexpected because Veneziano postulated his formula to explain the strong nuclear force, not gravity. Nevertheless, it began to appear that string theories might represent a quantum theory of gravity, which would be operative only near the big bang.

Unfortunately, early string theories did suffer from mathematical inconsistencies which made them less than attractive to most physicists. Suddenly in 1984, Michael Green of Queen Mary College, London, and John Schwarz, of the California Institute of Technology, discovered that these "anomalies" could be made to vanish if one formulated string theory in twenty-six dimensions.

At hearing this you halt the climb, walk to the edge of the Library ramparts, and prepare to jump. But the distance is large. Indeed, you can no longer see the ground.

Take heart, the idea that there should be more than three space dimensions and one time dimension is actually not new. It goes back to the work of Theodor Kaluza in 1921 and Oscar Klein in 1926. Just as special relativity unifies electricity and magnetism by including a fourth dimension (time), Kaluza had the remarkable idea of unifying electromagnetism with Einstein's gravitation by including a fifth spatial dimension, which is wrapped up on itself like a cylinder. Klein proposed that quantized nature of electricity could be explained if the cylinder were wrapped up on a scale of about 10^{-30} centimeters: a particle traveling in the fifth direction would actually be circling around a cylinder, but since the distance is so "compact" you couldn't notice.

Still, you are threatening to jump. Twenty-six dimensions is twenty-two too many for you. But really, is it any worse than David Bohm's declaration that quantum mechanics means space must be a multidimensional reality? Indeed, is it any *more* than David Bohm's declaration? "Hmm," you say, reconsidering. The New Agers present make plans to incorporate this intelligence into their program.

In any case, modern string theories incorporate the Kaluza-Klein idea. They may have ten dimensions, nine space and one time, but six of the spatial dimensions are rolled up, or "compactified."

But the reason for all the excitement after 1984 was not because of a multidimensional reality. The first reason was that Green and Schwarz's work appeared to predict a unique theory: only strings in twenty-six dimensions were allowed.

"Why twenty-six?" you ask.

The second reason is that theorists soon produced a consistent version of string theory in ten dimensions, which struck some as more economical, and it appeared that the standard model of particle physics fell out of this version at low energies—that is, as the universe cooled, as symmetry was broken, and as distinct forces fragmented out, one after another.

"Why ten?"

The third reason is that string theory did not merely predict that the carrier of the gravitational force, the graviton, existed; it predicted that when gravity froze out during the symmetry breaking it would be described by none other than Einstein's general theory of relativity, which everyone believes. Edward Witten, the world's leading string theorist, declares, "The fact that gravity is a consequence of string theory, to me is one of the greatest theoretical insights ever."

For all these reasons, in the mid-1980s, strings began to be hailed as the Theory of Everything.

"This is a typical end-of-the-millennium phenomenon," mutters Anonymous, on the way down.

A decade ago Paul Ginsparg and Sheldon Glashow expressed similar sentiments. Noting that string theory would operate only at energies forever unattainable to experiment, they warned, "Contemplation of superstrings may evolve into an activity as remote from conventional particle physics as particle physics is from chemistry, to be conducted at schools of divinity by future equivalents of medieval theologians.... For the first time since the Dark Ages we can see how our noble search may end, with faith replacing science once again."

Witten disparages such nattering naybobs of negativity. "Good wrong ideas are extremely scarce, and good wrong ideas that even remotely rival the majesty of string theory have never been seen."

Stephen Hawking, on the other hand, has evidently sided with Ginsparg and Glashow. In 1994, referring to the fact that quantum gravity with no strings attached predicted the fluctuations in space necessary for the inflationary universe scenario (scheduled for tomorrow) and allowed Hawking's own prediction of black-hole radiation, he said:

Neither of these predictions will be changed even if string theory is the ultimate theory of nature. But string theory, at least at its current state of development, is quite incapable of making these predictions except by appealing to general relativity as the low energy effective theory. I suspect this may always be the case and there may not be any observable predictions of string theory that cannot also be predicted from general relativity or supergravity. If this is true, it raises the question of whether string theory is a genuine scientific theory. Is mathematical beauty and completeness enough in the absence of distinctive observationally tested predictions? Not that string theory in its present form is either beautiful or complete.

"Verily," adds an anti-string hedgehog, "by the early 1990s it was known that there are literally thousands of string theories consistent the way Green and Schwarz's were consistent, and in four dimensions. So what's the big deal? Why choose any one of them?"

Well, this is true and perhaps partly for that reason after the glory days of the mid-'80s, string theory lapsed into semiobscurity for half a decade. But Hawking and the hedgehog may have spoken too soon. Since 1994, string fever has risen again. First, one no longer refers to string theory. At the moment the thousands of theories, practitioners claim, have been reduced to five consistent, ten-dimensional versions of string theory. *Au courant* academicians tend to refer to the five theories collectively as "brane" theories. One-dimensional strings are now just part of a set of "membranes"—called D-branes—that can have any number of dimensions.* A string is a one-brane, a sheet is a two-brane, and so on. A brane with p dimensions is termed a p-brane.

"A pea-brain?" your friend asks.

"My brain has been braned by pea-branes," you conclude, preparing to hurl yourself off the Library once again. "Hawking was right. Harp not on that string."

Harp then on a brane, imagining it to be something like a p-dimensional glob of jelly, jiggling through space.

"It is apparent that this system of the world contains many unnecessary epicycles," remarks Copernicus, passing the group as he walks downward along the spiral.

Actually, Copernicus's theory contained more epicycles than Ptolemy's, but it was Copernicus's theory that turned out to be correct in its essential idea. Similarly, branes have recently produced a spectacular result. In 1996 Andrew Strominger and Cumrun Vafa used D-brane technology to derive the Hawking black-hole entropy *exactly*! The question we posed before, How does one count the internal states of a black hole?, now has an answer. In the context of string theory, the number of internal configurations of a black hole *is* the number of D-brane states within the hole. Moreover, following Strominger and Vafa's discovery, colleagues went on to derive Hawking radiation from D-brane theory.

So Hawking may have hoisted himself on his own petard. Nonetheless, he may have a point. It appears to us that the situation is reminiscent of the situation after 1900 when Planck discovered blackbody radiation. What was oscillating? In terms of field theory we know the electromagnetic field is oscillating, but Planck did not have to go that far for a derivation of blackbody radiation. It may be possible to find an intermediate theory that does not require branes to reproduce Hawking's results. But one can hardly deny that the fact that Strominger and Vafa got the *exact* answer is remarkable.

---

* D refers to Peter Gustav Lejeune-Dirichlet, a nineteenth-century mathematician.

The black-hole results are only one reason strings are again in vogue. Perhaps the more important one is *duality*.

"Is this related to harmonic convergence?" a lost tourist asks.

Conceivably. As a general rule, it is easier to make calculations in what is called the "weak-field limit" of a theory than in the "strong-field limit." Hawking's calculations of black-hole radiation were weak-field approximations, possible because he confined himself to considering relatively weak gravitational fields (he did not try to quantize gravity itself). Had he considered the full interactions between gravity and the other fields (the strong-field limit) the problem would have proven intractable.

However, it has been known for many years that certain theories exhibit a peculiar property known as duality. It turns out that the strong-field limit of one theory can become the weak-field limit of another theory. In 1916 Debye was trying to formulate a theory of the thermal properties of solids. A direct assault on the problem is very difficult because a solid is the strongest-field limit one can imagine: all the atoms in a crystal are coupled by complicated, long-range forces to every other atom; if you move one atom, all atoms move. Debye's solution was to realize that when you tap a solid it begins to vibrate, producing sound waves that propagate through the solid almost as if the atoms weren't there at all. Thus the strongly coupled behavior of the atoms produces the weakly coupled behavior of the sound waves, which was enough to allow him to calculate the heat capacity of solids.

What string theorists have discovered in the past few years is that the five versions of string theory, as well as supergravity,* are connected by a web of similar dualities. Where the approximations of one theory break down because the fields are too strong, the weak limit of another theory takes over. The existence of D-branes is connected with duality and therefore the black-hole calculations have depended on it.

That the five string theories and supergravity, which were once all regarded as distinct, seem to be related by "duality transformations" has led Ed Witten to propose that they are all just various manifestations of a yet unexplored M-theory, a name he has coined to stand for "mystery," "magic," "matrix," or "membrane," as the mood strikes.

And so, if Witten is correct, then Jnaneshwar was truly prophetic when he wrote

In the same way, O Arjuna, when duality comes to an end, they and I live in one place, with one name.

A Practical Exercise—The Reader and the Authors Contemplate String Theory

By the time the discussion of string theory came to an end, the party had reached a height so vast that only stars were visible and it seemed as if the Earth itself had disappeared. But shivering in the vacuum of outer space, you and your friend remained troubled. Many questions seemed unanswered.

* Actually, supergravity formulated in eleven dimensions.

FRIEND: "Something has been bothering me. You said it was the introduction of an extended object—the string—that cured field theory by preventing forces from becoming infinite. This seems to me a very simple idea, so simple that I'm surprised that nobody tried it before."

GEORGE: You are correct, and it was tried before. People, particularly Yukawa and his group, spent lots of energy on "extended field theories." The idea was to regard an electron as an extended object, so that forces would not become infinite when two electrons touched. However, such theories could never be made consistent with special relativity. Researchers hoped that the contradictions were due to approximations, but careful work showed that the exact theory would be worse than the approximate theory.

Now we assert that string theory solves the problem by including extended objects. Question: How come? Part of the answer is that strings, unlike electrons, vibrate and produce an infinite number of particles. A finite number of point particles gives more infinities. But with an infinite number it is possible to get a nicer result. Of course I am speaking vaguely, like a good prophet. It is much like poison: poison helps poison to remove poison.

TONY: The homeopathic remedy for infinities. I have another problem. String theorists claim that the five theories are consistent, but recently I ran into a well-known anti-string hedgehog who maintained that the theories are known to be finite only to the lowest approximation.

GEORGE: This is true. By "renormalizable" we conventionally mean that the infinities can be removed at any level of approximation. No such demonstration of renormalizability to all orders exists for string theories.

FRIEND: This reminds me, I am a bit confused about compactification. Do the string theorists have a method to roll up the extra dimensions?

GEORGE: Do you mean a mechanism? No, it is merely assumed that this is the way the cookie crumbles. This is the ordained nature of the world.

YOU: I read an article in the *New York Times* by the science writer John Horgan. He called string theory the best example he could think of, of "ironic" science—science set free from truth. It's become like literary criticism, just opinion. If I understand what you've said, there seems to be no way to test string theory directly by experiment because the energies are unimaginably high.

TONY: Well, there are the predictions of black-hole entropy.

GEORGE: But have radiating black holes been observed? This is just another calculation.

TONY: You're just being stubborn.

GEORGE: No, they merely reproduced Hawking's result. Can they make any *new* predictions? You remember what Shelly Glashow says in his article—

TONY: Yes, I have it right here.

GINSPARG AND GLASHOW: A naive comparison of length scales suggests that to calculate the electron mass from superstrings would be a trillion times more difficult than to explain human behavior in terms of atomic physics.

GEORGE: Yes, this is Philip Anderson's point—you not only have to have the laws, you have to know how to get from one level to the next, from one discipline to the other. If they can't do this, is it science?

TONY: I would be interested to find out how much of that Glashow still believes. Remember, Ed Witten points out that string theory has predicted gravity.

WITTEN: Even though it is, properly speaking, a postprediction, in the sense that the experiment was made before the theory.

TONY: Witten regards it as a mere historical accident that general relativity predated string theory. On another planet, it might have been the reverse. We may well have reached the point where the only way to test a theory is to make contact with a previous theory—like relativity—that we already believe. In that sense I'm not sure strings are as "ironic" as Horgan believes.

GEORGE: But how do you know that many theories won't give you general relativity?

YOU: Do I detect a generation gap here?

TONY: This is true. The older generation can never accept the next generation's theories. But general relativity does fall out of string theory in a rather natural fashion—

GEORGE: Well, they've chosen the action they want to effect this miracle—

TONY: On the other hand, the black-hole calculations give you only the entropy, they don't give you the second law of thermodynamics. String theories, like all field theories, are time reversible, and so they cannot explain any irreversible phenomenon.

GEORGE: Correct.

FRIEND: How can this be a theory of everything?

YOU: Why ten?

GEORGE: Ah, this is the question. Why ten? Merely because it happens to work.

GINSPARG AND GLASHOW: The theory depends for its existence upon magical coincidences, miraculous cancellations, and relations among seemingly unrelated (and possibly undiscovered) fields of mathematics.

GEORGE: Agreed. There is no underlying principle involved with regard to the thing. The theory cannot answer the question Why ten? And earlier we swept under the rug the fact that the ten dimensions comes about because it is a supersymmetric theory.

TONY: Hence "superstrings."

FRIEND: More sparticles?

GEORGE: Unbelievable numbers. For all the assertions that string theory predicts gravity and supersymmetry, none of the sparticles have been found.

TONY: If they are found, though, it will be a big point in favor of strings.

FRIEND: That seems reasonable.

GEORGE: You should ask your string-theorist friends how long they are willing to look for sparticles before they give up.

YOU: I will if we ever get down to Earth.

The Discussion Widens into an Argument over the Point of Theories of Everything

By general consensus, the only thing for it was to jump. At that point the entire debating party leapt off the Library and continued arguing on the way down, while in free fall. Untune that string and, hark! what discord follows.

As the party fell, some foxes argued that because M-theory cannot answer all questions there will be yet another M-theory ("M" for "meta") lurking in the background, and another level of the onion will present itself for peeling. D-branes, for example, are undoubtedly composed of pre-branes, just as quarks are unquestionably composed of preons. The existence of a Meta-theory is possible but not necessarily the case. It may turn out that M-theory unifies the four forces, is logically consistent, and leaves no room for adjustment. It is in this sense that Steven Weinberg, who believes that a final theory is probably attainable and that something like M-theory may be it, compares the last theory to a piece of fine porcelain "that cannot be warped without shattering." Then, "although we may still not know why the final theory is true, we would know on the basis of pure mathematics and logic why the truth is not slightly different." Consequently, once the final theory is attained, "the old question Why? will still be with us." Why ten dimensions? Why does quantum mechanics work? Ultimately, Why is there something rather than nothing?

Weinberg is almost certainly correct about Why? There is no logical reason that space-time must have ten dimensions as opposed to thirty-seven; there are only Ginsparg and Glashow's "magical coincidences and miraculous cancellations" that allow string theory to work in ten and no other dimensions. Why? will still be with us.

Quantum mechanics itself gives some feeling for the frustration a logically complete theory will inevitably leave. There are no experiments that contradict quantum mechanics but, as has probably became clear by now, this does not imply we understand why quantum mechanics works. And thus as the party plummets toward Earth not a few of the participants ask:

What's the point?

Hermann Bondi, the English cosmologist, has expressed the view that there is no point whatsoever; the idea of a complete theory is a mirage. Of unified field theories, he said, "A vast number of hours and indeed years of the time of these towering intellects have been spent on this enterprise, with the end result (measured as one should measure science, by the lasting influence on others) of precisely zero. . . . In my view it is by no means fortuitous that all this endeavor was in vain, for I think that to aim at such completeness of description is meaningless in principle." He echoes your earlier concern that without experimental verification "unification is virtually bound to be sterile."

After this declaration, Bondi is made an honorary fox, but Diracian hedgehogs reply by reminding us that mathematical beauty will guide us to unification. Bondi will have none of it. "Experience indicates that while an individual theoretician may perhaps find such a concept heuristically helpful, it is not one on which different people can agree."

Weinberg disagrees. "The history of physics in this century," he maintains, "has been marked by the gradual growth of a realization that it is principles of symmetry that dictate the dramatis personae of the drama we observe on the quantum level."

At these words, telescopic observation of the still-distant ground shows that the symmetry battle has just flared up yet again; you suspect it will never end. In this state of free fall, Weinberg's position strikes the authors as both true and false. If we examine even atomic physics, the properties of simple systems—the hydrogen atom—*have* yielded to a description by group symmetries. However, even hydrogen with one additional electron—the negative hydrogen ion—cannot be described by groups. You perhaps recall from the Third Debates that certain conserved quantities, like momentum or energy, correspond to symmetries and determine how an object is allowed to move. But a hydrogen ion is a three-body system and it is known that its motion is chaotic; apart from energy and momentum nothing more is conserved and thus no new symmetries can be brought to bear on the problem. In a similar way, one might have thought the periodic table of the elements—on which chemical elements with similar properties are clustered together—would have pointed the way to a group description of chemical properties. But to work out the properties of similar elements required detailed calculations that had nothing to do with the periodic table or symmetry.[†] To say that the structure of the rare earth elements represents "broken symmetry" would be a statement without content. Even the neutral pion and the Omega-minus, whose celebrated predictions were made by symmetry considerations, represent only two particles. The vast number of discoveries in particle physics were first made by experimentalists and the notion of symmetry was applied afterward. Symmetry, as an experimentalist declared, is a language. To say that symmetry dictates the progress of physics is a little like saying file folders are important for physics because all physics papers end up in file folders.

For all these reasons the authors tend to side with Bondi in the resurgent subargument over mathematical beauty and physics.

But academician Bondi makes another interesting point regarding theories of everything. Not only is completeness a mirage, but it is not what science aims for. "The very power and elegance of the thermodynamic appraisal," he says, "lies in its essential incompleteness. . . . The fact that we can say a great deal about such a system without knowing about it in detail is a source of pride rather than of regret at the incompleteness of our knowledge. . . . No understanding whatever is needed of anatomy, physiology or the properties of leather to establish that one cannot pull oneself up by one's bootstraps. Indeed, one can argue that science is only possible because one can say *something* without knowing *everything*."

Looking back, virtually the entire downward dispute had thus far been among hedgehogs while the foxes remained unhappy. But now, as the outlines of the Academy gradually become visible, they raised another stark question:

So what?

It is a good question. Suppose we found a final theory in the sense that it unified the four forces, was mathematically consistent, and (somehow) was verified by experiment. How satisfied would we be, scientifically, aesthetically, emotionally?

In terms of the scientific satisfaction, it is true that physics has become a house divided. String theorists pursue M-theory, while the remainder of the physics community goes about its business unaffected. This is a fundamentally different situation than the one that existed in 1925 when the discovery of quantum mechanics changed everything. Even if M-theory should be proven correct, it will almost certainly have no impact on how the remainder of physics is carried out, except through the introduction of new mathematical techniques (which has already taken place). No one in his or her right mind would attempt to calculate the behavior of the atom directly from string theory. String theory's theological potential is large.

Impact, or lack of it, is not unrelated to the question of aesthetics. Suppose M-theory is accepted as correct. Will physics have ended? True, a few aesthetes might regard physics as complete and turn away from the field, and it would unarguably be pleasing to know that strings are related to quantum mechanics through a web of dualities. But, if while in free fall you lie back, close your eyes, and imagine M-theory to be true and whole before you, not much would change, in physics or anything else. The entire range of irreversible, chemical, biological and emotional phenomena would still be present in all their mystery. For this reason, it is hard to imagine that the extension of symmetry principles to achieve M-theory will have anything approaching the impact of, say, the Copernican revolution.

One must continually bear in mind how little of experience science as a whole encompasses, let alone one branch of physics. Hardly anyone has put it better than Arthur Eddington, to whom we cede the final word on this topic:

> One day I happened to be occupied with the subject of "Generation of Waves by Wind." I took down the standard treatise on hydrodynamics, and under that heading I read—
>
> > The equations (12) and (13) of the preceding Art. enable one to examine a related question of some interest, viz. the generation and maintenance of waves against viscosity, by suitable forces applied to the surface.
> >
> > If the external forces p_{yy}, p_{xy} be given multiples of $e^{ikx + at}$, where k and a are prescribed, the equations in question determine A and C, and thence, by (9) the value of η. Thus we find:
> >
> > [There follow several long equations.]
> >
> > where σ^2 has been written for $gk + T'k^2$ as before . . .
>
> And so on for two pages. At the end it is made clear that a wind of less than half a mile an hour will leave the surface unruffled. At a mile an hour the surface is covered with minute corrugations due to capillary waves which decay immediately the disturbing cause ceases. At two miles an hour the gravity waves appear. As the author modestly concludes, "Our theoretical investigations give considerable insight into the incipient stages of wave formation."
>
> On another occasion the same subject of "Generation of Waves by Wind" was in my mind; but this time another book was more appropriate and I read—

There are waters blown by changing winds to laughter
And lit by the rich skies, all day. And after,
    Frost, with a gesture, stays the waves that dance
And wandering loveliness. He leaves a white
Unbroken glory, a gathered radiance,
A width, a shining peace, under the night.

The magic words bring back the scene. Again we feel Nature drawing close to us, uniting with us, till we are filled with the gladness of the waves dancing in the sunshine, with the awe of the moonlight on the frozen lake. These were not moments when we fell below ourselves. We do not look back on them and say, "It was disgraceful for a man with six sober senses and a scientific understanding to let himself be deluded in that way. I will take Lamb's Hydrodynamics with me next time." It is good that there should be such moments for us. Life would be stunted and narrow if we could feel no significance in the world around us beyond that which can be weighed and measured with the tools of the physicist or described by the metrical symbols of the mathematician.

The Existence of Other Theories of Everything Is Revealed

As the downward journey continued, it struck you that the New Age had been almost entirely absent from today's discussion. When you queried the authors about the oversight they replied that, yes, for all its emphasis on the unity and inseparableness of the universe, the New Age seems to have ended with quantum mechanics (the Supersymmetric Field of Invincibility excepted). Either New Age advocates view the development of supersymmetry, strings, etc. as mere embellishments on the theme of symmetry, which they regard as a nondynamic relic of the West's Hellenistic past, or they have somehow missed the latest developments. Unless Deepak Chopra's remark that "you are like a wiggle, a wave, a fluctuation, a convolution, a whirlpool . . . " refers to D-branes.

Thus it was with considerable surprise that, as the levels of the Great Library flashed by, the party spied a man sitting cross-legged on a carpet that slowly glided in their direction. Although the fellow was thin to the point of malnourishment and clad in nothing more than a loincloth, he smiled to himself as if he knew something that others did not. The foxes greeted him, sensing a new prophet in their midst, and asked if there really could be a theory of everything and did it make any sense.

The fellow, whose name turned out to be Shankara, replied in astonishment, "You have not read my commentary on the Brahma Sutra in language accessible to everyone, without diagrams or math?"

"What is this BS?" asked one of the hedgehogs.

Shankara answered, "It is a superstring of formulae which describes the essence of the theory of everything."

The foxes admitted they hadn't read it and Shankara offered to give a brief outline of his commentary to the Brahma Sutra: "*Now we shall deliberate on the Brahman.*"

The foxes gathered around. "*From the Brahman is derived everything,*" Shankara began.

"This seems rather substantial," said a fox.

"This seems incomprehensible," replied a hedgehog.

"*Light is Brahman,*" continued Shankara.

"It is important to include light in a theory of everything," one of the foxes nodded. But another asked, "Does he mean light from the Sun or does he mean the Supreme Self? I never liked the use of technical terms."

"*From the Brahman is born Space and all things certainly originate from Space.*"

"I think he is talking about compactification," remarked a fox.

However, a sonic hedgehog objected. "This is not a theory. Where are its predictions?"

"*The entity possessed of the qualities of not being seen is Brahman,*" Shankara said.

"This is indeed a very inclusive theory," agreed the foxes.

The hedgehogs, on the other hand, remained dissatisfied. "What are entities that cannot be seen? Of what use is an entity that cannot be detected?"

"*The measured One is the Supreme Brahman.*"

"He must mean the Superbrahman," sneered a hedgehog.

"Or perhaps this is the pre-Brahman," suggested a fox.

"*The size of Brahman depends on the state of the heart.*"

A sonic hedgehog shook its head sadly. "What kind of measurement depends on the state of the measuring apparatus?"

"*The Brahman is the source of bliss,*" Shankara said.

"This is certainly a more inclusive theory of everything than any theory from physics," one of the foxes observed.

"This is truly BS," concluded a hedgehog.

"*The vital essence of the universe is vibration,*" Shankara went on.

"Hmm . . . " reflected the hedgehogs, hesitantly.

The fall continued . . .

In Which the Eighth Debates Are Summarized

Feet firmly planted on the ground, you and your friend reviewed the day's findings.

The morning began with a discussion over what a theory of everything could possibly mean. The authors suggested that to physicists a theory of everything was merely a theory that unified the four forces of nature into one "superforce." Since all natural processes are presumed to result from the interplay of the four forces, such a theory would explain everything.

The program of unification was reviewed. The world consists of fermions, or matter particles, and bosons, or particles that transmit forces. Unification of the two permits new interactions among fermions, which in turn requires new force particles— bosons. Unification has produced the standard model of particle physics, but because the forces are really kept distinct within their "unification," a more natural scheme was desired. In the 1970s Grand Unified Theories (GUTs) that truly united all the forces with the exception of gravity rode a wave of popularity. However, due to a total lack of experimental evidence, GUTs have since fallen out of favor.

Unification can be extended to an attempt to treat fermions and bosons on an equal footing, or abolish the distinction between forces and matter. Theories that do so are known as supersymmetric theories. All supersymmetric theories predict many new particles, known as "sparticles," none of which has ever been discovered.

Well up on the Library spiral, you initiated a discussion of symmetry breaking. The fundamental forces are unified only at high energies, such as that which would be found at the very beginning of the universe. As energies drop, the forces fragment out in a process of "spontaneous symmetry breaking," very much as left- and right-handed crystals form from a sodium chlorate solution. The world we live in is not a world of unified forces.

As the debate ascended to great heights, gravity remained the odd man out. Various attempts to unify quantum mechanics and general relativity into "quantum gravity" have been made, but because quantum field theories assume space is flat and general theory of relativity assumes space is curved, success has been modest.

The debate achieved unbelievable heights. Hedgehogs introduced string theory as the best candidate for a theory of everything. In string theory the fundamental objects are not point objects but either one-dimensional strings or p-dimensional D-branes. Proponents of string theory point out that it is consistent, predicts general relativity and supersymmetry, and has been used to derive black-hole entropy and radiation. Opponents object that no supersymmetric particles have ever been found and that the claims for consistency are based only on the lowest approximation. Edward Witten proposes that the five known versions of string theory are all related to each other by "duality transformations" and are just manifestations of an underlying M-theory, which has yet to be discovered.

Then the party jumped.

On the way down, the debate broadened into a discussion over whether a search for a theory of everything was meaningful. Opinion was sharply divided. Eddington juxtaposed two descriptions of waves and questioned which of them had more content. Finally, Shankara showed up with his theory of everything, which seemed more inclusive than string theory and which, strangely, seemed to suffer from many of the same defects.

Then a miracle occurred.

HOW DID WE GET HERE?
Cosmology

I: *General Relativity and the Standard Model*

All science is cosmology, I believe.
　—Karl Popper

And which of the gods was it
that set them on to quarrel?
　—The Iliad

The Question

If you have ever strolled among the Academy's excavations at Hissarlik, about six kilometers from the Aegean Sea in northwest Turkey, you have, as many before you, imagined that in a previous cycle of existence you were a member of the Greek army besieging ancient Troy. A popular incarnation among tourists is Odysseus, hiding in the Trojan Horse, waiting for nightfall in order to signal the destruction of the city. Or perhaps your previous avatar was Achilles, with his shield forged by no less than Hephaestus himself, pictured on it earth, heaven and sea, unwearied sun, moon waxing, all the stars that heaven bears for garland. . . . Or, if you are of a more literary turn of mind, perhaps Homer, recording the epic for posterity. Helen has of late lost popularity among past-life regressees.

But of course as you strolled through the ruins, your tour guide reminded you that these personages are more legendary than historical. The Greeks themselves were unsure of many fine details, such as whether Homer witnessed the immortal conflict, or whether he just missed it, or whether, as Herodotus claimed, the bard came along only four centuries after Aeneas fled the burning city with his father on his back. With such uncertainties as to Homer's dates, it is not surprising that there is no consensus as to his birthplace. (Semonides of Amorgos prefers Chios, while Pindar opts for Smyrna.) And with such disputes about his whereabouts there arises inevitably the question of whether Homer actually existed and whether one person or two (why not three?) wrote both the *Iliad* and the *Odyssey*. Computer studies have undoubtedly concluded something.

The war itself has provoked exciting controversies. Was there a Trojan War? When? Was it really caused by Paris, who, with the help of Aphrodite, abducted Helen, whose beauty not only launched one thousand ships but provided the first scientific unit of measurement (the milliHelen, mH = beauty required to launch one ship)? Or were there more mundane factors, such as economic expansion and Troy's strategic location on the Hellespont?

These questions might strike you as secondary if scholars are not even certain the Trojan War took place. As for Troy's strategic location, you might have wondered, strolling through the ruins, how are we sure that Troy existed?

All is not lost, however. Archaeology has provided some answers where legend and literary scholarship have failed. The statue of Athena in the sixth book of the *Iliad* cannot, evidently, be older than the eighth century and the use of the phalanx, Trojans massed and running, may be even later. Perhaps these are additions from a subsequent age, you argue. Even so, your guide replies, the *Iliad* probably achieved its final form before 700 B.C., when reports of its recitations begin to surface. In any case, we are certain the *Iliad* and the *Odyssey* exist. That is a fact.

It is also a fact that Troy existed. Heinrich Schliemann, who originally plundered the mound at Hissarlik in the nineteenth century, unearthed nine. Nine Troys, that is. Due to the vast quantities of gold he was able to steal from the second oldest city, and the fact that it had been destroyed by a devastating fire, Schliemann did not hesitate to declare Troy II to be Priam's Troy, which fell to the Greeks. Nowadays, with better dating methods, archaeologists tend to think Troy VIIa, destroyed by a fire in the thirteenth century B.C., was the Troy of legend, although some texts advance Troy VI as a candidate. Geomagnetic surveys indicate that the city may have been far larger than previously thought.

And, as the excavations began near the Great Library on the dawn of the Ninth Debates, it was probably no coincidence that one of the archaeologists present remarked, "Troy is much like history itself. It is difficult to assign causes to events that occurred only once, if they occurred at all, and there are many gaps in the historical record. But just because we don't know everything, doesn't mean we know nothing."

"In that cosmology is much like history," mused one of the cosmologists, taking pickax in hand. "The extraordinary thing is not that we don't know everything, but that we know anything at all."

"No," dissented a second academician, rolling up her sleeves. "Cosmology is more like religion. Just as the beginnings of Troy and the Trojan War are bound up with mythology and legend, cosmology cannot avoid essentially mythological answers to its own questions. The extraordinary thing is not that cosmology can't answer all the questions it poses, but that people even try."

Thus, as the debating party broke ground, everyone present was asking not merely, "What is cosmology?" but "Can it be distinguished from religion and philosophy?"

A Contemplation on the Meanings and Difficulties of Cosmology

At the disquieting quantum sunrise a few days past, academicians kept to themselves and sported dueling pistols. By contrast, today's atmosphere seemed quiet, even placid; no factions of any sort were to be found. Once more a large contingent of visitors from southern California was evident (cosmology has been called "the Hollywood science"), but many amateur astronomers, theologians, philosophers, historians and closet Einsteins also appeared at the Academy gates. Trading on academic ideas had been resumed on the World Exchange and due to anticipation of the outcome of the deliberations, the Knowledge Marker opened at a record high.

As ever more people were deposited by tour buses at the excavation site, they asked what the company could possibly be digging for. "The Oblivion File," replied one academician, "the ultimate resting place for most theories, almost all artistic creations and every job application." Relishing the opportunity to help confirm the existence of the legendary Oblivion File, many of the tourists immediately volunteered, and the number of debate participants increased threefold.

Early on you noticed that one of the tourists found herself next to a cosmologist and immediately asked him for advice on eye shadow and lipstick. To your everlasting surprise, the cosmologist took her at her word and produced a selection of various shades and textures. He later explained that having heard the request at least a dozen times a year, he enrolled in a school of cosmetology in order to be able to provide expert service.

To a scientific cosmologist, cosmology and cosmetology have much in common. Cosmetology is the study of the artificial enhancement of the whole person; cosmology is the study of the evolution of the universe as a whole. Some say both deal in illusions. A cosmologist these days is any scientist, be it a theoretician or an observational astronomer, who is concerned with the large-scale evolution of the universe, in particular with the origin of the universe.

But scientists, you discover as the dig gets under way, are not the only ones who adorn themselves with this mantle. Some historians who study the world constructs and creation myths of ancient civilizations also refer to themselves as cosmologists.

"They should all be called theologians," quips the same skeptical academician, as the first rays of the sun glance across the horizon.

There is truth here. Cosmology suffers from the difficulty that it attempts to deal with the origin of *everything*. "Why is there something rather than nothing?" is the ultimate cosmological question. It is also the ultimate theological question and any answer you give inevitably boils down to the assertion that either something always existed, or that something arose from nothing, or that something was created by something else, which always existed. None of these responses is terribly comprehensible to the human mind, but if you tend toward the first two you are classified as a cosmologist, while if you tend toward the last you are termed a theologian.

Academy visitors have sometimes found these distinctions less than obvious, and for good reason. Bookstores at the millennium are filled with titles by reputable scientists such as *God and the New Physics, The God Particle, Genesis and the Big Bang* and *The Physics of Immortality*. Stephen Hawking has referred to the task of uncovering the laws of nature as revealing "the mind of God." After the discovery of the ripples in the cosmic microwave background radiation (to which the debaters are eager to turn), mission head George Smoot spoke of viewing "the face of God," and the media reworked the metaphor into the "handwriting of God."

Theologian Paul Tillich once observed that physicists are the only scientists capable of using the word "God" without embarrassment. He may have been on to something, and it is also true that to the general public the perennial philosophical aspects of cosmology are its most alluring feature. If you doubt this, observe closely the approaching tourist. He is meeting someone; she is an engineer. Listen. The first thing he asks her is for help in repairing his car. Notice the second tourist. She is meeting someone; it is a cosmologist. She is immediately plunged into an existential crisis. What is the meaning of life? What are we doing here? Where did it all come from? Without waiting for a reply she forthwith schedules a banquet in the cosmologist's honor at which the guests will discuss the great mysteries. The cosmologist asks whether she needs advice on eye shadow.

You see, despite Tillich's observation, not all cosmologists are comfortable discussing God. Many, including the hapless cosmologist/cosmetologist just cornered, would prefer to make a clean distinction between cosmology and theology and leave the latter to the theologians and philosophers. Just as archaeologists needn't ask how Troy I originated in order to study Troy VIIa, so too can cosmologists ignore the instant of creation and the philosophical dilemmas it brings. The usual procedure in day-to-day research is to assume that the universe has already started off somehow (typically with a bang) and then ask how it evolved into its present state. One's tools in this task are mathematics, computers, brain cells and telescopes; religious commentaries simply don't figure. Only when confronted by journalists do cosmologists go theological.

The displeasure with such a nonholistic approach is visible on the faces of the crowd. In any case, you want to know, is a distinction between cosmology and theology possible? It may be feasible to study Troy VIIa without knowing where Troy I came from, but if you wanted to know where the Earth came from, you must place it within the larger context of the solar system, and if you want to know where the entire universe came from, you are stuck. There is no Court of Appeal.

The point cannot be denied and is exactly why cosmological and theological territory eventually merge. It is also natural that people want to know what lessons can be drawn from contemporary cosmology. But as Richard Tolman said, over sixty years ago: "It is appropriate to approach the problems of cosmology with feelings of respect for their importance, of awe for their vastness and of exultation for the temerity of the human mind in attempting to solve them. They must be treated, however, by the detailed, critical, and dispassionate methods of the scientist."

So be it.

Yet.

Even if one puts aside theological issues, one immediately runs aground against two other significant dilemmas associated with cosmology, dilemmas that distinguish it from most laboratory sciences.

The first is that cosmology relies on astronomical observations for its evidence and astronomers are restricted to making observations from the vicinity of Earth. It is one of the major achievements of twentieth-century science that astronomers have been able to learn so much about the universe solely by collecting electromagnetic radiation in Earthbound telescopes, but this handicap does force scientists to make assumptions about distant regions of the universe, which are often difficult, if not impossible, to verify. Is the average number of galaxies the same everywhere? Are distant galaxies composed of matter or antimatter? Such questions have no obvious answers. As a result cosmology has been rife with principles, philosophical if not exactly theological. One has been so prominent that it has effectively guided the history of the subject.

We are referring to the famous "cosmological principle," which states "merely" that the universe should be the same everywhere. The cosmological principle was introduced by Einstein before astronomical observations told scientists much about the large-scale structure of the universe. But the irregular clustering of galaxies across the heavens shows that the real universe is simply not the same everywhere. The cosmological principle provided the basis for what has become known as the "standard," or FLRW, model of the universe,* and one question that those gathered here have already begun asking is, How reasonable is the cosmological principle as the basis for cosmology?

To which we return.

But cosmology's main difficulty is that there exists only a single universe, which is evolving. Not only is this the ultimate source of the theological question, but it presents practical dilemmas as well. Each day on the radio, for instance, the weather forecaster says something like, "The probability of rain is thirty percent." What the forecasters have done (more or less) is searched historical records for past days with similar atmospheric conditions and discovered that 30 percent of the time it rains. But what would it mean to say, "The odds of the Trojan War breaking out were 30 percent"? Historical events take place only once, making it difficult to calculate odds. The hurdle rarely stops policy analysts from making oracular pronouncements. How often have you heard the obligatory retired colonel on a late-night news program claim, "I estimate the odds of this crisis escalating into a Balkan war to be about 50 percent"?

You are asking, "If a war doesn't break out, was he correct?"

Precisely. The notion of probability is difficult, if not impossible, to apply to unique events. If you demand of a science that its results be reproducible, then any discipline dealing in unique events, such as history and perhaps even economics, should not be called science.

At this remark, to no one's surprise, the economists and historians present have taken umbrage, and begin pelting the authors with stones from the excavation. Economic results are predictable, the Fed's chairman would maintain. That is why each time the Fed raises interest rates the people scream bloody murder. Yes, it is true that economists can predict that raising interest rates will lower inflation (most of the time), but would they predict that when the Fed chairman says the stock market is doing well, stocks plummet? Could you predict the behavior of the economy for next year? In gen-

---

* FLRW stands for Friedmann, Lemaître, Robertson, Walker, after the Russian meteorologist Aleksandr Friedmann, who first worked out the mathematical details in the 1920s, and after the Belgian Abbé George Lemaître, the American H. P. Robertson and the Englishman A. G. Walker, who all independently developed essentially the same model in the 1920s and 1930s.

eral the Fed waits to see how things are going and then applies a corrective maneuver. It is a very experimental approach.

"Is that any different from most sciences, biology for instance? What kind of predictions does biology make?"

Now the biologists have angered, responding that at least the results are generally reproducible, if not always predictable.

Watching the argument they have touched off, the authors retreat a step; they had merely intended to point out that cosmology suffers from the ultimate uniqueness problem: there is only one universe and we are in it. With current technology we cannot generate multiple universes in a laboratory and observe the results. In this sense cosmology is not an experimental science.

Because of the universe's uniqueness and because answers to the Ultimate Questions elude us, cosmology has often been viewed with skepticism. Until fairly recently, the *New York Times* would launch periodic assaults on the big bang theory and from time to time books appear claiming that it never happened.

But to return to the cosmetologist/cosmologist's position, just because archaeologists cannot accurately date the Trojan War does not necessarily imply it never took place. Cosmology is in a similar position. Because light from distant astronomical objects takes longer to reach us, then just as digging downward carries archaeologists to earlier levels of Troy, observing galaxies at ever larger distances automatically allows astronomers to view most of the universe's history. As for the missing details, cosmologists create models—with computers or without—that are meant to reconstruct the past that would have produced the present. Cosmology is less a predictive than a retrodictive science.

But it is a science.

Before the Big Bang Is Reached, an Argument Erupts over General Relativity. Is Space Curved? And Other Matters

Despite the authors' intent to concentrate on cosmology's scientific aspects, you perhaps remained perplexed about the discussion thus far. The authors asserted that cosmology exists and makes reasonable statements about the universe's past, but they failed to prove anything. Why shouldn't one trust the *New York Times* when it says the big bang never happened?

The authors now reply. First, consider Einstein's theory of gravitation, general relativity, with which he extended Newton's own gravitational theory. Because, for reasons explained in the Third Debates, only feeble gravity is thought to determine the large-scale evolution of the universe, modern theoretical cosmology is basically the study of how general relativity can be applied to various universe models.

There are few doubts that general relativity, GR in Academy corridors, is essentially correct. All its predictions have been confirmed, for example "gravitational lensing," in which light from a distant object (a quasar) is bent around an intervening galaxy such that we see several images of it. The Hubble telescope's discovery of black holes in galactic centers is another direct confirmation of general relativity.

There are others. Just as accelerating electric charges emit electromagnetic radiation, according to GR accelerating masses should give off gravitational radiation. Because gravitational radiation is incredibly weak it has not yet been directly detected, but it has been indirectly. Two stars orbiting each other are accelerating and so should emit gravitational waves. In the process the stars should slowly spiral into each other and their orbital period decrease. Observations of the binary pulsar (which can be used as a precise clock) have confirmed GR's predictions to fourteen decimal places, making it the most precise test of a theory in history.

Still, scientists would prefer a smoking gun and for that reason teams at Caltech and MIT are currently constructing LIGO, the Laser Interferometer Gravitational Wave Observatory, which is expected to go online in 2002. (Similar devices are being built by the French and the Italians, the Germans and the Scots, and the Japanese.) Everyone expects LIGO to make a direct detection of gravitational radiation. If it doesn't, relativists will have some very hard thinking to do.*

At this moment the Academy mail carrier arrived at the excavation with a sack overflowing with gravitational theories from enthusiasts worldwide, but the authors did not have time to read them because they wanted to make another point about general relativity.

GR differs from Newtonian gravity not only in its predictions but in its description of gravity. Newton regarded gravity as a force between two objects. In GR, as in the other modern theories, forces are replaced by fields. Here is the origin of Combs and Holland's assertion, "Beginning with Einstein the cosmos became an undivided field. Objects such as atoms and stars are viewed as properties of this field. They are seen as local concentrations of it. . . . They have no separate existence in and of themselves. . . ." Combs and Holland, following Fritjof Capra, wish to argue that the concept of a field is holistic, implying that "material objects are not distinct entities, but are inseparably linked to their environment; that their properties can only be understood in terms of their interaction with the rest of the world."

But the Zen master has already hit you over the head with a field once or twice. No one would deny, even in Newtonian physics, that objects are often inseparably linked to their environment. A collection of Newtonian particles *does* manifest a gravitational field (though Newton would not have pictured it this way) whose configuration depends on all of them; if one object is removed, the entire configuration changes, much as would be the case in general relativity. By the same token, neither in GR is the existence of objects such as neutron stars or galaxies denied. Rather these objects tend to be regarded as sources of the gravitational field.

Indeed, in Einstein's equations, one side describes the distribution of matter and this is equated to quantities on the other side that describe the field. What is interest-

* More about LIGO can be found at the LIGO web site, http://www.caltech.edu.

ing about Einstein's equations is that the mathematical quantities (known as tensors) describing the field can be viewed as describing how spacetime is curved. In the Second Debates we briefly discussed non-Euclidean geometries, in which parallel lines may meet or diverge and in which triangles may contain more than 180 degrees or fewer. GR takes over the mathematical apparatus of non-Euclidean geometries lock, stock and tensor. For that reason Einstein and his successors have always viewed general relativity as a *geometric* theory. When light is bent around a black hole it is no longer traveling on a straight line in Euclidean space; it is traveling along the straightest possible path—a geodesic—in curved spacetime, analogous to a great-circle route on the globe.

The idea of spacetime curvature is probably the single idea from general relativity most deeply ingrained in the minds of scientists and public alike. As Capra exclaims, "We must not think that the field fills the space and 'curves' it. The two cannot be distinguished; the field is the curved space!"

But at hearing this, one of the cosmic archaeologists present exclaims, "All this talk about curvature of space is meaningless."

The workers cease digging, a hush falls over the crowd, and the relativists in the ditch turn their pickaxes on the heretic, ready to impale him. No spacetime curvature! Inconceivable! Why, the computer graphics in Disney's *Black Hole* showed the curvature during the opening credits.

Actually, it's not inconceivable. In the Second Debates, a traveling Texan surveyed the triangle made by Dallas, Austin and Houston and found that it contained more than 180 degrees. For this reason he concluded that the surface he was traveling on—the Earth—was curved like a sphere. Beings restricted to life on Texas can determine the geometry of this space only by measurement. Beings restricted to living in the universe—all of us—can determine its geometry only by measurement. If measurements reveal that light is deflected around a black hole, then space is curved.

"Well then, that settles the matter," you say, relieved.

Does it? At this juncture Henri Poincaré appears above the excavation pit, wondering what everyone is doing down there, and declares, "Experiment tells us not what is the truest, but what is the most convenient geometry."

He admonishes us to survey a triangle in space with light rays. Let us agree that the result is more than 180 degrees. "Aha!" exclaim the relativists, "space is curved."

"No," answers Poincaré, "we may interpret the result in two ways. We may say that light traveled along a geodesic, the straightest possible path in a curved space. Or, contrary to our usual custom, we may say that space is flat but that light traveled along a more complicated, curved path. My memory is bad, but if you reflect on the shipping lanes of the Fourth Debates, you will agree that the situation is exactly analogous. On a flat map the path of the ships appeared curved rather than straight. Ancient seafarers naturally attributed the bending of the shipping lanes to an Arctic demon sucking on the ships and pulling the trade routes northward. Scientists of our own age could say either that the Earth is flat and a complex force is acting on the ships, or that the Earth is curved and that the ships are following great-circle routes, the straightest paths on a globe, along which no forces are acting."

"Wait a minute," you interrupt, feeling swindled. "There is no question that the Earth is curved, like an orange. Astronauts observe the curvature from outer space."

"But, unlike astronauts viewing the Earth from above, we are not at liberty to view the universe from the outside," Poincaré replies. "We are restricted to making measurements from within the universe."

Poincaré's position is not so easily dismissed. As recently as 1972 Steven Weinberg advocated a similar view:

> We saw that Gauss was led to introduce . . . curvature as the true measure of the departure of a two-dimensional geometry from that of Euclid, and that Riemann subsequently introduced the curvature tensor to generalize the concept of curvature to three or more dimensions. It is therefore not surprising that Einstein and his successors have regarded the effects of a gravitational field as producing a change in the geometry of space and time. At one time it was even hoped that the rest of physics could be brought into a geometric formulation, but this hope has met with disappointment, and the geometric interpretation of the theory of gravitation has dwindled to a mere analogy, which lingers on in our language in terms like "metric". . . and "curvature," but is otherwise not very useful. The important thing is to be able to make predictions about images on the astronomers' photographic plates, frequencies of spectral lines, and so on, and it simply doesn't matter whether we ascribe these predictions to the physical effect of gravitational fields on the motion of planets and photons or to a curvature of space and time.

Weinberg's view was admittedly heterodox; nevertheless, in recent discussions with him it appears that he continues to hold this position.

Weinberg is wrong in a practical sense when he says that the geometric interpretation of gravitation is not very useful. Useful, if nothing else, is exactly what the geometric picture has proved to be. It allows relativists to perform calculations that would otherwise be virtually impossible. Since the 1980s string theorists have also taken over geometric language in order to make their calculations tractable. But the real question is whether the geometric description of relativity is merely a language or something intrinsic to nature. If, then, someone maintains that geometry is merely a language, cannot we reply with the question of the Third Debates: Shouldn't the same be said about symmetry?

While you ask yourself once more whether the symmetry argument will ever end, the symmetrists at the dig turn on the authors with the same hostility that relativists once greeted Weinberg's heresy. The authors stand their ground. Surely symmetry can be no less a language than geometry. The geometric "analogy" is, in fact, exact. There is simply no difference between the mathematical formulation of general relativity and the mathematics of non-Euclidean geometry. In which case, how can one distinguish gravity from geometry? Moreover, in string theory, the geometry of spacetime itself is viewed as being derived from the geometry of strings, and so the analogy now extends to an even deeper level. Given that it appears impossible to distinguish geometry from the formulation of these theories it is reasonable to ask, Is Weinberg's position metaphysical?

Nevertheless, even if one accepts the geometric interpretation of gravitation, there remains Poincaré's question, Which geometry? Poincaré, certain of his position, boldly declares, "I challenge anyone to give me a concrete experiment which can be interpreted in the Euclidean system and which cannot be interpreted in [a non-Euclidean system]. As I am well aware that this challenge will never be accepted, I may conclude that no experiment will ever be in contradiction with Euclid's postulate; but, on the other hand, no experiment will ever be in contradiction with [the postulates of non-Euclidean geometry]."

Poincaré waits for an adversary, but neither modern academicians nor visitors dare reply while they remain confused by his language. He merely means to say that any experimental result can be interpreted in a simple way in curved space or in a complicated way in flat space, but both are consistent. Academician Ian Roxburgh, siding with Poincaré, puts it forcefully: "The curvature of space is in the mind of the scientist. . . . Is space curved? The answer is yes or no depending on the whim of the answerer. It is therefore a question without empirical content, and has no place in physical inquiry."

Now those present find themselves persuaded by the reasoning and begin to harangue the relativists for misleading the public over the past decades. You find yourself depressed, ready to go home and burn all the popular expositions of relativity you have collected.

But at this moment the authors bravely step forward to take up the thrown gauntlet. Have not Poincaré and Roxburgh overstated the case? Imagine that you live in a universe shaped like a bagel. You can always cover a tiny piece of the bagel in your immediate vicinity with a flat sheet of paper. Geometers say any reasonable space can be made locally flat. Locally Poincaré is correct; an observer can always regard space as flat but adorned with a complicated gravitational field that produces complex light paths.

Globally this is not the case. A bagel-shaped space is intrinsically different from flat space, meaning that you cannot wrap a flat piece of paper smoothly around the entire bagel without cutting or folding the paper somewhere. In a bagel, or toroidal, universe light could travel around, eventually returning to its point of origin. And so our proposal to Poincaré: Take an extremely powerful telescope. Wait for a long time. If you eventually see the back of your head, then you do not live in a flat universe. Rather you live in some more complicated spacetime, likely a bagel, or perhaps a pretzel.[†]

Poincaré objects that this is not a practical experiment. Ah, we reply, but it is. If we are living in a bagel universe, then we will see many images of the same galaxy over and over again. All that is required is to identify the images as arising from the same galaxy.

Poincaré continues to voice reservations as to the outcome of such an experiment but honorably pays up.

You sigh in relief, glad to know something of geometry remains. The universe always struck you as like a bagel anyway.

However, one of the authors now has second thoughts. Perhaps the experiment just suggested is not definitive after all. "Why are we making the matter so complicated? Instead of a bagel-shaped universe, imagine ourselves to be in a tailor's dressing room, surrounded by mirrors on four sides. You will see many images of yourself, exactly as before. My assertion is that to produce multiple galactic images does not require a bagel-shaped universe. The universe could be flat, surrounded by mirrors—"

"—or what is equivalent," adds Poincaré, immediately grasping the argument, "merely a flat universe, resembling a wall map, on which you match the eastern and western borders, such that each time one passes off the eastern edge of the map, one reappears on the western edge. Of course, we must also assume that each time one passes off the northern border, one reappears on the southern. No one will doubt that this is equivalent to the bagel universe you proposed. I humbly thank you for correcting my oversight. How stupid of me. It is now clear that Euclidean geometry can never be contradicted by experiment."

The authors prepare to hand back their winnings to Poincaré, but now the other author has second thoughts. "A bagel universe was too simple. Consider instead a spherical spacetime, like an orange. Suppose we are located at the North Pole. The image of a star located exactly at the South Pole, or in general at our antipode, would be seen everywhere, very much as the image of an object at the focus of a concave mirror spreads out over the entire mirror. I don't believe you can account for this in any other way than by assuming the universe is intrinsically curved, globally."

"Let me think about this for a moment," replies the first author.

"Why don't we call it a draw," you suggest, unearthing some Tylenol from an ancient medicine cabinet.

Your friend nods assent. "This doesn't seem to be a trivial question."

We agree.

As the excavation continued downward, try as you might, you could not put aside your feeling that the authors have cheated. In previous debates when general relativity was mentioned, they gave no hint of any ambiguities in the description of gravity, but blithely tossed about words such as "curvature" just as everyone else does. We don't apologize for the pedagogical maneuver. One should introduce confusion a little bit at a time, like Chinese water torture. Nor should you be angered or depressed, for the discussion provided yet another illustration of the famous dictum "The map is not the territory."

The geometric language employed by general relativity, in fact, turns out not to be its only ambiguity. Another perplexing feature of the theory has been emphasized by cosmologist George Ellis over the years. Listen attentively, for the conundrum is tricky.

As we have said, Einstein's equations equate a tensor describing the curvature of space (if you accept this language) to another tensor describing the distribution of matter in space. For example, in the vicinity of the solar system, the matter distribution is very uneven—a few planets in otherwise empty space—and the tensor describing the geometry will reflect that, showing highly irregular curvature. Now, suppose we step back to a greater scale, one that includes the entire galaxy. On this scale, matter is more evenly distributed and the curvature tensor will be smoother. Nevertheless, cosmologists assume that GR is correct for both the solar system and the galaxy and do not hesitate to equate the matter tensor to the curvature tensor at both scales.

Is this allowable?

Suppose we assume that GR is correct at the scale of the solar system. We equate the two tensors. Next, imagine passing from the solar-system scale to larger and larger scales. In doing so, we smooth out the matter distribution by performing some sort of averaging operation on the matter tensor until we arrive at the galactic scale. (We might do this with a cosmic bulldozer.) Having smoothed out the matter tensor so that it is identical to the one describing the original (unaveraged) matter tensor at the galactic scale, we can equate them. Next, apply the same averaging operation to the curvature tensor. Once having arrived at the galactic scale we now ask, Can we equate this smoothed curvature tensor to the original (unaveraged) curvature tensor at the galactic scale? If not, then Einstein's equations do not hold at this scale.

Unfortunately, there is simply no reason the two curvature tensors should be the same. To see this a little more clearly, consider two numbers, 2 and 4. Their average is just $(2 + 4)/2$, or 3. The square of the average is $3^2 = 9$. But the average of the squares is $(2^2 + 4^2)/2$, or 10. Notice that 10 does not equal 9. In words, the square of the average does not equal the average of the squares.

In the same way, the curvature tensors of the average matter distribution (the smooth, galactic-scale distribution) are in general not the same as the averaged tensors (those obtained by passing from the solar-system scale to the galactic scale).

"I think I get it," says your friend. "The tensors of the average are not the average of the tensors."*

Correct. Although cosmologists do not seriously doubt the validity of general relativity, cosmologists are still working to give a convincing explanation of Ellis's conundrum: why should GR hold at all scales?

"But this problem seems to be of a different nature than the previous one," you object. "The question about the geometrical interpretation of relativity seemed almost philosophical, a matter of definition. Ellis's puzzle seems to indicate either that GR does not hold at all scales or that cosmologists do not know how to perform averages."

True enough, but as in the case of quantum field theory, it does indicate there are important features of a fundamental theory that we do not entirely understand. In fact, one of the most basic questions general relativity was invented to answer remains the subject of debate: where do accelerations come from?

When you round a corner in a car you feel a "centrifugal" force that throws you to one side of the car. The fact that you experience a force means you are accelerating. But accelerating with respect to what? Astronauts feel the same force in the space shuttle far from any planet. Thus the acceleration cannot be with respect to the Earth. To Newton there can be only one answer: "With respect to absolute space."

"Nonsense," dissents Ernst Mach, the nineteenth-century physicist and philosopher. "One cannot measure absolute space. For me only relative motions exist. You have a right to say only that you are accelerating with respect to something you can observe— the entire universe, the fixed stars."

The somewhat vague idea that the distant stars provide the reference frame against which accelerations are measured has become known as Mach's principle. One

---

* Technically this comes about because the tensors of relativity are nonlinear functions.

consequence of Mach's principle is that were the fixed stars absent you would feel no acceleration when rounding a corner. Mach provided an inspiration to Einstein, who had hoped to incorporate Mach's principle into GR and thereby show precisely how the distribution of matter in the universe gave rise to accelerations ("inertial forces" to be precise). Unfortunately, Einstein's hope was not fully realized. General relativity admits cosmological models that are empty of matter, but which nonetheless give rise to inertial forces, in contradiction to Mach's principle. On the other hand, there are respects in which GR does support Mach's principle, suitably interpreted.[†]

Thus the question, How do accelerations arise? is still without a definitive answer. One might agree with Abraham Pais, who says, "The origin of inertia is and remains *the* most obscure subject in the theory of particles and fields."

As the sun reached the zenith, illuminating the excavation pit, your sense of unease over relativity not only lingered but grew, very much as the Marchioness's legendary confusion after her dispute with the King of the Moon. At this moment, though, your thoughts were interrupted when one of the archaeologists cried out, "We've found it!" Everyone gathered around, expecting that the Oblivion File had been discovered. However, the excitement subsided somewhat when it became clear that the expedition had unearthed only a large crypt from an earlier age of the Academy. When the doors were opened, two well-preserved academicians were revealed, seated at a table and arguing over the origin of the universe. But exposed to sunlight they quickly dissolved into dust before resolving their differences.

"Well, we shall now resolve the matter for them," said one of the cosmologists. "The universe was created in a big bang."

"You seem much surer than the *New York Times*."

"There is no doubt about it."

"No doubt?"

Without hesitation the cosmologist went on to give

Four Reasons to Believe the Big Bang

In his book *Before the Beginning,* the eminent astrophysicist Martin Rees declares he would offer ten-to-one odds that the big bang took place. Rees is a modest man. Despite the *New York Times* most of his colleagues would probably offer hundred-to-one odds. Given the difficulties with cosmology discussed earlier, many people find it surprising that cosmologists can be so certain about the creation of the cosmos.

Here we interject a note of caution: as indicated by the cosmetologist/cosmologist at the start of the morning, the standard big bang model makes no statement about the moment of creation. To be sure, in the FLRW universe, time zero, the moment of creation, is an impossibility, a singularity, where physical quantities such as temperature and pressure all become infinite. One reason that investigators search for a theory of quantum gravity is that quantum effects might rid GR of the big bang singularity. The existence of the singularity is also why when discussing the origin of the universe cos-

Figure 9.1

mologists typically start their models off a few instants after the big bang, when the equations are well behaved.

Nevertheless, there is much evidence that the observable universe arose from a highly compressed state. The first and still one of the most important pieces of evidence is that the universe is expanding. During the early decades of the century a number of astronomers, notably Vesto Slipher and Edwin Hubble, recognized that most galaxies appeared to be receding from us. With Hubble's announcement of his famous law[†] in 1929, it became reasonably clear that not only were distant galaxies receding from our own Milky Way, but that virtually all galaxies were receding from each other. The standard crutch to help visualize this is to think of coins pasted onto the surface of a balloon. As the balloon expands, all the coins move away from each other. The universe is expanding. Running the cosmic motion picture backward, it is a fair presumption that at one time in the past all the galaxies were compressed together—a big bang.

Now you interrupt, "But where is the center of the universe?"

"There is no center of the universe," the authors reply. "As the balloon model shows, no galaxy is any more central than another; they are all receding from each other.

Unconvinced, you shake your head. "There is obviously a center to the balloon. Not on the surface but in the middle."

Ah, we cannot deny it; you are correct. Here is where the balloon model breaks down. A balloon is a two-dimensional surface residing in a three-dimensional room. Standing in the room we can view the surface from the outside. But as Poincaré pointed out, we are constrained to view the universe from the inside. Unlike balloons or bagels, the universe must be thought of as a four-dimensional spacetime complex, which is admittedly difficult to visualize. You can't look off the surface of the universe in the way you can look off the surface of the balloon. In that sense the famous balloon model is very misleading.

"Hmm," you reply, not entirely satisfied, "but what is the universe expanding into?"

Ah, the most frequent question put to cosmologists and not unrelated to the other. It most likely arises either because we are conditioned by the blasted balloon model or because the term big bang misleadingly connotes an ordinary explosion, which must take place within something else. The universe is not inside anything else and the big bang should not be viewed as a bomb going off inside an empty room which was already present. Spacetime itself begins expanding at the big bang. Perhaps it helps to visualize a rubber sheet stretching infinitely in all directions, with galaxies on the grid marks. At the big bang the grid marks begin moving apart. But the sheet is not expanding into anything; it is everywhere.

"I am not satisfied," you interrupt again. "You've pasted coins on the balloon. But galaxies aren't pasted on the universe. They're in the universe. Are galaxies themselves expanding?"

Ah, difficult. The usual assumption in GR is that measuring rods—anything held together by gravity or other forces—are not expanding. But it is not entirely clear at what scale the expansion starts. Is the space inside a galaxy expanding? No, because the galaxy is gravitationally bound. But on the scale of superclusters. . .? Moreover, imagine a giant, spherical region of the universe in which there is no appreciable matter. Is space within this cavity expanding? Well, no, the boundary of the sphere is expanding, but two objects within the cavity will actually not get farther apart. So, you see, apparently simple questions about where space is expanding and where it isn't do not always have obvious answers.

To continue.

In 1964 a discovery was made that convinced most cosmologists that the big bang theory was correct. That year radio astronomers Arno Penzias and Robert Wilson stumbled across what has since become known as the cosmic microwave background radiation, or CMBR, which represents nothing less than the heat left over from the big bang itself.[†] Near the singularity, the temperature of the universe approached infinity, but as the cosmos expanded and cooled, the temperature dropped until today it is a crisp 2.7 degrees above absolute zero. Any doubts about this interpretation were dispelled in January 1990 when, after about ten minutes of observing time, the COBE (Cosmic Background Explorer) satellite confirmed that, to the limits of its precision, the CMBR had a perfect blackbody spectrum.

You do not see the implications.

In the distant First Debates we discussed the blackbody radiation that is given off by many objects. It is, in fact, the type of radiation emitted by any uniformly hot body.

According to the standard model, the heat left over from the big bang should be precisely that of black body. COBE, by producing the most perfect blackbody spectrum ever measured, unequivocally confirmed the prediction.

"There is no question in anyone's mind that the CMBR is from the big bang?"

No. No one has been able to produce a remotely plausible alternate explanation.

Closely related to the CMBR is the process of "primordial nucleosynthesis." In the first few minutes after the big bang, when the CMBR was at a temperature of not 2.7 but 1 billion degrees, the available neutrons and protons underwent fusion reactions similar to those that take place in a hydrogen bomb and were "cooked" into helium, deuterium and a few other light isotopes. The abundances predicted by the FLRW model are about 76 percent (by mass) hydrogen nuclei (protons), 24 percent helium, with the other isotopes such as deuterium and lithium accounting for parts in 100,000 or less. The truly extraordinary thing is, all of these abundances are close to what astronomers actually measure in the universe. No other explanation gets all the abundances remotely correct. Many cosmologists regard the primordial light-element abundances as the most important evidence for the big bang, and this leads to a convincing picture of the universe as early as a tenth or one one-hundredth of a second after the big bang.

There is more. A prominent feature of the universe is the existence of galaxies. Galaxies are not spread uniformly across the cosmos but are lumped in clusters, which contain dozens to thousands of galaxies; and the clusters are arranged in superclusters, which constitute enormous filamentary structures strung across the universe. The existence of these large-scale structures is rather strange because the standard cosmological model assumes that the universe began with perfect uniformity, or to use the technical term, *homogeneity*. Galactic structures represent obvious irregularities—*inhomogeneities*—in the density of the universe.

The question is, How did the large-scale structure of the universe emerge? If the universe began homogeneously, then it is a reasonable assumption that gravity caused matter to clump, eventually producing high-density galaxies. In the most general sense that is certainly what did happen, but the details have proven extremely difficult to work out because any successful theory of structure formation must account for the observed shapes of the structures, their mass ranges, their velocities, and it must do so within the current age of the universe. It is not easy to meet all these demands.

Nevertheless, cosmologists have devised numerous models in an ongoing attempt to account for the observed structure of the universe. You have perhaps encountered some of these models, which have names like Cold Dark Matter (CDM) and Hot Dark Matter (HDM). Because many details of such models are under dispute, we do not go into them.* For our purposes the important thing is that all models require that gravity produce clumps that are of a certain size at about 300,000 years after the big bang, when the CMBR first became visible and the structure-formation process got under way.† You may recall in 1992 the extraordinary press coverage of the "handwriting of God" and Stephen Hawking exclaiming, "The greatest discovery of the century, if not of all time."

---

* See references in the Commentaries.

It was neither the handwriting of God, nor the greatest discovery of the century, but in its first two years of mapping the universe, COBE had detected primordial ripples in the microwave background of the required size to evolve into galaxies.

"Why don't you think it was the handwriting of God or the greatest discovery of the century?"

We've corresponded with God; his handwriting is different. Seriously, to claim that the COBE finding was greater than the discovery of the atomic nucleus or the expansion of the universe itself strikes us as an exercise in sophistry. Moreover, as magnificent as the COBE achievement was, it basically confirmed what theoreticians had long insisted must be there. Paradoxically, COBE would have achieved a more revolutionary result had it not discovered anything. In any case, now you have four reasons to believe that the big bang took place.

The Standard Model Is Not the Universe

In retrospect, at this stage time remained real and the debate, which concerned mostly relativity and the standard model, remained well defined. Hourly updates from the World Idea Exchange were optimistic. During the exegesis, the party reached a yet older cultural level of the Academy. At this stratum, the excavation revealed much less paper than at the higher level, as well as fewer of the modern structural devices known as overhead. Ever more layers of soil were removed, the discussion of the big bang progressed, and your friend reported a widely circulated rumor that, depending on the matter density, the universe was destined either to expand forever or recollapse, and both of you requested the authors' opinion on the subject. The authors' reply was unexpected.

"The strictly correct statement is that the standard model of the universe will expand forever or recollapse, depending on the density."

Occasional detractors notwithstanding, the standard cosmological model is by now so ingrained in the collective unconscious of both the public and the scientific community that even most cosmologists have forgotten that it is a tremendously simplified model, not the universe itself. A few moments ago we said that the FLRW model is based on the assumption that the universe is perfectly homogeneous—has the same properties at any given point. Actually, it assumes an even stronger condition: not only is the universe identical at every point, but from each observation point it is identical in all *directions*, a condition known as *isotropy*.* An isotropic universe must also be homogeneous, but not necessarily vice versa. As a simple example, take an infinite field of tiny arrows all pointed north. Since this universe is the same at all places, it is homogeneous, but because one can distinguish north from south, it is *anisotropic*.

Precisely put, the cosmological principle is the statement that the universe is isotropic (and hence homogeneous). Given that anything with the slightest irregularities—and thus anything one encounters in the real world—is not isotropic, this is an extraordinary simplification. What is true of the rumor you report is that in an isotropic universe—the FLRW model—there exists a "critical density." If the actual matter density falls below the

---

* The term arose in the First and Third Debates.

famous critical value, the universe will expand forever. If the matter density is above the critical value, the gravitational attraction of all the galaxies will eventually halt the expansion and cause the universe to recollapse. If the density is exactly equal to the critical value, the universe will just barely expand forever.* You have perhaps heard that one of the Hubble telescope's missions is to determine once and for all which outcome Fate has store for us.

"But it can't," a renegade academician now objects. "You have forgotten the horizon."

True. Because light travels faster than anything else, no signal can reach us from any point in the universe that lies farther than the distance light has traveled since the big bang. Just as on Earth one cannot see ships beyond the Earth's horizon, our knowledge of the universe is absolutely restricted to events that have taken place within the cosmological horizon, which lies at a distance of some tens of billions of light-years.†
On larger-than-horizon scales, there is no obvious reason why the universe can't be expanding in one place and collapsing in another, much as the sun itself can oscillate in a way such that its outer layers are moving in one direction while its inner layers are moving in another. In this case, on larger-than-horizon scales the universe would be very inhomogeneous.

"Do you mean, we are spending billions of dollars to answer a question that cannot in principle be answered?" you want to know.

The Buddha smiles silently and the authors get down to digging.

Now, while time is still real, academician Bryce DeWitt reminds the expedition that it has overlooked another important issue. All this talk about "open" and "closed" universes, he insists, is nonsense.

What DeWitt means is that if you have taken one of the Academy's introductory astronomy courses, the later chapters of the textbook have undoubtedly contained figures much like those inscribed on page 31 of these debates. If the universe is destined to recollapse, the geometry is usually described as similar to that of a sphere: it is closed in the sense of being finite, but unbounded in the sense of having no edges, and is said to possess "positive" curvature. On the other hand, if the universe is destined to expand forever, its geometry is usually described as open, like a saddle, and is said to possess "negative" curvature.

Once again, this description refers to the standard model and the standard model has been confused with the universe. In general these distinctions between open and closed turn out to be untrue. We have already discussed a bagel, or toroidal, universe in which light can travel round many times. In the past few decades mathematicians led by William Thurston have shown that there exist an *infinite* number of universes similar to the bagel in the sense that light can travel many times around along closed paths. What is extremely strange about them is that, unlike the bagel or the sphere, they have everywhere negative curvature. They are closed in the sense that they are finite (like the

---

* There are anisotropic models that collapse regardless of the density, but they tend, for example, to make the CMBR more anisotropic than observations permit.

sphere or bagel), but they are open in the sense that the density is less than critical and they will expand forever (like the saddle).

Such universes are terrifically difficult to visualize. The simplest one dates from 1933 and is known as the Seifert-Weber dodecahedron. If you can accept that one can construct a dodecahedron (a twelve-sided solid) in a negatively curved space, then twist each face and identify it with the opposite side (in the same way Poincaré identified the opposite sides of the map to make a bagel), then you have a closed universe where the curvature is everywhere negative. The more complicated ones resemble multidimensional knots or pretzels.*

The conventional astronomers, astrophysicists and cosmologists present collectively answer DeWitt with a huge yawn. "So what?" is their reaction; Ockham's razor requires that no one think about anything other than the simplest model.

This raises an interesting philosophical issue. If one is going to restrict one's attention to the simplest possible model, the FLRW model, then one is stuck with the dilemmas the standard model brings, for instance the big bang singularity. Also, the famous inflationary universe scenario (which the debaters are readying to take up) is largely superfluous if the universe is assumed to be standard from the outset. Then there is De-Witt's reply to Ockham's razor: "The Lord in His wisdom would not have created the universe as the standard model."

You find yourself unsure of these things but are open-minded enough to want to know whether it is possible to determine whether we live in a standard-model or a negatively curved dodecahedron.

Excellent question. One possible way to determine that we lived in the Seifert-Weber dodecahedron or, to make it simpler, a bagel, would be to carry out an experiment such as the one we proposed to Poincaré: if astronomers are able to observe multiple images of the same galaxy, then we live in a bagel or some more complicated universe. However, the experiment would be far from trivial. If it takes ten billion years for light to circumnavigate such a universe, then the images will be snapshots of the source galaxy at ten-billion-year intervals. Because galaxies age considerably in that amount of time, it would be virtually impossible to recognize the images as originating from the same object.

"This does not seem to be a feasible experiment after all," interjects Poincaré.

A perhaps more realistic version would be to observe the patterns of lumps in the microwave background. If the "handwriting of God" is identical in two regions of the sky, this might be the signature of a universe connected in a topologically strange fashion. A few scientists are already hoping that when precise data on the CMBR is available from the MAP and PLANCK satellites in the next few years (to be discussed at a lower cultural level), such a determination might actually be made.

It occurs to you that it would be scientifically convenient to have many universes at our disposal. Yes, here the problems introduced by the uniqueness of the universe are becoming manifest. As for the feasibility of the microwave background experiment, well, for now we agree to deposit the bet money in an escrow account.

---

* The situation is made more confusing by the addition of a cosmological constant (see pp. 243–244), which can cause an "open" saddle universe to recollapse, or a "closed" spherical universe to expand forever.

A Perplexed Meditation on Two Celebrated Cosmological Conundrums

As the downward trend continued amid reports that the Knowledge Marker was falling, someone pressed the authors to elucidate their provocative statement of a few moments ago, that if one insisted on describing the universe by the standard model all the way to the big bang, then one would be stuck with the problems the standard model brings. The authors evidently had more in mind than the initial singularity, an acknowledged impossibility of the model. Yes, apart from the big bang itself, there are several other conundrums associated with the FLRW universe, which were emphasized during the 1970s by Robert Dicke and James Peebles and which became famous in 1981 with the introduction of Alan Guth's inflationary universe hypothesis.

The first of these is the so-called flatness problem. The flatness problem can be stated in several ways. An FLRW universe, in which the density of matter and energy is just equal to the critical density, is termed flat because its geometry *is* flat, hovering between the closed spherical geometry and the "open" saddle geometry. In principle the density of the real universe could be anything, a million times greater than the critical value or a million times less. But astronomers over the past decade have consistently measured the matter density to be 20 or 30 percent of the critical value, that is, to within a factor of five of critical.*

Thus the universe is reasonably flat. Which is exceedingly peculiar. According to the equations governing the FLRW model, the universe becomes more curved as it ages—it grows away from flatness.† Consequently, if the universe is reasonably flat now, it must have been *extremely* flat early on. Indeed, it is simple to run the equations of relativity backward and show that if the universe is within a factor of five of flatness now, then at 10^{-36} seconds after the big bang, when the natural forces were to have been unified, it would have had to have been flat to one part in 10^{55}. This is very flat.

Cosmologists find it extremely improbable that the universe could have begun with such extraordinary "fine tuning" in order to keep it "reasonably" flat now.

"Excuse me," your friend interrupts. "I do not understand the use of the term 'improbable' here. As you have stressed, there is a single universe. It must have started out in some fashion, with some initial conditions. Perhaps they were exactly the initial conditions needed to keep the density within a factor of five of the critical value at the present time."

We confess that the use of "improbable" is ill-defined here. The uniqueness of the universe is hitting us with full force. But look at it this way. If the universe starts out *exactly* flat, then it remains flat for eternity. Otherwise it rapidly becomes curved. Suppose the universe were just a tiny bit more curved than by one part in 10^{55} at 10^{-36} seconds. Then if it were curved in the closed direction, it would have recollapsed after an incredibly short amount of time and galaxies never would have formed. If it were curved in the open direction, the matter density would quickly have dropped to zero and the

---

* All these numbers are approximate, if for no other reason than the critical density itself depends on the Hubble constant, which is uncertain. We do not enter into the debate.

universe would be dark. So you can rephrase the flatness problem as asking why there are galaxies or why there are enough photons to see.

"Let there be light," interjects one of the theologians.

"I see your position," your friend responds. "Nevertheless, since you claim these are all ways of phrasing the same conundrum, it boils down to the same thing: how the universe started. And it must have started *some* way—here we are. To say a particular beginning of the universe is too 'improbable' is much like saying the odds of my grandparents' meeting were infinitesimal and therefore I should not exist—a phrase you have used yourself. The flatness problem seems to me very metaphysical."

At this moment another cosmologist interrupts. "You have not mentioned the anthropic resolution to this dilemma."

True. We encountered the anthropic principle in the Fifth Debates when Boltzmann argued that unless the universe began in a low-entropy state, life could not exist. The weakest form of the anthropic principle merely states that the properties of the universe must be consistent with the existence of life, otherwise we would not be present to observe the universe. Dicke's famous anthropic argument was that we must observe the universe to be at least ten billion years old; otherwise the elements necessary for life—in particular carbon—would not have had time to be cooked in stars, and we wouldn't be here asking the question.

Not surprisingly, at hearing the celebrated argument, the academicians launch into a side debate on validity of the anthropic principle, a debate which we can safely predict will be eternal and vicious. A perennial objection to the anthropic principle is that it is incapable of making testable predictions and is therefore unscientific. We dimly recall from the First Debates that many principles enjoying widespread support among scientists—reductionism, mechanism, beauty, symmetry—do not directly result in predictions and yet are regarded by practitioners as even more sacred than laws. In part because they do lead to testable theories. The anthropic principle, certainly, is the only one that explicitly invokes life and is consequently anathema to many scientists. However, one might argue that the principle of beauty implicitly invokes life (life being a prerequisite for taste) and, in any case, anthropic reasoning *has* led to some concrete predictions. Therefore it is somewhat peculiar that the anthropic principle among all the principles of science exercises some physicists the most. Another objection is that the anthropic principle reinjects teleology into science. It is not a large leap from the contention that the universe must be consistent with life to the contention that the universe was made for the purpose of life—the argument from design. The Aristotelians return once again.

The authors, meanwhile, recognize that the anthropic dispute now erupting is a battle without exit and accept discretion as the better part. We content ourselves by clarifying a few of the above remarks in the Commentaries. In terms of the present discussion, one might invoke the anthropic principle to argue that the existence of life requires the universe to be nearly flat; otherwise galaxies and stars would not exist.

"It seems to me," objects an academician, overhearing the authors, "pushing anthropic reasoning to extremes to explain the flatness of the universe to one part in 10^{55}."

"That is not clear," answers a pro-anthropic cosmologist. "The statement that the universe must be flat enough to form galaxies *is* the statement that the universe must have been fine-tuned to one part in 10^{55}."

As we said, discretion is the better part. We turn now to the second famous conundrum inherent in the standard model: the "horizon" problem. The horizon problem is best phrased in terms of the cosmic microwave background. When the COBE satellite detected the first stirrings of galaxies at 300,000 years after the big bang, it did not directly observe the clumping matter. Rather, it detected slight variations in the temperature of the CMBR, which would accompany gravitational clustering. These temperature variations were very slight—only several one-100,000ths of a degree above or below the average temperature of 2.735 Kelvins. Apart from the COBE ripples, the temperature of the CMBR is remarkably constant over the entire sky—2.735 degrees above absolute zero. In the language of cosmology, the CMBR is extremely isotropic. The horizon problem asks, What made it that way?

The horizon problem was clarified by Wolfgang Rindler as far back as 1956. To understand why the conundrum is so striking, recall that the cosmological horizon is the farthest distance we can see in the universe—the distance light has traveled since the big bang. When two objects (say galaxies) lie beyond each other's horizons, physicists say they have no causal contact; no signals can be exchanged between them and consequently there is no possibility of physically influencing one another. The extremely peculiar thing about the standard-model big bang is that near the big bang itself, *all* particles lay beyond one another's horizons. How then could the universe become so isotropic? How could the CMBR become so free of irregularities?

"Excuse me," you say, interrupting. "I thought you said the standard model *was* isotropic. You don't have to explain its isotropy, you've assumed it from the outset."

True enough. But cosmologists would rather not assume isotropy; they would rather explain it by some mechanism.

"This seems like another metaphysical dilemma," your friend adds. "But I'm still puzzled about the nature of the horizon problem. I don't understand why all the particles at the big bang lay outside each other's horizons."

Yes, we have not explained this strange phenomenon; however, a picture is worth a thousand words, and so we refer you to Figure 9.2, which shows that no matter how close together two particles are at the present epoch, as you go back to the big bang, they will *always* lie outside each other's horizons and hence could not physically influence one another. By the same token, one can show that any two galaxies today separated by more than about 60 degrees on the sky are only just now entering each other's horizons and hence could have had no causal contact before the present.

We know the universe was highly isotropic as early as one second after the big bang. (Models anisotropic at this time produce more helium in primordial nucleosynthesis than is allowed by observations.) If widely separated regions of the sky are coming into causal contact only now, fifteen billion years after the big bang, how could the universe become isotropic as early as one second?

Figure 9.2
The farthest distance from a given galaxy one can observe in the universe is called the horizon: the distance light has traveled since the big bang. Suppose the horizon (h) is smaller than the "radius" of the universe (R), as in (a). Since h is increasing faster than R, the horizon gradually catches up to R, encompassing ever more galaxies, as in (b). Conversely, suppose h and R are equal at the present time, as in (b). Running the clock backward, h will shrink faster than R. At some time in the past, (a), all galaxies will lie outside the horizon, making any communication impossible. If different parts of the sky could not communicate, how did the universe become so isotropic at early times? This is the horizon problem.

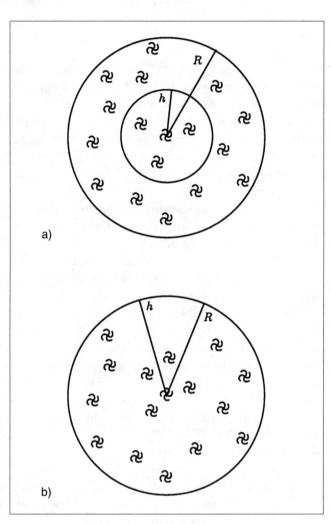

a)

b)

That is the horizon problem.

"I think I see it now," replies your friend, "but I repeat, in the standard model there is no problem—you have assumed the universe is isotropic from the outset."

One can't deny this, but if you imagine all possible ways the universe could have started out—with great anisotropies or inhomogeneities, then it does seem extraordinarily implausible that it should have begun with no irregularities and nearly exact flatness.

"Again, words like 'implausible' applied to the entire universe. What does this mean?"

The people are silent.

Cosmology

II: *Inflation, Quantum Cosmology and Quantum Theology*

For deriving all from nothing
there suffices a single principle.
    —*Leibniz*

Try as they may to savor the taste of eternity, their thoughts still twist and turn upon the ebb and flow of things in past and future time. But if only their minds could be seized and held steady, they would be still for a while and, for that short moment, they would glimpse the splendor of eternity which is for ever still.
    —*St. Augustine*

In Which Inflation Is Described and Vigorously Debated

In retrospect, by this time it seemed that the minutes and seconds had been stretching, causing the debate to continue for aeons. Truthfully, the excavation had reached a very deep stratum of the Academy. No sign of the Oblivion File was to be found, but at such an age, close to the very founding of the institution, there seemed little evidence to make any concrete statement whatsoever about academic life of that period. The few documents and potsherds unearthed split the archaeologists into two camps. The first claimed that the similarity of the academic alphabet and artwork to outside alphabets and artwork proved that the Academy was originally settled by barbarians from the north. The second claimed the opposite: that the Academy was the earlier institution and its influence gradually overwhelmed the barbarians. Better dating methods were called for.

The debate turned once more to the flatness and horizon problems. Before 1981 few cosmologists actually paid much attention to Dicke's conundrums. But in that year Alan Guth announced a solution—his celebrated "inflationary" universe scenario. Guth posited that at approximately 10^{-36} seconds after the big bang, the universe underwent a period of exponential expansion (lasting, say, 10^{-5} seconds), during which its size increased by at least 25 orders of magnitude, from a radius of roughly 10^{-25} centimeters to 1 centimeter in the balloon model.

Such a period of inflation had two immediate consequences. The universe became extraordinarily flat, just as the surface of an enormous balloon appears essentially flat to a mite on its surface. In this way inflation solved the flatness problem.

Second, under inflation, the horizon also increased by at least 25 orders of magnitude. Thus widely separated regions of the universe were brought into causal contact—not after ten billion years as in the standard model, but in the first 10^{-30} seconds or so. Once different regions were able to interact, any irregularities in density or temperature could be smoothed out, say by particle collisions, thus making the universe highly isotropic. In this way, inflation solved the horizon problem.

Ah, but we see you are wondering why inflation occurred, and why it occurred at the particular time. Listen attentively.*

Guth based inflation on the premise that at the time in question the temperature of the universe would be so high (10^{29} degrees, roughly) that the electroweak and strong forces would be unified, according to Grand Unified Theories. As the universe cooled, it underwent the same spontaneous symmetry breaking that has been discussed several times and the forces became distinct. Inflation is not only meant to have taken place during the process of symmetry breaking but it relies on symmetry breaking to take place at all.

In the FLRW model, the expansion rate of the universe depends on the energy density of its contents. The more matter or radiation in the model, the faster it expands. At late times, such as one second after the big bang, the contents of the universe is assumed to be ordinary matter, such as electrons, photons, neutrinos.** Such fields (to identify particles with fields) are termed "thermal" because the density of particles is linked to their temperature—knowing the temperature tells you the density and vice versa. Just as the density of, say, marbles inside an expanding balloon decreases, the universal expansion caused the density of thermal particles to decrease, with a concomitant decrease of their energy density and temperature.

At Grand Unified times, however, there existed other, "nonthermal" fields, fields whose energy does *not* decrease in the usual way as the universe expands. One of these is the famous Higgs field.

In the Sixth Debates we encountered an achiral sodium chlorate solution, which upon precipitation produced crystals that were mirror images of one another. A solution that showed no preference for direction produced crystals that distinguished left from right. Symmetry was broken. In the case of the sodium chlorate, the mechanism behind the symmetry breaking is unknown. In the case of GUTs, scientists postulate a new field—the Higgs field—that is responsible for creating from one force, many. One way of thinking about the situation is that above unification temperatures there was a single electroweakstrong force. The respective bosons of the electroweak and strong forces were indistinguishable. As the universe's temperature dropped, the Higgs field caused the electroweakstrong force to fragment into the electroweak and strong forces and give separate identities to the electroweak bosons (the photons, W and Z) and the strong-force bosons (gluons).

---

* The description that follows does not correspond exactly either to Guth's original inflationary model or to any of the many versions that have come after. It might be called "generic" inflation.

** People tend to think of electrons as matter and photons as energy, but by $E = mc^2$ one can speak of matter density or energy density interchangeably.

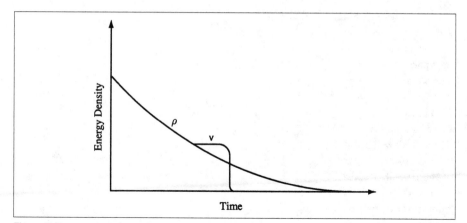

Figure 9.3

The energy density (ρ) of ordinary, thermal particles decreases rapidly as the universe expands and the density of particles is diluted. Inflation posits that at very early times the universe suddenly became dominated by the energy of the Higgs field (V). The energy density of the nonthermal Higgs field remains nearly constant for the duration of inflation. Because the higher the energy density, the faster the universe expands, the large Higgs field resulted in a vastly increased expansion rate of the universe—inflation.

Because the Higgs field *is* a field, it has an energy associated with it, and because energies at the time were so huge, the energy density associated with the Higgs field was also great. But as the universe expanded, the energy of the nonthermal Higgs field did not drop as fast as the energy of thermal particles. Suddenly the universe got "hung up" in a state in which the energy density was much greater than in the standard model. As a consequence, the expansion rate of the universe also became much larger. This was the onset of inflation. Eventually, even the energy of the Higgs field dropped to zero, and once again the universe's expansion was governed by that of ordinary matter; the expansion rate fell to that of the standard model. This transition marked the end of inflation.

The "hung-up" state of the universe is referred to as metastable and many examples of metastable states exist in nature. The supersaturated sugar solution from which children make rock candy is one of the most common. Another is supercooled water. Normally, lowering the temperature of water causes it to freeze, because ice is in a lower (free) energy state than liquid water and all systems "gravitate" toward the lowest possible energy. But if water is cooled slowly enough, it can be cooled to as much as 20°C below zero without freezing. The supercooled state is metastable, meaning that it is a local low-energy state with a small energy barrier that prevents it from falling into the lowest-energy state of ice. But the slightest perturbation will knock the system over the barrier and it will freeze, releasing much heat. The water is the symmetric state; the ice crystals represent the state of broken symmetry and, like rock candy or the sodium chlorate, distinguish directions. The transition from water to ice is known as a phase transition.

However, the metastable state of the universe differs somewhat from supercooled water. A less familiar but more accurate comparison would be to an ammonia molecule.

Figure 9.4
A close-up of the Higgs potential shows that it resembles one side of a sombrero hat. Inflation is said to occur when the universe (represented by a ball) is stuck in the small well at the top and cannot roll over (top). Eventually, the universe tunnels through the barrier and falls to a lower energy state. The potential drops to zero and inflation ends. Most inflationary models do not rely on tunneling. Rather, they posit a very flat potential (bottom); as the universe slowly rolls down this potential, inflation takes place. As the universe falls into the well at right, the Higgs energy is released and converted into the matter we observe today.

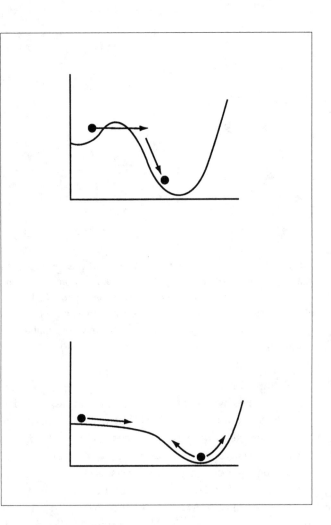

An ammonia molecule consists of three hydrogen atoms and a nitrogen atom in the form of a pyramid with the nitrogen at the apex. Now, the ammonia molecule constantly oscillates between two states, one with the nitrogen atom up, and the inverted state, with the nitrogen atom down. Ammonia performs this transition about thirty billion times per second. The two pyramid states are metastable with a small energy barrier between them. Nevertheless, due to quantum mechanical tunneling, the nitrogen atom actually "leaks" through the barrier presented by the hydrogen atoms to the other side and inverts the pyramid.

In this respect the universe is more like the ammonia molecule than water. As the universe expanded it got hung up in the symmetric, metastable state dominated by the enormous Higgs energy, a state commonly termed the "false vacuum." Inflation took place during at this time. Eventually, though, the universe tunneled through the energy barrier created by the Higgs field until it fell into the lower-energy state of the "true vacuum" and inflation ended. In this way inflation depends on symmetry-breaking phase transitions taking place in the early universe.

Listening to the exposition, you are struck by the terrific elegance of the idea of inflation. Yes, in fact not only has inflation survived fifteen years—rare these days for a theory—one might say it has prospered. Apart from solving the horizon and flatness problems, inflationists have also claimed for it at least two other major successes. The first (in fact one of the original problems Guth set out to solve) is the monopole problem. In the Third Debates we pointed out an evident asymmetry of nature: that magnetic monopoles do not exist. However, according to GUTs, they should exist, in great numbers, numbers perhaps as great as those of neutrons and protons. This is a grave difficulty since, given that the magnetic monopoles should have a mass 10^{16} that of the proton, we undoubtedly would have detected them by now on laboratory balances. The exponential expansion of inflation cures this problem by dilution; the volume of the universe increases so much that the number of monopoles is reduced to, say, one per observable universe, and we haven't found it yet.

"Excuse me," you interrupt, catching the authors off-guard. "If inflation dilutes the number of monopoles so much, and the number of monopoles equals the number of ordinary protons and neutrons, why do we observe any particles at all in the universe? Why are we here?"

Excellent question. It is true that the exponential expansion of inflation essentially empties the universe of *all* particles. What we have not mentioned is that during the phase transition, when the universe tunnels from the false vacuum to the true vacuum and the Higgs field drops away, its energy must go into something. It does. Just as heat is liberated by freezing water, so is heat liberated by the decaying Higgs field. This heat—in a process known as "reheating"—goes into replenishing the universe with the matter and energy observed today. Inflation first depletes the observable universe of particles then refills it. But in the process the universe has first been made flat and all irregularities have been smoothed out.

"Clever," you respond.

Yes. Nowadays, though, many consider inflation's most significant success to be its prediction of the spectrum of primordial ripples detected by COBE. At first glance this might admittedly seem a strange collection of words. Did we not just say that inflation wiped out any trace of irregularities? Where then did the COBE ripples come from?

A few of the extremists present are agitating to move on to quantum cosmology, which will inevitably happen soon enough. We pacify them by previewing one feature. A classical swing that is not moving is said to have zero energy. It is at rest, so it has zero kinetic energy, and it is at its lowest point, so it has zero potential energy, and that's all the energy there is. In quantum mechanics, though, the situation changes somewhat. According to the Heisenberg uncertainty principle, a system's momentum and position cannot both simultaneously be zero. Since potential energy is the energy of position, and kinetic energy is the energy of motion (velocity or momentum), Heisenberg's principle states that even at a temperature of absolute zero, a quantum swing must have some energy. This energy is called the zero-point energy, and because the swing must always be regarded as being in motion, we can regard it as executing "zero-point fluctuations."

Any quantum field exhibits zero-point fluctuations, or vacuum fluctuations, as they are also known, including the Higgs field of the early universe. Inflation's exponential expansion subsequently enlarged these microscopic perturbations to macroscopic amplitude; it is these magnified ripples that were observed by COBE. But the important thing is not even the ripples themselves, it is the distribution of their amplitudes, or spectrum.* Because inflation does not predict exactly how long the universe inflates, it does not predict the amplitude of the perturbations, but it does predict their spectrum.** Although COBE did not exactly confirm the inflationary prediction, the COBE results were certainly consistent with inflation; many would argue that inflation gave the best fit to the COBE data.

"Yeah," grouses an unregenerate cosmologist of the old school. "On the other hand almost any idea you can think of will give a similar spectrum."

Perhaps, but not so naturally. In any event, the fact that the ripples are magnified quantum fluctuations dating from times even earlier than inflation was the source of Hawking's hyperbolic, "The greatest discovery of the century, if not of all time."

Unregenerates aside, all those present have been extremely impressed with inflation's successes and, as the exposition ends, fall silent. The only sound is that of spades and pickaxes taking the expedition ever downward, toward bedrock.

Eventually, your friend gets up the courage to say, "May I ask a stupid question? In yesterday's debates everyone agreed that there is no evidence for GUTs, and that the original version has been ruled out because no one detected the decay of the proton. If that is the case, then is not the monopole problem also a metaphysical problem?"

"Well," replies a right-wing cosmologist, "it may not be metaphysical, but one is indeed invoking inflation to solve a problem arising from the possibly incorrect theory it is based on."

"May I ask an even dumber question?" you say. "If inflation has solved the flatness problem, then as I understand it, the density should be exactly equal to the critical value. But earlier you claimed that astronomers measure the density to be only 20 to 30 percent of critical. This does not sound flat to me."

Yes, dumb questions are always the hardest to answer. Like superstrings, this situation has resulted in another fragmentation of the scientific community. Listen to the conversation among the observational astronomers on the dig: "The value of Omega is .22, believe me." "You are off, way off; the large-scale measurements imply Omega is .54." "My value of Omega is the only true Omega: .38."

---

* We have been loosely speaking about the "size" of the density perturbations. It is important to realize that when cosmologists speak of "sizes" in regard to the density fluctuations, they are not talking about physical size. Rather, they are talking about the change in density compared to the average. A "large" density fluctuation means that a certain region has a much greater or lesser density than average, whereas a "small" fluctuation means that the density in that region is about average. For this reason we have chosen the word "amplitude" here, where precision is necessary.

** According to inflation, the percentage variation in density in every (spatial) size range should be the same, a spectrum termed "scale invariant."

Now tune in to the theoreticians: "The value of Omega is exactly 1." "The value of Omega is exactly 1." "The value of Omega is exactly 1." "The value of Omega is exactly 1."

The mantra continues. Omega (Ω) has become the most popular word in astronomy and cosmology. It merely denotes the ratio of the observed density of the universe to the critical density. If inflation is correct and the universe is critical, then Omega must be 1. The problem you have identified is that observational astronomers have consistently concluded that Omega lies below 1 and the universe is "open," whereas theoreticians have insisted, "The missing mass will be found."

We bring our microphones in closer to a press conference held in early 1998. Five separate research groups stand at the podium. "Omega is between 20 and 30 percent. The result is unequivocal. The universe is destined to expand forever."

"Bah, humbug," mutters an inflationist, "the cosmological constant will save us."

The inflationist is overheard and, suddenly, amid great cries, the scientists present begin dueling with shovels. The cosmological constant is an arbitrary number—the one Einstein called his "biggest blunder"—which for mathematical reasons can always be consistently added to the equation governing the FLRW model. Like all other fields it has an energy density associated with it and, depending on whether it is positive or negative, the constant can act like extra gravity, slowing down the universe's expansion, or like a pressure that speeds up the expansion. Recently it has become a popular "hypothetical" to concede that the matter density of the universe may be below critical and make up the difference by adding a cosmological constant—

"I won't have it!" cries an anti-inflationist who, face nearly as red as Suleiman the Magnificent's after his defeat at Malta, proceeds to tick off three reasons why any reasonable human should oppose the cosmological constant. "One: the cosmological constant is a constant."

In certain theories, true enough. The anti-inflationist means to say that

Action and Reaction

has proven itself to be a good motto in physics. All dynamical fields act on other fields and are acted upon. But the cosmological constant is unique in that, *if* it is truly a constant, it is able to act on other fields but cannot be acted upon. This is strange and disturbing.

"Two: to introduce the cosmological constant is to essentially reintroduce the flatness problem."

This also can hardly be denied. In the equation governing the FLRW model there is matter and there is curvature. The flatness problem asks why the curvature is small compared to the matter; inflation answers by eliminating curvature altogether. But if we introduce a cosmological constant, we are just adding a new term to the same equation. The question now becomes, Why is the cosmological constant the size that it is? We have merely swapped the flatness problem for the cosmological constant problem.

"This is aesthetically displeasing," muses one of the philosophers.

"That's not all it is," interjects the previous anti-inflationist, still livid.

The renegade is evidently referring to the size of the constant. Like Einstein, one may simply regard the cosmological constant as a fudge factor, an arbitrary number inserted into the equations. But one would like to believe it arose from some physical mechanism in the early universe—presumably from the still unknown quantum theory of gravity. However, in that case, any reasonable estimate of its value shows that it should be about 125 orders of magnitude larger than is necessary to bring Omega to 1. A cosmological constant that large would cause the universe to expand (or contract) about 62 orders of magnitude faster than any astronomer claims is the case, and so is suspect.

"This is the third reason why any sensible person must believe the cosmological constant is strictly zero."

In which case we are back to Omega being below 1.

A flat universe brings with it a further difficulty, which has become known as the age problem. The age of the universe depends on both the nature of its contents and, not surprisingly, its rate of expansion, which is given by nothing other than the famous Hubble constant. Recent measurements of the Hubble constant suggest that the universe is younger than previously believed. With the older value of the Hubble constant, the age was thought to be about fifteen billion years for an open universe. With the newer values, the age may come down to ten or twelve billion years. This is not necessarily bad, except that the children are generally born after their parents and astronomers have long held that the oldest stars in the galaxy are approximately fifteen billion years old. Inflation exacerbates the problem by bringing up the density to critical, which turns out to make the universe younger still; with the newer values of the Hubble constant, the age would go down to about eight billion years.[†]

Although the age dilemma has received a fair share of publicity, it may not be so serious. There are substantial errors in determining both the Hubble constant and the age of the oldest stars. With the latest data from the *Hipparcos* satellite, the age of the universe is going up and the age of the oldest stars is coming down. Our microphones at the 1998 news conference also detect a claim that the value of the Hubble constant has decreased again and the age of the universe has risen once again to fifteen billion years. If all else fails one can also solve the age problem by adding an appropriate value of the cosmological constant to slow the expansion—

"Don't you dare!"

In any case, it is premature to say whether the problem is resolved and whether the inflationary age of the universe is now excluded. Stay tuned.

After arguing for what seems the age of the universe, you desire a coffee break, but the academicians are just warming up. At this moment a heterodox cosmologist throws down her shovel and states, "I don't believe inflation has solved the flatness problem at all."

She goes on to explain that although inflation makes the universe extremely flat, even extraordinarily flat, it does not make it *exactly* flat. And because the equations say that curvature *always* wins eventually, eventually the universe will become dominated by curvature again and the density will depart from critical. Admittedly, this will take a long time, 10^nth ages of the universe, where n is a *large* number.

"But in that case," an inflationist ripostes, "all the stars will have long since ceased to burn, galaxies will have ceased to exist, the universe will consist of nothing but black holes and there will assuredly be no life."

"Agreed. But then you are using an anthropic argument to explain why we observe the universe to be flat. Has inflation solved the flatness problem or has the anthropic principle solved the flatness problem?"

Cosmologist Andrei Linde, with his "chaotic" inflationary scenario, has recognized the point. According to chaotic inflation, some regions of the "megauniverse" inflate, some don't, some inflate enough to become flat, some don't. We happen to live in one region of the megauniverse in which life has evolved. Only in such a region will cosmologists ask whether inflation or the anthropic principle has solved the flatness problem.

By now time was behaving in the strangest fashion. Whether day had become night or life still existed on the Earth above was impossible to determine. Dazed and confused by the course of the argument, your only desire was to find the Oblivion File and cast yourself in. The academicians were not done. At this juncture one remarked, "We have basically been arguing about the flatness problem. There are difficulties with inflation's solution to the horizon problem as well."

It cannot be denied. Inflation solves the horizon problem by smoothing out any initial cosmological irregularities. But the issue may be stated more broadly. Must we posit very specific initial conditions to explain the universe we observe today? Or can we create a theory that frees us from specifying the initial conditions? Inflation's great attraction is that it claims to do the latter. Regardless of how the universe might have begun—with large anisotropies and inhomogeneities—inflation produced the universe we observe.

"But this is impossible," claims an unrepentant anti-inflationist. "Inflation is a time-reversible theory."

Whoever tells the truth is chased out of nine villages. Inflation, because it relies on GR and quantum mechanics, is time-reversible, as are all the fundamental theories that came under scrutiny in the Fifth Debates. That being the case, it is easy to see that inflation is incapable of smoothing out all initial irregularities. Imagine a universe in which the COBE fluctuations are, say, ten times larger than the real ones. Here is a universe more anisotropic than observations permit. Now we merely run the equations describing this universe backward to a time before inflation. That gives us a set of initial conditions that inflation will not isotropize.

Theorists typically respond that the number of initial conditions that inflation can handle is much larger than the number of initial conditions that it can't. But we do not have any idea of the number of possible initial conditions of the universe, nor do we have any idea of how to count them. Again we face an attempt to apply probability to a unique event, the birth of the universe.

The debate has also raised the interesting issue: if inflation wipes out all traces of preinflationary times (quantum fluctuations excepted), does that mean that theories such as quantum gravity and superstrings, which operate at those times, are exercises in metaphysics?

The visitors are especially eager to attack such broader questions, and we can assure them that a discussion will take place later in the day or night.

Finally, what is the consensus on inflation? Probably the most judicious remark is by Martin Rees, who writes:

> Most theorists, especially those who have come to cosmology from a background in high-energy particle physics, regard inflation as a beautiful idea, offering a compelling insight into why our universe has its distinctive properties. But some are less enraptured, particularly those who prefer a more geometrical approach. The most distinguished of these is Roger Penrose. For him, inflation is a "fashion the high-energy physicists have visited on the cosmologists"; he notes that "even aardvarks think their offspring are beautiful." Despite such discordant voices, and some innovative ideas along alternative lines, most thinking about the ultraearly universe incorporates the concept of inflation.

It may be that the controversy will soon be laid to rest. Early next millennium, two satellites, named MAP (Microwave Anisotropy Probe) and PLANCK, will remap the microwave background with about one thousand times the sensitivity as COBE, and a much higher resolution. If all goes as planned, they will be able to determine the Hubble constant, Omega, and the cosmological constant to within 1 or 2 percent and tell us once and for all what sort of universe we live in and whether or not inflation occurred.

This is an exciting prospect but a perplexing one. If we imagine for a moment that inflation is confirmed, there will remain the questions, Why were the initial conditions of the universe such that it took place? and Why are we here to view the fruits of its success?

At Last the Epoch of Quantum Gravity Is Reached and the Universe Is Created from Nothing

By now zero time was approaching, at least you thought so, but all watches had broken, The expedition had reached a cultural layer so deep that there was not only no sign of the Oblivion File, but no artifacts of any kind whatsoever. At last, when the archaeologists had despaired of finding any evidence for the founding of the Academy, someone cried out, "Om!" or perhaps "Um!" For when everyone gathered around the faint inscription chiseled into the ancient rock, no one could say exactly whether they faced the Russian word "um" for mind, the Sanskrit "uma" for wisdom, or the first syllable of the French "ombilic," for navel.

After much discussion two possibilities alone remained. Half the camp saw in the message from the dawn of time "um," the cosmic shrug, the universal pause, the final, honorable admission of defeat in the face of overwhelming mystery. Half the camp saw in it the sacred "Om," standing for supreme Reality, the symbol of what was, what is, and what shall be. What lies beyond, past, present, and future. The Brahman, the whole world in one syllable.

Both interpretations were reasonable, for the excavation had attained the epoch of quantum gravity, which does not exist. To be precise, a complete theory of quantum gravity does not exist. What does exist, as the Eighth Debates revealed, is the conviction that any marriage of general relativity and quantum mechanics will operate only in the first 10^{-43} seconds after the big bang, a time known as the Planck time. It is at this time, before inflation, where the expedition now finds itself, that cosmologists find themselves confronting the Question of Questions, How did it begin?, which is an altogether different class of question than asking how Troy began. Asking for the origin of a series is not the same as asking for the origin of a particular member of the series.

Do cosmologists have an answer to the Question of Questions? We have said that insofar as cosmology goes there are only two possible answers: either the universe was not created and existed for all time, in one form or another, or it came into being at a finite time in the past.

In 1982 Alexander Vilenkin of Tufts published a paper called "Creation of Universes from Nothing," in which he claimed that the universe could spring into existence from literally nothing. By nothing we do not mean the quantum mechanical vacuum of spacetime, Deepak Chopra's quantum field of potentiality, we mean nothing, a state of zero space and no time. A state that does not even noth.

"What do you mean, 'noth'?"

"Well, the verb to do nothing, to create nothing, to be nothing."

"But can nothing noth?"

"This is the question."

"I don't understand."

That is the problem. It is exceedingly difficult to conceive of what absolute nothing is, or does. Vilenkin himself wrote, "The concept of the universe being created from nothing is a crazy one." Many people over the centuries have had difficulty with the concept that there was a moment before which there were no moments. If there was no time, what picked out the instant of Creation? If there was a time zero, must not there have been a time minus one second? Here is St. Augustine posing the same questions in the fifth century :

> Those who ask, "What was God doing before he made heaven and earth?" are still steeped in error, which they should have discarded. "If he was at rest," they will say, "and doing nothing, why did he not continue to do nothing for ever more, just as he had always done in the past? If the will to create something which he had never created before was new in him—if it was some new motion stirring in him—how can we say that his is true eternity, when a new will, which had never been there before, could arise in it...?"

If one substitutes "cosmos" for "him" and "big bang" for "will," Augustine's questions could have been asked today. No one has ever provided Augustine an answer that can be visualized in terms of our experience, but Vilenkin argued that mathematically the creation of the universe from nothing made sense. To be sure, his idea has since received something like cult status. Similar proposals have been put forward by Andrei Linde, and Jim Hartle and Stephen Hawking, and the chapter of Hawking's

A Brief History of Time visitors most often ask about is the one on quantum cosmology and "imaginary time," both of which figure prominently in creations of universes from nothing.

"I never did understand that section of his book," someone says as the visitors gather round.

Let us see what sense can be made of it.

First consider a marble in the center of the sombrero hat, rocking back and forth a little bit (see Figure 9.4). Classically, if it does not have an energy greater than the height of the central barrier, it cannot climb out, exactly like a roller coaster that is moving too slowly. But we know that according to quantum mechanics there is a small probability that it will tunnel through to the other side. This is exactly what happens in the radioactive decay of uranium, which emits a daughter particle, known as an alpha.* One can envision the alpha particle as something like a loose tooth knocking around the inside of the nucleus until it leaks out. The tunneling rate was first calculated by George Gamow in 1928 and is now a textbook exercise. The important thing in this calculation is that you have two states—the central dent in the hat, and the lower "ground" state outside—between which the particle can tunnel. Knowing this, Gamow could calculate the rate at which the alpha particle can leak through.

Roughly speaking, Vilenkin took Gamow's formula for the tunneling rate of an alpha particle and applied it to the entire universe. Actually, he used a more modern version due to Julian Schwinger that gave the rate at which particles are created by an electric field from the quantum mechanical vacuum. (This is Chopra's quantum field of potentiality, and the electric field literally yanks particles from the vacuum.) Schwinger employed a particular mathematical device of multiplying ordinary time by the imaginary number $i = \sqrt{-1}$, which makes time imaginary. Of course you are now scratching your head, wondering what this could possibly mean, but for the moment let us just regard it as a mathematical trick to facilitate computation. Despite the trick, two essential elements remain in Schwinger's version of Gamow's calculation: there is an initial state and there is a final state between which tunneling takes place, and so many particles are created per second, per real second. Time may be imaginary for the purposes of calculation, but time is real for interpreting the result. As Groucho said, time flies like an arrow; fruit flies like a banana.

Vilenkin took the Einstein equation governing the size of an inflating universe and, like Schwinger, let the time become imaginary. The advantage to this is that the big bang singularity disappears. Instead, Vilenkin found that at a (real) time equal to zero, the universe abruptly popped into existence with a finite size and thereafter continued to expand. He interpreted this as a universe tunneling from nothing into reality. The size of the universe, according to his estimates, would be very small—of subatomic size; inflation would then enlarge it. This did not explain inflation, however; an inflating universe was assumed from the outset.

But now a perturbed academician interrupts, "By the same argument the universe could suddenly disappear into nothing."

---

* Alpha particle is another term for a helium nucleus—two neutrons and two protons.

True. The equations are totally reversible and also describe a universe that contracts to a minimum size and suddenly vanishes—Nirvana, the annihilation of the universe into nothing. In terms of the Gamow formula this would correspond to the "undecay" of uranium, which is not observed in nature.

All this is weird enough, but you are categorically dying to know that if Vilenkin just borrowed the Gamow formula, what was the previous state from which the universe emerged? According to Vilenkin, it's pure nothing. In fact the words "previous" and "from" also lose meaning because there was no time before and no place from.

"But this is absurd," interrupts a nuclear physicist. "As you have stressed, in the old days, radioactive decay took place in real time, and the alpha particle really was someplace before—inside the nucleus." At these words, the debate on the relationship between mathematics and reality erupts once more. Watching the academicians go at it, it strikes you that this is probably what Sartre really had in mind for hell, to be damned to argue the same questions over and over and over and . . .

One cannot deny that something has changed over the years. Nowadays, quantum cosmologists routinely assert, with a certain disdain for reality, that "tunneling must be thought of as taking place in imaginary time." Evidently a mathematical formalism has taken on a life of its own. As the nuclear physicist just indicated, to Gamow and Schwinger tunneling took place in real time and therefore a previous state must have existed to tunnel from. As with supercooled water, the ammonia molecule, and the inflationary scenario, the previous state must have been metastable; if it were absolutely stable, it would by definition exist for an eternity and never decay. But if it existed for only a finite time, then a state previous to *that* one must have existed, and so on, as St. Augustine understood.

By going to imaginary time, one eliminates previous states, but the question remains whether the appropriation of a formula for the creation of particles in an electric field and applying it to the creation of the universe means anything. Schwinger's formula gave a rate of particle production. Should we then regard Vilenkin's formula as predicting the creation of so many universes per second? Also, there is another class of solutions in which the radius of the universe was negative. Does this have less meaning than imaginary time?

But does imaginary time have less meaning than negative probabilities? Both require interpretations. Some members of the expedition insist that there is indeed an interpretation of imaginary time. Physically? Well, mathematically, there is. In relativity, imaginary time is more like a distance; in the Commentaries we provide a few details. Converting to imaginary time near the big bang does eliminate the singularity and frees one from worrying about what came before time equals zero. As Hawking has said, we do not ask what is north of the North Pole. In a similar way, there is no time beyond the singularity. It just vanishes.

Vilenkin's model of the universe was very simple; indeed he considered only the behavior of its radius. About a year later, however, James Hartle and Stephen Hawking published a paper that gave a very similar picture of how a universe might spring from nothing. Closely following the methods of Bryce DeWitt, Hartle and Hawking attempted to construct a quantum mechanical wave function of the universe itself.

As you by now suspect, they did not actually consider the real universe; as DeWitt did, they considered a simple model—in fact an isotropic and homogeneous model. But to apply quantum mechanics to any entire universe brings with it certain conceptual difficulties over and above the problems of a quantum theory of gravity discussed a few days ago. In the Third and Fourth Debates, Laplacians stressed how one must always externally supply the initial conditions (in classical physics the initial positions and velocities of the particles involved) for a given problem. Since the time of Newton, equations of motion and boundary conditions have been distinct; only after the latter have been supplied by hand can the equations of the theory take over and predict the evolution of the system.

But to assign an initial position and velocity to a swing requires standing outside the swing. To assign the boundary conditions to any system requires standing outside the system. Clearly this is impossible when considering the universe as a whole and a specification of boundary conditions must somehow become part of the theory. The Newtonian distinction between laws and boundary conditions dissolves in any theory of the entire universe.

Hartle and Hawking did not start with the Gamow-Schwinger expression for tunneling. Rather they began with what might be regarded as a more appropriate springboard, the Wheeler-DeWitt equation, which can be thought of as the Schrödinger equation for the universe. We know that according to quantum mechanics any system is described by its wave function and the behavior of the wave function is governed by the Schrödinger equation. Moreover, a sphinx's wave function consisted of a superposition of all her aspects. The Sphinx was not one-third woman, one-third bird and one-third lion, but all of them; a measurement reduced her to one or the other. The wave function of a particle with spin consisted of one part spin up and another part spin down, although only a single orientation was observed with each measurement.

The universe has neither wings nor spin, but it does have geometry (at least in most people's minds). The wave function of the universe can be thought of as consisting of a superposition of all allowable geometries: mildly curved, seriously curved, etc. As the universal wave function evolves, you can imagine it sampling all possible worlds. The question is, Which of these worlds is actually realized? Which of these world histories becomes ours?

In ordinary quantum mechanics, the outcome is decided by a measurement, which collapses the wave function and selects a spin or an aspect of the Sphinx. And here the New Age members of the expedition perk up, for they cling to the notion that an external observer is required to decide the outcome of an experiment and there is nothing external to the universe.

"Cosmic consciousness collapses the wave function of the universe," maintains one of the visitors from southern California. "It must be."

"But what collapses the wave function of cosmic consciousness?" asks another.

"Synchronicity," the first replies. "It's obvious."

Probably not synchronicity. Perhaps decoherence.

"I never thought I'd hear that word again," you say, amazed.

Ah, you have forgotten the prime mantra, "Use it until it breaks." To refresh your memory, modern physicists have turned from the idea that consciousness or observers are necessary to collapse the wave function. Now they lean toward decoherence, which refers to any natural, environmental disturbance that changes a quantum system to a probabilistic classical system. Decoherence does not decide the exact outcome of an experiment, but it does get you halfway. A quantum moviegoer potentially sees all films. Once decoherence sets in, you know that he sees four-star films with a 90 percent probability, sees three-star films with a 70 percent probability and avoids anything called "the greatest movie of the year," but you never know exactly which film he will decide to see tonight. A quantum photon passing by a double slit passes though both slits. Once the system has decohered, it has a fifty-fifty probability of going through either.

Decoherence thus provides a natural "out" to quantum cosmologists who cannot invoke an external apparatus to collapse the wave function of the universe. Environmental influence will determine the probability that the universe passed through one series of geometries or another. The ultimate goal is to determine which histories have a high probability and which don't. One would like, presumably, to show that the history of producing our universe was high.

At this stage you are a little intrigued and a little impatient. "Much of this sounds suspiciously like the many-worlds interpretation of quantum mechanics that was debated the other day. Are you sure the universe isn't just splitting? And what does this have to do with creating the universe from nothing?"

To answer the first question, people do couch quantum cosmology in terms of the many-worlds interpretation. But notice that today we have been using the term "history" as a potential history of the universe, one that may still be going on, and not in the common sense of the single, unadulterated chronology of events that have come and gone. You certainly recollect the confusion when Murray Gell-Mann introduced multiple interpretations of the many-worlds interpretation, rejecting the idea that the universe truly splits. Now the source of the confusion is at hand. Gell-Mann was using the term "history" as we use it now. Not as anything that really happened, but as a possibility. Of many such possibilities one was made manifest. You would not call the others alternate universes. Just paths not taken.

"But you have already emphasized that we have no way of knowing which initial configurations of the universe are probable and which are not. How can a program of calculating the probability of universe trajectories make any sense whatsoever?"

Before tackling that question, we can say the following: In the Seventh Debates we mentioned that it is easy to make a system decohere. Water dripping before one of the slits, microbes applauding, will make a photon lose coherence. But these are influences external to the photon. Nothing is external to the universe. Therefore, as pointed out by Roland Omnes, to cause the universal wave function to decohere into histories with classical probabilities can be accomplished only by internal disturbances. Perhaps these debates would perturb the universe enough to do the trick. Nevertheless, to date no one has actually demonstrated that internal disturbances are sufficient to cause the universal wave function to decohere into the various classical histories.

Back to the creation of universe from nothing. At about the time Vilenkin published his paper, Stephen Hawking was also grappling with the problem of boundary conditions to the universe. We have already seen that going to imaginary time removes the big bang singularity, since time simply vanishes, or perhaps more accurately, gets rotated into a spatial distance. So we now stop thinking in terms of spacetime and begin thinking in terms of more ordinary geometries like globes and wall maps. If one were going to write down the wave function for the universe on a flat map, one would have to specify how it behaved on the edges—the boundary conditions. But on a globe, there are no edges and one is saved from having to specify boundary conditions.

Thus, in 1983, Hartle and Hawking made the "no-boundary" proposal, which Hawking discusses in *A Brief History of Time*. In formulating quantum mechanics for the whole universe, one should first go to imaginary time. Then, recalling that the wave function of the universe samples all possible histories, and that a history is a sequence of possible geometries, we restrict our attention to geometries, like the globe, that have no boundaries. This saves us from having to specify how the wave function behaves at the edge of the universe. As Hawking puts it, "The boundary condition of the universe is that it has no boundary." However, one should realize that he is not talking about the spacetime of relativity, he is talking about "Euclidean" spacetime, in which time is imaginary. Furthermore, the no-boundary condition *is* a boundary condition—a choice that nature has made of allowable geometries.*

One should also bear in mind that to calculate the behavior of the universal wave function in any realistic situation would be impossible. For that reason Hartle and Hawking restricted their attention to the simplest conceivable geometries— standard-model geometries. The wave function, then, can be regarded as a superposition of all possible standard models (which are really all the same except for differing sizes). In such a simple model they found, as Vilenkin did, that the universe could spontaneously appear from nothing in an inflating state. They were also able to go one step beyond Vilenkin and calculate quantum fluctuations—the ones that inflation magnifies into primordial galaxies observed by COBE. But again they chose the simplest possible form of the fluctuations, and a full calculation, to our knowledge, has not been carried out.

Hawking has made many claims for the no-boundary proposal. For instance, he has stated that "the no-boundary proposal seems to account for the universe we live in."

"But if they were working with the universal wave function," your friend asks, "which samples all possible worlds, how did one world—ours—emerge from the calculations? By decoherence?"

---

* Because Hartle and Hawking have chosen to sample over only closed (globelike) geometries in imaginary time, and because these must be matched onto the "real" universe when time becomes real, the real universe must also be topologically closed. However, if the universe is ever-expanding, as observations suggest, then it cannot be the standard model but must be one of the negatively curved, topologically closed universes discussed in the previous debates!

The precise question—and one they simply did not address. Although it is possible to calculate the spectrum of quantum fluctuations observed by COBE without worrying about the collapse of the universal wave function, nothing they did has explained why when you go out at night and gaze at the stars you see one universe and not many.

Hawking has also implied that the no-boundary proposal explains the arrow of time. It is, for instance, these tiny quantum fluctuations, representing the first inhomogeneities of the universe, that drive Penrose's Weyl tensor of the Fifth Debates away from zero and get the arrow of time pointed in the right direction. But as Hawking himself concedes, there are two "directions" to the equations (as there must be, since the theory is time-symmetric), and he is selecting one of them. Apart from this issue, Prigogine would undoubtedly ask, "What do quantum fluctuations at the Planck time have to do with the increase of entropy in my room?" A side debate on time irreversibility takes off.

Hawking has also said, "Any measurement made in cosmology can be formulated in terms of the wave function. Thus, the no-boundary proposal makes cosmology into a science, because one can predict the result of any observation."

It is extraordinary that in light of chaos theory and so on, someone would claim that anything (yet alone everything) could be predicted in terms of the universal wave function. Bear in mind that in a fully developed theory, the universal wave function truly accounts for Everything—even the cackling of chickens and the swarming of mosquitoes. Not to mention the fact that, as we just argued, no one has calculated even in the simplest case how the classical universe emerges from the quantum superposition. In this regard Hawking seems the ultimate reductionist.

One's first reaction to Hawking's "the no-boundary proposal makes cosmology into a science" is naturally to ask how he would have classified cosmology before the no-boundary proposal. Would he have called it a religion? This is difficult to say because Hawking's views on cosmology and religion sometimes seem contradictory. On the one hand, he concluded his book with the statement that a complete theory would allow us to "know the mind of God." On the other hand, he also mentions that after the advent of Newtonian physics, God's role had shrunk but he was nevertheless still allowed to wind up the clockwork and choose how to start it off. In light of the no-boundary proposal even that role is denied him. Hawking asks, "What place, then, for a creator?"

Before examining the theological meditations of contemporary scientists, we might risk answering Hawking's question with a question: Who chooses the no-boundary boundary condition?

Having reached imaginary time, the expedition prepared to ascend, but since time no longer existed motion was impossible. You now realized that you were trapped in these debates for eternity and that you too would join the academicians, frozen in Sartre's stasis, or was it Dante's . . .? At the moment of your greatest despair—moments? What moments could there be?—one of the visitors approached the authors and asked, "Excuse me, there is something I didn't understand."

What is that?

"How did the universe begin?"

Om.

Um?

A Practical Exercise—In Which the Authors Ponder the Nature of Time

Aeons passed in debate, or perhaps only seconds; in the region of imaginary time you had no way of knowing. Whether the expedition was ascending homeward or still descending was impossible to say. No trace of the legendary Oblivion File had been found, but after some—or no—time it occurred to you that you were in it. In the zone of timelessness, the only thing manifest was a general sense of frustration among the archaeologists. Although some believed they had answered the Question of Questions the majority felt, as Steven Weinberg had foretold, that Why? was still with us. When you overheard the authors engaged in earnest conversation, you also began—how could anything begin?—to suspect that neither were they entirely satisfied. One of them, at least, seemed to be perplexed about the very idea of motion. In the current—perpetual?— situation there was much time—distance?—for confusion.

The first author remarked that he couldn't really understand how the universe could be expanding at all. "Question: how can there be an equation of motion for the entire universe?"

At this the second author tried to move without time, replying, "Einstein's equations tell the universe how to expand."

But you could see the first author didn't buy this pat response. "For there to be motion an object must be in a larger system. A train is seen to be moving only from the outside. A swing is in motion only with respect to a larger system. What framework is the entire universe changing in?"

As the stasis continued, his colleague thought and thought and thought . . . and you began to believe the question actually made sense. The second author, however, did eventually answer. "Einstein's equations tell the universe how to move from one geometric configuration to the next. You don't need anything from the outside. You can tell a balloon is expanding if you are on the surface—the galaxies pasted on get farther apart. That's all there is to it."

The reply appeared to be reasonable, but to your surprise the first author riposted, "This doesn't solve the problem. The assertion is that one cannot measure motion from inside a system. One should then view the different geometric configurations as different universes with some external time connecting them."

The other author didn't see where that got you. "You have all these different snapshots. Each one shows a universe at a different time and a different size. These snapshots are ordered by labels, A, B, C . . . We call these labels different times. Now, just as you can measure temperature in Fahrenheit or Centigrade it doesn't matter what numerical values of time you assign these labels. But regardless of how you label time, the equations always get you safely from one configuration to the next. So the geometry is changing in a well-defined way and nothing is gained by looking at each snapshot as a separate universe."

You were certain that ended the argument, but the first author rejoined, "But there were no clocks near the big bang at all. Question: What was measuring time?"

Now, you wondered, how could such a fundamental issue have slipped through the cracks thus far? Do all the researchers investigating the early universe assume that something is measuring time?

The second author conceded, essentially yes, and this a conceptual difficulty. "To a physicist time is what a clock measures, whether the clock is a mechanical contrivance or a natural process. As early as one second after the big bang, physical processes should behave in a conventional fashion, and so cosmologists assume that time can be measured. But people usually say one needs periodic motion, like pendulums, to measure time, and earlier on it is not 100 percent clear what periodic processes would serve as clocks. On the other hand, it is difficult to even conceive of a clock without the notion of prior (or at least coexisting) time—just as it is impossible to conceive of a meter stick without the idea of distance. All these questions are great mysteries. At the earliest moments after the big bang one is merely guided by blind faith that time behaves the same as it does today."

The author further conceded that at the Planck time, when researchers believe spacetime itself must be quantized, the last assumption almost certainly breaks down. Cosmologists often talk about "spacetime foam," which is meant to conjure up a picture of a fuzzy sort of time, but to be honest no one has a clear idea of what this means. Trapped here in imaginary time, you had no idea of what that meant either.

What was clear to the authors is that none of the calculations discussed in the Ninth Debates deal with this issue. The authors also agreed on one other topic and advised you, if you ever returned to the Real World, to check the last chapter or two of the nearest modern astronomy text. It is quite likely you will find that the chapter on cosmology has been given a provocative title such as "Toward the Beginning of Time." This is unfortunate. For reasons that are not clear, unless it is the influence of John Wheeler or Stephen Hawking, writers and cosmologists alike have indeed fallen into the habit of calling the big bang the "beginning of time" and the big crunch—if there will be one—the "end of time." Yet it is not beyond dispute that time's beginning coincided with the big bang and that both arose *ex nihilo*. A theoretical trend has crystallized with neither evidence for the conjecture nor a clear understanding of what the words mean.

Here in imaginary time, where science was without form, and speculation and fact were as yet undivided, the expedition at last entered into a

Meditation on Cosmologist-Theologians

In the sublime eleventh chapter of the Bhagavad Gita, Prince Arjuna is granted a vision of Krishna in his full nature, the universe itself. "There, within the body of the God of gods, Arjuna saw all the manifold forms of the universe united as one." Arjuna falls before the unbearable vision, prostrate, and exclaims, "O Lord, I see within your body all the gods and every kind of living creature. I see Brahma, the Creator, seated on a lotus . . . . I see infinite mouths and arms, stomachs and eyes, and you are embodied in every form. I see you everywhere, without beginning, middle or end . . . . I see you, who are so difficult to behold, shining like a fiery sun blazing in every direction. You are the supreme, changeless Reality, the one thing to be known . . . ." And, after Arjuna has enumerated many of Krishna's attributes, Krishna adds a few more: "I am time, the destroyer of all; I have come to consume the world . . . ."

Arjuna's cosmic vision is one of the most exalted passages of all literature and if its writers intended that the listener be humbled they succeeded beyond qualification. The Bhagavad Gita expresses better than any scientist can something of the awe and humility one experiences when faced with the beauty and order of the universe. It is to partake of this beauty, if only in the smallest degree, that many embark along the scientific path, and few at some stage along that path fail to encounter a timeless moment when one *sees*, when ideas join in almost concrete form, when the order of the universe becomes manifest. Yet, when attempting to convey this sense of beauty, the conviction of order immanent, we fall back on the banality of the expressible.

In their experience of the mysterious, scientists are far from alone. Artists know such moments, when a structure feels true and complete. They know when (rarely) they achieve economy of means and perfection of form. Yet to demonstrate by word or brush stroke how it came about or why it is true is no easy task. The creators of the *sangaku*—merchants, samurai, women, children—had an artist's sense of form. When beauty of design is united with the unexpected mathematical relationships arising in those extraordinary problems, you have a fair sense of the order a scientist perceives in the enveloping world.

And so it is natural that when the expedition encountered a watch embedded in the eighteenth-century cultural layer, philosopher William Paley would conclude that its "several parts are framed and put together for a purpose." It is natural that Paley, as many before and after him, would view the apparent design of the universe as evidence for a Designer. It is natural that cosmologists, confronting creation as they do, resort to religious analogies and metaphors, speaking of the "mind of God," the "face of God," and less exaltedly, "the handwriting of God." It is perhaps understandable, if less natural, that textbooks should reinforce the theological overtones of cosmology with their chapters, "Toward the Beginning of Time," which inevitably conjure up visions of creation and creator. And in light of all this it would be surprising if visitors, faced with the profusion of cosmological-religious book titles that have recently overwhelmed bookstores, were not convinced that theology and cosmology were one and the same.

Many scientists and cosmologists share this belief. In their loose references to God, one of course hears the persistent echo of Einstein, who frequently invoked the Lord (though any Lord he may have believed in was not the Judeo-Christian one). But apart from coy stylistic references there has been a genuine upsurge of scientists who profess to find evidence for a divine plan in the coincidences of nature that make the universe habitable. Twenty years ago Freeman Dyson wrote:

I conclude from these accidents of physics and astronomy that the universe is an un-expectedly hospitable place for living creatures to make their home in. Being a scientist, trained in the habits of thought and language of the twentieth century rather than the eighteenth, I do not claim that the architecture of the universe proves the existence of God, I claim only that the architecture of the universe is consistent with the hypothesis that mind plays an essential role in its functioning.

Few have argued from design as eloquently as Dyson. At the time Dyson was a lone figure, rebelling against biologist Jacques Monod's injunction, "Any mingling of knowledge with values is unlawful, forbidden," and Steven Weinberg's soon-to-be notorious, "The more the universe seems comprehensible, the more it also seems pointless." As Dyson said, "If Monod and Weinberg are truly speaking for the twentieth century, then I prefer the eighteenth."

In the two decades since Dyson declared his faith, large numbers of other scientists have also come out of the closet. If they sometimes hesitate to refer to the anthropic principle by name, cosmologists, biologists, geologists and astrophysicists no longer hesitate to avow belief in a divine plan. "The exquisite order displayed by our scientific understanding of the physical world calls for the divine." "A common-sense interpretation of the facts suggests that a superintellect has monkeyed with physics, as well as with chemistry and biology, and that there are no blind forces worth speaking about in nature." "We know that nature is described by the best of all possible mathematics because God created it." Weinberg sometimes seems a lone, lonely figure in maintaining that "no one has ever discovered any correlation between the importance of *anything* to us and its importance to the laws of nature."

But reality is not always determined by referendum. Martin Rees reminds us that "scientists' incursions into theology or philosophy can be embarrassingly naive or dogmatic," and he for one refuses to venture into those realms regardless of how profound cosmology's implications may be for them. There is wisdom in Rees's diffidence. Much of cosmology is amenable to theoretical analysis and increasingly to observational test. More to the point, Rees reminds us that for all the close historical connection between physics and philosophy, physicists tend to make terrible philosophers. We cannot decide the existence of God in these debates. It might be more profitable to ask how much of the recent surge in religiosity is genuine, how much is fashion, and what of substance astrophysicists and cosmologists have added to the age-old clash between science and religion.

A certain collective amnesia is evident in the current resurgence of design arguments. One reason that physicists tend to make terrible philosophers is that their outlook is as a rule ahistorical and they are forever rediscovering the wheel. The debates over science and religion, it goes without saying, have a long pedigree and it is interesting that the similar discussions at close of the nineteenth century over the theological implications of the second law of thermodynamics have been forgotten. As Bishop Ellicott once wrote to Maxwell, "Theologians are a great deal too fond of using up the last scientific hypothesis they can get hold of." Today's theologian-scientists have also forgotten that William Paley himself, after conceiving of his watchmaker, said, "Astronomy is not the best medium through which to prove the agency of an intelligent Creator."

The "philosophical naivity" of scientists Rees notes is aided and abetted by the fact that the day-to-day activity of science is concerned with models and experiments. Only afterward do scientists attempt to fit their activities into a larger framework. As a result, scientists' beliefs can appear surprisingly contradictory. When Stephen Hawking states on the one hand that all one can ask of theory is that its predictions correspond to observations, and on the other hand invokes the anthropic principle to explain certain features of the universe, one must wonder how seriously his position is thought out. The famous Russian astrophysicist Yakov Zel'dovich, a proponent of the inflationary scenario, once exclaimed in a fit of pique, "I believe the universe started isotropically!" Several other well-known supporters of inflation have made similar remarks in print, citing considerations of simplicity and symmetry; Ockham's razor. Yet, an advertised purpose of inflation is to make the universe isotropic; if one believes it began that way, then many of the arguments for inflation are superfluous and one is not shaving with Ockham's razor.

The early universe has become fertile territory for cosmologists to reveal their philosophical inclinations and contradictions. In his book *The Inflationary Universe,* Alan Guth describes a variant known as eternal inflation. According to eternal inflation, the universe continues to inflate forever, all the while producing an infinite number of small "pocket" universes, in one of which we live. Because our tiny pocket may be the one thousandth or one millionth created, Guth can argue that "although eternal inflation apparently does not avoid the question of how the universe originated . . . any hypotheses about cosmic origins become totally divorced from observational cosmology." Yet he goes on to say, "Given the plausibility of eternal inflation, I believe that soon any cosmological theory that does not lead to the eternal reproduction of universes will be considered as unimaginable as a species of bacteria that cannot reproduce."

One must ask whether such statements are any different from the old Indian myth that tells of the Earth supported on a tortoise, which stands on a second tortoise, which stands on a third, which stands on . . . There may be some unarticulated consistency in Guth's position, but without a breath he rejects the old oscillating universe model, in which the universe goes from big bang to big crunch eternally, on the grounds that the "theory relies solely on speculation."

It is against this background that one should judge scientists' assertions about God, creation and the origin of time. To physicists, certainly, the issues cannot be disentangled. Here, in a dialogue between a physicist and a skeptic, Paul Davies attempts to dispel a few common notions about time:

PHYSICIST: In the external world there is no past, present and future. How could the present ever be determined with instruments? It's a purely psychological concept.

SKEPTIC: Oh, come now, you can't be serious. Everybody knows the future hasn't happened yet, whereas the past is gone—we remember it happening. How can you confuse yesterday with tomorrow, or today for that matter?

PHYSICIST: Of course you must make a distinction between various days in sequence, but it's the labels you use that I object to. Even you would agree that tomorrow never comes.

SKEPTIC: That's just a play on words. Tomorrow does come, only when it does we call it today.

PHYSICIST: Precisely.

And so on. To Davies the flow of time does not exist. He would reasonably ask if time does flow, how do we measure its velocity? Seconds per what? Hence the physicist's claim that in the external world there is no past, present and future. But sidestepping the debate on time's arrow now reerupting, we may again ask how far understanding has progressed. Here is another part of St. Augustine's famous meditation:

What then is time? I know well enough what it is, provided that nobody asks me: but if I am asked what it is and try to explain, I am baffled. All the same, I can confidently say that if nothing passed, there would be no past time; if nothing were going to happen, there would be no future time; and if nothing were, there would be no present time.

Of these three divisions of time, then, how can two, the past and the future, *be,* when the past no longer is and the future is not yet? As for the present, if it were always present and never moved on to become the past, it would not be time but eternity. If, therefore, the present is time only by reason of the fact that it moves on to become the past, how can we say that even the present *is* when the reason *it is* is that it *is not to be?* In other words, we cannot rightly say that time is, except by reason of its impending state of not being.

Einstein has taught us that there is no absolute past, present or future, but mere sequences of events whose order may depend on the motion of the observer. But to each of us, measuring time, Augustine's questions still stand and no one since has put them better.

With such basic, unanswered issues about time surfacing and surfacing again, one might hesitate to make categorical assertions about creation. Yet we have already seen that cosmologists do not shy away from deriving the world. Physical chemist P. W. Atkins also gamely takes a whack at it:

Now we go back in time beyond the moment of creation, to when there was no time, and to where there was no space. From this nothing came spacetime, and with spacetime there came things. In due course there came consciousness too, and the universe, initially nonexistent, grew aware.

Atkins, after Hartle and Hawking, has assumed the universe was created from nothing. This is fashion. The oscillating universe, with its infinite cycle of bangs and crunches, is an example of a cosmos that has existed and will exist forever. The oscillating universe is speculative and no longer popular, but neither has any observation ruled it out, and it is difficult to see how any observation can. We simply do not know what happened before the present big bang, if there was a before, and it is best to regard talk of creation that pretends to certainty of these things as poetic license.

There is less doubt about the social intent of such exercises. Atkins again:

FIRST, THERE IS THE BEGINNING.

In the beginning there was nothing. Absolute void, not merely empty space. There was no space; nor was there time, for this was before time. The universe was without form and void.

By chance there was a fluctuation, and a set of points, emerging from nothing and taking their existence from the pattern they formed, defined a time. The chance formation of a pattern resulted in the emergence of time from coalesced opposites, its emergence from nothing. From absolute nothing, absolutely without intervention, there came into being rudimentary existence.

Atkins's professed aim is to show that there is no need to invoke the supreme being to account for the creation of the universe. Fair enough. But it is also fair to ask what has been gained from the substitution of scientific terminology for Old Testament art. Having read Atkins's reworking of Genesis, do we feel satisfied that an understanding of the universe's creation is in hand? Has he answered Augustine's question about what God was doing before creation?

Like Atkins, Paul Davies at one time argued that physics renders religion obsolete. Contemplating whether God exists within time or is free from time's shackles, Davies reasoned, "Clearly, God cannot be omniscient if he is subject to the physics of time, nor can he be considered the creator of the universe if he did not create time." Davies goes on to say that once spacetime existed, everything else follows "as the result of perfectly natural activity. Thus, many would argue that God is not really needed as a creator at all *except* to create time (strictly, spacetime)."

Again we ask whether such speculations add anything new. Two hundred years ago Laplacians contended that once God started off the universal clockwork, there was no further need for him. Hawking's provocation, "What place, then, for a creator?" is the latest extension of the argument; the no-boundary condition proscribes him from interfering even in the creation of spacetime. Actually, Epicurus, wanting happiness even for God, put it more genially several thousand years before: "And do not let the divine nature be introduced at any point into these considerations but let it be preserved free from burdensome duties and in entire blessedness."

For those who take the Bible literally, Epicurean arguments pose a genuine threat. Biology and physics have progressively circumscribed the role of a Judeo-Christian God until today some scientists have felt cornered enough to make spectacular efforts to ensure the compatibility of Genesis not only with evolution but with general relativity. And yet most scientists, Einstein included, do not hold a firm vision of God. For many "God" is merely a poetic synonym for "universal order" or "laws of nature," in which case their pronouncements become tautological or meaningless. Other scientists would probably best be described as pantheists; they see god everywhere. To all but the most literal-minded the conflicts between Bible and cosmology seem irrelevant and evaporate.

Because of advances in cosmology, the debate at the millennium has been characterized as a new turn in the age-old contest between science and religion: "As science advances, religion recedes." But to many participating in this expedition, today's dispute has seemed comically Occidental. Christianity has been substituted for religion. It should go without saying that many of the world's religions, Buddhism primary among them, do not recognize a supreme being. Neither do the Jains in India worship God. Even in Hinduism the various gods are regarded as, ultimately, an extension of oneself—Thou art That—and so are not viewed as omnipotent in the same sense as the Christian God. Brahma, the creator, is sometimes viewed within the universe, sometimes without. The ambiguity is evident in the creation hymn from the oldest of the Indian spiritual texts, the Rig Veda:

> There was neither non-existence nor existence then; there was neither the realm of space nor the sky which is beyond. What stirred? Where? In whose protection? Was there water, bottomlessly deep?

The author goes on to describe the lack of any distinguishing sign, the emptiness, the desire that came upon "that one in the beginning," which was the first seed of mind. Then the ancient bard asks:

> Was there below? Was there above? There were seed-placers; there were powers. There was impulse beneath; there was giving forth above. Who really knows? Who will here proclaim it? Whence was it produced? Whence is this creation? The gods came afterwards, with the creation of this universe. Who then knows whence it has arisen?
>
> Whence this creation has arisen—perhaps it formed itself, or perhaps it did not— the one who looks down on it, in the highest heaven, only he knows—or perhaps he does not know.

The refreshing thing about this hymn is the perplexity with which the narrator retells events. There is no mention of God. In the final stanza the narrator even admits the possibility that the universe was created from nothing.

There are lessons here for both philosophers and scientists. One of the questions perpetually asked by historians of science and which came up early on in these debates is why the Chinese, with their obsession with legalistic law, never invented modern science. To the extent that the question has an answer Joseph Needham seems to opt for a theological one. Without faith in a rational creator, "there was no conviction that rational personal beings would be able to spell out in their lesser earthly languages the divine code of laws which [the creator] had decreed aforetime." There is plenty of irony here and maybe parity. The West got both God and physics, the East got neither, and for the same reason.

The irony produces a smile at the New Age attempt to blend Oriental mysticism and modern physics, which, if Needham is correct, is an attempt to blend historical incompatibilities. Nevertheless, even if the New Age has not entirely grasped either the physics or the philosophy it is hoping to combine, one can sympathize with the impetus.

Underlying the current revival of religiosity is a dissatisfaction with what science and technology has given us: a comfortable life by history's standards but one devoid of spirituality. In their return to design arguments the cosmologists participate in the general dissatisfaction, but in restricting the debate about religion and science to Western religion, they have set up a straw man. Ask the Zen master what God was doing before the big bang, and he is likely to reply, "A frog jumps into the water; hear the sound!" New Agers have seen the wisdom of the response. Their attempts to escape the shackles of a personal God are perhaps closer to the spirit of Einstein than the cosmologists who have given false or empty answers to the perennial questions. It is fair to ask whether a scientist who proclaims that "a superintellect has monkeyed with physics, as well as with chemistry and biology" should be deferred to more than Deepak Chopra and his quantum wiggles.

Ultimately physics is what we experience or, as Clausewitz might put it, an extension of experience by other means. Can we conceive of time before spacetime? If not, then we can't translate the question into categories of physics. Questions like "What happened before the big bang?" may be legitimate or they may turn out to be meaningless unless we have a supertheory, a theory in which the universe is viewed as being in something else. It is a bit like the woman who went to the opera wearing curlers. "How can you do that?" the other operagoers exclaimed at her impertinence. "But I'm going somewhere else afterward," she replied.

It is disingenuous to pretend that by introducing mathematics which may be untranslatable to the realm of experience one has answered the ultimate questions. Richard Tolman's warning of sixty years ago is just as relevant today: "The discovery of models, which start expansion from a singular state of zero volume, must not be confused with a proof that the actual universe was created at a finite time in the past." Tolman's words are for scientists.[†] Cosmologist Joseph Silk, to whom Martin Rees refers for his position, puts it more generally: "Humility in the face of the persistent great unknowns is the true philosophy that modern physics has to offer."

The Tao that can be expressed is not the true Tao.

In Which the Ninth Debates Are Summarized

And suddenly, without warning, the expedition popped out of nothing and found itself on the Academy grounds again. No one could say how much time had passed. As the academicians dispersed for the night, there was speculation that the safe landing of the Eighth Debates must have been due to a rotation into imaginary time, which had prevented a singular catastrophe. The Buddhists were disappointed that timelessness had an end and searched for a universe that would vanish into nothing. You and your friend on the other hand were relieved that eternity had proven so short lived—had it?—and that you might live to see the Real World yet. As you walked out toward the beach to sleep, you found sufficient time to review the excavation's main discoveries.

Perhaps the main lesson drawn from the Ninth Debates was that cosmology differs from most other sciences in that the system under observation is unique; it cannot be reproduced. The uniqueness of the universe has forced astronomers to make many assumptions about unobserved and unobservable regions of the universe, the most famous being Einstein's "cosmological principle," which states that the universe should be absolutely uniform everywhere. The standard, or FLRW, model is based on this assumption, and so is the simplest cosmological model. Popular statements by both astronomers and the press about the fate of the universe invariably refer to the FLRW model, not the universe. The FLRW model does have many features in common with the observable universe: the Cosmic Microwave Background Radiation, the correct abundances of helium and deuterium, the ripples in the CMBR, and for this reason is considered by most astronomers close enough for jazz.

Nevertheless one must bear in mind that the standard model is a model. Debaters pointed out that other cosmologies exist (the ever-expanding, topologically closed models) that would be difficult or impossible to distinguish from the standard case. Moreover, no model can describe regions of the universe lying beyond the cosmological horizon, the distance light has traveled since the big bang.

Along the way to earlier cultural levels, several arguments erupted over the interpretation of general relativity. Is space truly curved? Or is this merely a convenient language? Henri Poincaré and Steven Weinberg made a strong case that the language of curvature is merely a convention and not the underlying reality. Is relativity valid at all scales? Does relativity support Mach's principle? None of these questions received definitive answers.

At a very deep cultural level, while the Oblivion File had still not been found, the discussion took an even more skeptical turn. Alan Guth introduced inflation to solve two "cosmological conundrums." The first was the so-called flatness problem, which asks why the universe appears so nearly flat when the generic behavior of the model is to become dominated by curvature. The second was the "horizon problem," which asked how the CMBR came to be so nearly uniform in temperature, when according to the FLRW model, different areas of the sky could not have influenced each other earlier than the present epoch.

Inflation touched off a long debate, which raised interesting issues: Are such questions physical or metaphysical? The universe is unique and must have begun in some fashion. Do the observations support inflation, which requires the density of the universe to be at the critical value (Omega = 1)? Does one not require the anthropic principle for a full explanation of the apparent flatness of the universe? Since inflation is time-reversible and one can thereby construct initial conditions that inflation cannot transform into the observed universe, why were the initial conditions of the universe such that inflation worked?

This question led us to the deepest cultural level attained by expedition, the level of quantum gravity. No trace of the Oblivion File was found, but Alexander Vilenkin created the universe out of nothing. The authors examined what sense could be made out of the proposal and the similar one by Hartle and Hawking. The main lessons here were that the introduction into the calculations of imaginary time made the physical interpretation of the ideas difficult. Furthermore, the Hartle and Hawking calculation was for a quantum superposition of universes and no one has yet explained how the quantum superposition of universes evolves into the universe we actually observe.

As the party became trapped in imaginary time, the authors entered a cautionary meditation on the meaning of motion in a system that included everything, and

Figure 9.5

whether it was reasonable to expect everyday concepts, such as time, to have the same interpretation near the big bang when clocks were unavailable.

At roughly the same time, Hawking asked whether the no-boundary proposal (which determined which geometries should figure in the quantum superposition of universes) ruled out the need for a creator. However, the authors asked who chose the no-boundary boundary condition. If it takes the status of a law of nature, as it must, then the no-boundary proposal brings cosmology closer to religion because it represents a step toward teleology. Had Hartle and Hawking incorporated both initial and final boundary conditions into their model, its entire evolution would have been preordained. Mention of the "creator" led to a contemplation of the recent theological speculations of scientists and found them wanting. They seemed to add little to the age-old controversy between science and religion, nor did they consider non-Christian worldviews.

Then the expedition was recreated from nothing, and time restarted. The Knowledge Marker, however, appeared permanently frozen. Popping back into existence a few moments after everyone else, the hapless cosmologist/cosmetologist caught up with the party and claimed that at a level even below that of quantum gravity he had discovered the remains of a Cosmic Blow Dryer. He proposed this as a new explanation for the big bang, but you met his suggestion with skepticism and continued strolling toward the beach.

WHAT DO YOU MEAN?
Metaphors, Analogies, Culture Wars

The folly of mistaking a paradox for a discovery, a metaphor for a proof, a torrent of verbiage for a spring of capital truths, and oneself for an oracle, is inborn in us.
—*Paul Valéry*

Analogies prove nothing, that is quite true, but they can make one feel more at home.
—*Sigmund Freud*

The Question

At some point in your life it is likely that you have attended a concert given by a large symphony orchestra. The lights dimmed, the audience fell silent, the conductor gave the downbeat and the ensemble launched into the music. You did not know the composer, or particularly care, but as the music continued, it seemed to you to mean something. Wildly shifting emotions, beyond anything in pure music, convinced you that a story was unfolding. The anguished chords of the opening, the soft, tender, almost nostalgic second theme peering into a deep past. A scherzo, nightmarish, wind whistling through cracks in a bleak, cold apartment; here something horrible had taken place. The triumphant marching songs of the finale, running straight into the arms of Death, the trajectory of a revolution gone wrong. All these pictures ran through your mind, but when you opened the program notes the annotator claimed that the composer was merely dealing with musical ideas, signifying nothing.

Strangely, at a gala concert to celebrate the expedition's return from imaginary time, you had the strongest sense of déjà vu. The Academy orchestra seemed to be playing the same piece you had once heard, and your reaction and the program notes were identical. As the concert ended, academicians began arguing about what the piece meant, an argument that followed you in your dreams and was continuing in full force when you awoke in the morning.

The first sight that greeted you at dawn was Igor Stravinsky proclaiming to a circle of admirers, "Music is, by its very nature, essentially powerless to express anything at all. The phenomenon of music is given to us with the sole purpose of establishing an order in things."

Others present emphatically agreed with Stravinsky. "There is no such thing as 'evil' or 'cruel' music. That makes as much sense as saying, 'Here is a bad electron.'"

You were beginning to wonder what it was you saw in the music, when other academicians joined the growing fray and claimed, "But anyone who listens to the Nocturne from Stenhammar's *Serenade* sees the same picture—a Swedish evening at summer's end. All the commentators have commented on it."

"And what about Dukas's *Sorcerer's Apprentice?* Surely no one can listen to it without seeing the apprentice chopping up thousands of marching brooms."

"But that is only because Dukas wrote the music to follow the poem and Disney made the film to follow the music. None of these pictures exist in the music itself. They are conjured up merely by association with preexisting images."

"Universal images."

"The collective unconscious."

"Speaking of which," chimes in another academician, "how is it possible that the New Age is able to mix up science and mysticism? This seems to be far worse than imparting images to music. Science has nothing to say about art, ethics, morality or mysticism."

"What a statement! Science has had an extraordinary impact on twentieth-century art."

"Impact, perhaps. But this is all by way of metaphor. There are no logical implications of science for anything except technology."

And thus as the sun rose on the morning of the Tenth and final Debates, the topic seemed to be metaphor and the connection between science and humanities.

A Stroll Through the Marketplace

Your friend had vanished overnight, leaving you with only a note that read, "Imaginary time was too much for me." So you headed alone toward the central plaza to discover it was market day and everyone present. Scientists, writers, literary critics, New Agers, politicians. At the stalls, everyone was hawking their ideas, haggling over the prices and plagiarizing. Someone had scrawled over the Knowledge Marker, "All Knowledge Is Relative." Across the way, at the Pennsylvania convention, John Adams and Benjamin Franklin are having it out over the form of the future legislature of the United States. The year is 1776. We tune in.

Delegates have just tabled a motion to add another assembly, under the name of senate or council, to the already proposed House of Representatives. Franklin, playing to the audience, appears diffident to the proposal. "Two assemblies," he recounts, "appears to me like a practice I have somewhere seen, of certain wagoneers who, when about to descend a steep hill with a heavy load, took one pair of oxen from before and, chaining them to the hinder part of the wagon, drove them up the hill; while the pair before and the weight of the load, overbalanced the strength of those behind. In this way, they drew the wagon slowly and moderately down the hill."

To Adams, Franklin is a man who "loves his ease, hates to offend and seldom gives any opinion till obliged to." Ol' Ben's anecdote causes him to lose patience and Adams, having studied physics at Harvard, replies to the Philadelphia sage: "The president of Pennsylvania might, upon such an occasion, have recollected one of Sir Isaac Newton's laws of motion, namely—'that reaction must always be equal and contrary to reaction,' or there can never be any *rest*."

"Mr. Adams," responds one of the natural philosophers observing the convention from the visitors' gallery, "a Harvard education is evidently not what it is reputed to be. Motion at constant velocity—which includes rest—is possible, verily common, when no force whatsoever is acting on a body. Rest would indeed seem more consistent with the inertia of our government, or as Mr. Franklin's colleague James Wilson has put it, the '*vis inertia* in publick bodies.'"

But Adams is far from alone in enlisting Newtonian physics for extrascientific purposes. Elsewhere on the market it is 1713, the year in which the second edition of Newton's *Principia* appears. Solipsist George Berkeley, addressing morning shoppers, sets forth a social system constructed in analogy with Newtonian gravitation. Just as there is a law of gravity in the material universe, there is "a principle of attraction" in the "Spirits or Minds of men." The attraction draws people together into "communities, clubs, families, friendships, and all the various aspects of society." What's more, "the attraction is strongest between those [bodies] which are placed nearest to each other." Berkeley goes on to elucidate a number of conclusions from his analogies regarding society, from the love of parents for their children, to the concern of one country for the affairs of another.

Leaving Berkeley behind, you soon stumble across Adam Smith, in the economic sector of the marketplace, who lectures those around him: "The natural price is the central price to which the prices of all commodities are continually gravitating." Afterward William Stanley Jevons steps up to argue for the introduction of differential calculus into economics on the grounds that it works for mechanics and that equations of the same form arise in both mechanics and economics. As your morning stroll continues, it becomes evident that Hume, Fourier, Montesquieu have all explicitly modeled their social theories on Newtonian physics.

"Hear ye! Hear ye!" a member of the iatrophysical school of medicine shouts, capturing your attention. "As you know, the bodily processes—locomotion, respiration and digestion—are purely mechanical. The body is a machine, powered by the life force, the élan vital. Allow me to sell you this bottle of élan lubricant, which enhances the circulation of the life force to the benefit of all extremities."

"Well, it doesn't sound *too* unreasonable," you say to yourself.

And here also in this Age of Reason, Jefferson: "We hold these truths to be self-evident." A reference to Locke or to Newton and the axioms of mathematics? Jefferson's library holds both the *Principia* and *Opticks,* and in his office as Secretary of State hang portraits of Locke, Francis Bacon and Newton.

At another quarter of the marketplace, in the waning half of the nineteenth century, the concept of social Darwinism is having enormous impact on all aspects of society. "Survival of the fittest" becomes not just an evolutionary principle but a battle cry. On the right, robber barons cite Darwin as a justification for economic exploitation of the lower classes and to buttress their opposition to child-labor laws. On the left, Karl Marx wishes to dedicate volume one of *Das Kapital* to Darwin.

"But which is the correct inference?" you wonder as George Bernard Shaw advocates his brand of socialism with the help of "creative evolutionism."

The year is 1880, more or less. Hermann Helmholtz, Charles Howard Hinton, H. G. Wells, Poincaré and theosophists begin to popularize the higher-dimensional geometries of Riemann, and the idea of a fourth spatial dimension captures the world's imagination, far outstripping the public fascination with black holes one hundred years later. In 1909 *Scientific American* offers a prize for the "best popular explanation of the Fourth Dimension." You do not enter. Now in 1915, the suprematists, led by Kasimir Malevich, are hanging their red and white rectangles on clotheslines strung from the Knowledge Marker to the palazzo. Intending to use higher dimensionality as a way of "liberating art from the ballast of the representational world," they have given their works subtitles like "Color Masses in Four Dimensions." You find the paintings more two-dimensional than four, but the Bauhaus architects are deeply impressed and incorporate the suprematist style into their designs, which propagate throughout the world.

Closer to the year 2000, yet another debate has erupted in the vicinity of the Oriental market. Physicist David Bohm attempts to make an analogy between thought processes and quantum mechanics. "If a person tries to observe what he is thinking about at the very moment that he is reflecting on a particular subject, it is generally agreed that he introduces unpredictable and uncontrollable changes in the way his thoughts proceed thereafter. . . . If we compare (1) the instantaneous state of a thought with the position of a particle and (2) the general direction of change of that thought with the particle's momentum, we have a strong analogy [with the Heisenberg uncertainty principle]."

Following Bohr, David Bohm speculates that the analogy may be deeper and that the neurons in the brain must be described in an essentially quantum mechanical way. Amit Goswami, not unexpectedly, pushes the model to extremes, attempting to sell A-mart shoppers the idea that the brain-mind must be deeply quantum mechanical. Among other things the "normal modes" of the mind (the fundamental vibrational states; according to Goswami, "mental quanta") are the Jungian archetypes.

The name Jung is also heard nearby, where Combs and Holland hawk their wares: "We live in a world in which, for the last fifty or sixty years, subatomic physics has described a universe founded at bottom on acausal connections, on paradoxical and seemingly illogical relationships and observations. Yet as a culture we still deny acausal, symbolic connections as part of our lives and the lives of our souls."

Deepak Chopra refuses to be outdone. "The coexistence of opposites—stillness and dynamism at the same time—makes you independent of situation, circumstances, people and things. When you quietly acknowledge this exquisite coexistence of opposites, you align yourself with the world of energy—the quantum soup, the non-material non-stuff that is the source of the material world."

A Recent Controversy

You weren't entirely certain you understood Chopra, or Goswami for that matter, but shortly afterward things took a turn for the worse when nearby, in a more staid quarter of the market, a whirlwind swept up everything in its path. Cultural critic and oracle Jacques Derrida had been deconstructing physics: "The Einsteinian constant is not a constant," he said, "is not a center. It is the very concept of variability—it is, finally, the concept of the game. In other words, it is not the concept of some*thing*—of a center starting from which an observer could master the field—but the very concept of the game. . . ."

Physicists turned to their colleagues and asked, "Is this supposed to mean something? What is the Einsteinian constant?" One physicist, Alan Sokal of New York University, struck back with a celebrated hoax in which he posed as a deconstructionist, extending the conclusions of Derrida and others to the ultraviolet: "In this way the infinite-dimensional invariance group erodes the distinction between observer and observed; the π of Euclid and the G of Newton, formerly thought to be constant and universal, are now perceived in their ineluctable historicity; and the putative observer becomes fatally de-centered, disconnected from any epistemic link to a space-time point that can no longer be defined by geometry alone."

Physicists and mathematicians—knowing that π and the gravitational constant, G, have been, are, and always will be constant—died of laughter.

For those who remember, Sokal's hoax made national headlines in 1996, was debated for a few months in *Time, Newsweek* and the *New York Review of Books,* then vanished from the public radar. But in the pages of *Physics Today* the discussion went on for a further year and at the Academy, as you might guess, culture wars are perpetual.

To be sure. Dazed by the cornucopia of verbiage issuing from the marketplace this morning, you decided to retire to a nearby café to contemplate all that you heard. But your plans for quiet mediation were thwarted by a group of academicians also just arriving from the plaza. Emotions at fever pitch, several sonic hedgehogs then and there proposed banning all deconstructions of physics from Academy grounds. "This is millennium madness!" Several others added New Age interpretations of quantum mechanics for censure. Everyone cited Steven Weinberg's essay on Sokal's hoax, in which he wrote: "Those who seek extrascientific messages in what they think they understand about modern physics are digging dry wells. In my view, with two large exceptions, the results of research in physics have no legitimate implications for culture or politics or philosophy."

But as you observed the carnival, you felt the strong urge to remind everyone that millennium madness and the digging of wells, dry or otherwise, had evidently gone on for some centuries. It also struck you that on the market you witnessed an entire spectrum of "scientific transpositions," from John Adams on Newton's third and government equilibrium, to David Bohm's analogy between quantum mechanics and thought processes, to Amit Goswami's farther-out speculations on the quantum mind, to the artistic community's appropriation of the fourth dimension. Was there validity to any of it?

To the critics on the left, who seemed to you mostly foxes, yes. At Weinberg's declaration they raised a great outcry, accusing him of promoting scientific/cultural dualism and they insisted that quantum mechanics is loaded with "telling cultural implications." To which Weinberg replied that quantum mechanics is nothing more than a mathematical formalism and it is simply not possible for it to have any cultural implications.

You could see that the response left the foxes highly dissatisfied; but Weinberg is a careful man. When he said "legitimate implications" he meant legitimate implications, not metaphorical transcriptions or inspiration. "I would agree that anyone is entitled to draw any inspiration they can from quantum mechanics, or anything else," he answered. " . . . But there is a difference between inspiration and implication. . . . There is simply no way that any cultural consequences can be *implied* by quantum mechanics."

As shouting engulfed the café and the mob spilled onto the streets, you were struck that nobody paid any attention to the physicist's willingness to grant two exceptions to his rule against implication and you inquired what these were. Weinberg's first exception, drowned out in the clamor, is "jurisdictional": sometimes discoveries in science reveal that topics (such as space and time) previously thought to be philosophical actually belong to the province of science. The second exception is the profound influence on culture of the discovery, dating to the time of Kepler, Galileo and Newton, that nature is governed by impersonal, mathematical laws. As Weinberg said elsewhere, "The product of our work [as scientists] is a world view that has led to the end of burning witches and . . . to the fostering of liberal democracy—or at least to an understanding that we are not living in a world with a nymph in every brook and a dryad in every tree. I feel that this, above all, is the thing about which we scientists can be most proud." But any advance Einstein might have made over Newton, or the difference between classical and quantum mechanics, "as far as culture or philosophy is concerned is immaterial."

Many visitors are unmoved at what scientists look on with pride and freely admit they would prefer a world populated by nymphs and dryads. Even those sympathetic to Weinberg find his last remark hard to swallow. Surely, relativity theory has forced an enormous rethinking on the part of philosophers about the nature of space and time. Chaos theory has forced a reexamination of such notions as determinism.

Philosophy be damned. It is really Weinberg's disassociation of modern physics from culture that has caused the foxes to go visibly nonlinear. You too remain perplexed, especially by his last statement, and ask, "Why Newton and not Einstein? Didn't Einstein's theories have enormous cultural impact? Cultural relativism and so on?"

Once again, Weinberg distinguishes between implication and impact. He might reply, "Imagine Einstein's theory suddenly turned out to be wrong. Would it make any difference in terms of what artists and politicians say about it?"

In light of John Adams one might ask the same regarding Newton. Whatever, a genuine question has emerged. The theory of relativity could have equally well been called "the theory of invariants." (In 1910 the mathematician Felix Klein did suggest that name.) Many fundamental quantities of the theory, in particular the speed of light, remain unchanged, or invariant, to all observers in special relativity. Relativity deals as much with what is absolute as with what is relative.

One can only wonder what the impact on twentieth-century art would have been had Einstein settled on "the theory of invariants."

Are Metaphors from Science Valuable?

And so, escaping again to the market, you began to wonder. Academicians had been legalistically haggling over the distinction between implication and inspiration. But is this the central issue? Is it not possible that inspiration ultimately proves more important than implication? Is not the more interesting exploration to learn by what means science and culture cross-fertilize each other?

And so we ask, Suppose relativity had been named "the theory of invariants." Would the novel have remained closer to its deterministic nineteenth-century form than its more chaotic twentieth-century equivalent? Would the phrase "everything's relative" have come into widespread use and would people have used it to advocate moral relativism? Not long ago, President Jiang of China cited Einstein to justify the imprisonment of political dissidents. "The theory of relativity worked out by Mr. (Albert) Einstein, which is in the domain of the natural sciences, I believe can also be applied to the political field. Both democracy and human rights are relative concepts and not absolute and general."

This is not merely harmful but reprehensible. In coopting the term "relativity," Jiang, like many before him, has attempted to put his beliefs on a scientific basis, just as the advocates of social Darwinism did a century ago. But all he has done is borrow a technical term out of context with no understanding.

Though the consequences of his statements are not nearly so dire, much the same can be said about Deepak Chopra and his morning remarks on the coexistence of opposites ("stillness and dynamism") and your alignment with the world of energy—the quantum soup. Chopra is evidently referring to the verses in the Bhagavad Gita that go

> The wise see that there is action in the midst
>   of inaction and inaction in the midst of action.
>   Their consciousness is unified, and every act is
>   done with complete awareness.

And

> Neither the sense of acting, nor actions, nor
>   the connection of cause and effect comes from
>   the Lord of this world. These arise from nature.

The essential idea is that when you are aware of things in entirety you do not consider yourself to be the actor, but you do the right action. Once you remove yourself from the agency and go with the flow, there is nothing you cannot do because what you want to do is the thing that is going to happen.

Wisdom is here in abundance, but what does it have to do with the quantum soup? Chopra's quantum soup is "the non-material non-stuff that is the source of the material world." This does not mean much. Although people routinely use the word "energy" to refer to something nonmaterial, like consciousness or the ur-stuff of the world, to a physicist energy is no more immaterial than matter; the quantum soup is as material as anything else. Moreover, Chopra uses the term "quantum" in exactly the opposite sense that physicists use the word. Quantum has been used since Planck to designate the smallest building blocks of a system, discrete units. Chopra plays at Humpty Dumpty by changing the meaning of the word to "integral" or "holistic."

Nevertheless, there is a sense in which Chopra is right. The most interesting property of quantum mechanics is not the discreteness of phenomena, but the wholeness of phenomena. Even though you cannot tell where a photon is, all of it is absorbed in one place on a photographic plate. The development of a Polaroid photo is gradual and occurs in grains, but the whole picture develops together. Two photons may be separated by light-years, but their measured spins may be bound together.

Chopra arrived somewhat late on the scene. Those who have read H. G. Wells's *Time Machine* and "The Plattner Story," Edwin Abbott's *Flatland* and even Dostoevsky's *The Brothers Karamazov* have asked themselves how such a concern with higher dimensions arose decades before Einstein. Art historian Linda Henderson has provided a compelling answer. As you have already discovered in your morning stroll through the marketplace, a great four-dimensional fervor swept across Europe and America in the late nineteenth and early twentieth centuries. The fascination with higher dimensions began in the 1860s, when several authors, especially Hermann von Helmholtz, began to popularize the concepts of Riemann's non-Euclidean geometry. Charles Dodgson (Lewis Carroll) and others took up the cause and directly inspired Edwin Abbott to pen *Flatland*, his 1884 *Romance of Many Dimensions*. Although the idea of time as the fourth dimension goes back at least to d'Alembert in 1754, among the late-nineteenth-century writers, only Wells seems to have conceived the fourth dimension as time. To everyone else, in particular Charles Howard Hinton, "higher dimensionality" referred to space—and spirituality.

Hinton styled himself "the first hyperspace philosopher." "I shall bring forward a complete system of four-dimensional thought—mechanics, science, and art," he wrote. "The necessary condition is, that the mind acquire the power of using four-dimensional space as it now does three-dimensional." He went on to say that "we can never see four-dimensional pictures with our bodily eyes, but we can with our mental and inner eye." Hinton believed that it is our inability to perceive the fourth dimension that limits our perception of reality. Perhaps our successive mental states are due to the passage of our four-dimensional beings through the three-dimensional states to which our consciousness seems confined: "We must really be four-dimensional creatures or we could not think about four dimensions."

Today it seems a little silly, not a description of any universe we accept. But Hinton's idea that life and thought can be freed only by going to a higher dimensionality had great impact on artistic movements of the early twentieth century—the cubists, futurists and suprematists. Marcel Duchamp explicitly tried to develop a "four-dimensional perspective" in his art—hence the "exploded" aspect of many of his paintings—four dimensions passing through three.

In Russia, Hinton's influence was felt through the hyperspace philosopher Piotr Ouspensky, whose ideas surfaced in the futurist and suprematist movements. "Ouspensky," writes Henderson, "posits a true reality that is immobile and constant, with the illusion of change resulting simply from our temporarily limited powers of perception. The very notion of time is transitional and will recede as our spatial understanding enlarges."

You cannot help but sense the resonance with Fritjof Capra's statement that "the space-time of relativistic physics is a similar timeless space of higher dimension. All events in it are interconnected but the connections are not causal."

What is most striking about works on the interaction of science and the humanities by art historians, artists or musicians, as opposed to essays on the same subjects by their scientific counterparts, is that writers in the humanities never question the validity of transposing science to other fields by metaphor. That is simply the way of the world.

It is true that the hyperspace philosophers and artists misconstrued the science, as do those of the New Age who misunderstand quantum mechanics today. But that it is easy to get the science wrong seems to be the minor lesson here. Contrary to Weinberg's remarks, it is difficult to see how the fostering of liberal democracy is a logical implication of even Newtonian physics. But that the scientific worldview has fostered liberal democracy is true nonetheless. In the 1980s, before the collapse of the Soviet Union, the names of political dissidents were often in the news worldwide. Heading the list was Andrei Sakharov, but also prominent were Anatoly Shcharansky, Yuri Orlov and Fang Lizhi. Not so well known in the West but also important were Valery Chalidze and Cronid Lubarsky. The most striking thing about this list is that everyone on it is a physicist.

Not all dissidents were physicists, of course, but a disproportionate number were, and this is no accident. A few, like Sakharov, became dissidents precisely because they came to regret their participation in weapons of mass destruction. However, there is more. The assumption of basic principles distinguishes scientists from almost all other professionals, call it their acceptance of natural laws, their understanding that the theory of relativity might have been called the theory of invariants. The habit of working by universally accepted laws and principles engenders a respect for one's colleagues, for one's rivals, and ultimately for human beings. Is this an implication of Newtonian mechanics? No. But it *is* a result and one that we should indeed be proud of. Before one dismisses all transpositions of physics into other domains, by the New Age and cubists, one should remember that John Adams from Harvard misconstrued Newton's laws. As a result the legislature of the United States has two houses and the government has a system of checks and balances.

The Use of Analogy in Science Comes Under Inspection

Try as you might, you cannot escape the mob of academicians. On the side streets and alleys, barricades have gone up, foxes on the one side, hedgehogs on the others and yourself caught in the middle. Representatives from both camps are calling for the partition of the Academy into two independent entities. Insults are hurled against the authors for discussing the influence of sciences on the humanities but not vice versa.

The authors appear with a white flag, promising to take up the topic in a few moments but protest that more remains to be said about metaphors. The foxes agree but, to show their resolve, shove forward a hostage; it is Alan Sokal, whom they have captured and bound. One of their camp unfurls Sokal's text and reads aloud his attack on Robert Markley, who has written somewhat incomprehensibly, "Quantum physics, hadron bootstrap theory, complex number theory, and chaos theory share the basic assumption that reality cannot be described in linear terms, that nonlinear—and unsolvable—equations are the only means possible to describe a complex, chaotic, and non-deterministic reality." Yet, buried amidst Markley's verbiage is the further contention, "These postmodern theories are—significantly—all metacritical in the sense that they foreground themselves as metaphors rather than as 'accurate' descriptions of reality."

"Repent!" cries one of the foxes, prodding Sokal with a sword. "Have you not gone too far? Have not the authors constantly emphasized that models are imperfect representations of reality? Are not models themselves metaphors?"

Sokal remains proudly silent, but the authors diplomatically step in, displaying unusual solidarity with the foxes. To be sure, hedgehogs are traditionally suspicious of metaphors, the equation of two unlike concepts. "Glory is like a circle in the water/Which never ceaseth to enlarge itself/Till by broad spreading it disperse to naught."

"That is a simile!" the foxes chorus from behind the barricades.

True enough, agree the authors, retreating a step, but the principle is about the same. The difficulty with regarding models as metaphors is that, unlike the case of Shakespeare's glory and water ring, it is not always clear what the model is a metaphor of.

"Reality," ripostes an ultrasonic hedgehog.

That does not help much. As candidates for metaphor, scientific models more closely resemble music than figures of speech. What was that piece of music at the concert representing? Who has knowledge of it? If one accepts the metaphorical content of music it is not too difficult to accept the metaphorical content of scientific modeling.

"Oh come on," remonstrates one of the scientists from beyond the hedgehog barricade. "You cannot possibly assert that science requires metaphors. Regardless of the *process* of model building, the result is expressed mathematically, with precision."

Philosopher Max Black, who considered this question in the 1960s, is nowhere to be found, but he might answer, "The mathematics may be precise, but it does not represent anything more than the model. Mathematics is merely a language; it can be used to model anything and the use of it does not imply the model is physically meaningful. In that sense the mathematics is the figure of speech—or the music." Not only is a mathematical model not necessarily physically meaningful, but the authors have often described instances in which the same mathematics could be used to model many physically distinct systems. This implies that something is missing in the mathematical description. Metaphors are always imperfect.

The hedgehogs remain riled and fire a few potshots into the air, but they do not produce a response.

"What about analogies?" you ask innocently from your position in no-man's-land.

"Always!"

"Never!"

It may be putting too fine a point on it to distinguish analogy from metaphor, but if a metaphor is the equation of two dissimilar things ("All the world's an academy") then an analogy is more specific, positing a similarity in form or function between two aspects of dissimilar concepts. Scientists use analogies all the time. Faraday searched for the relationship between electricity and magnetism, convinced that their reciprocal action was analogous to Newton's third law. The Rutherford model of the atom—a nucleus surrounded by orbiting electrons—was seen as an analogy to the solar system. The entire program of unification via symmetry is one of extending an analogy. The list is endless.

That science proceeds by analogy simply cannot be denied; the only question is whether the analogies are productive or infertile. And whether the analogies are mistaken for the real thing. We have often spoken about the spin of particles, and how particles can in some ways be viewed as tiny, spinning tops. And we have also cautioned against taking this analogy too literally—what does it mean to say a particle of zero breadth is spinning? Yet scientists have often failed to distinguish analogy from identity. Wolfgang Pauli himself objected to the idea of electron spin on the grounds that, if an electron were spinning, the velocity of the electron's surface would exceed the speed of light—an impossibility.

If scientists shun metaphors, misadventures like Pauli's have also left them with a lingering suspicion of analogy. Just how lingering is evident when the hedgehogs bring forward their own hostage, biologist Rupert Sheldrake, who in 1981 proposed the "morphic field." Fields are unquestionably the New Age's most coveted analogy and Sheldrake's field controls the development of an organism by influencing DNA and even brain activity. He has said, "Morphic resonance theory would lead to a radical reaffirmation of Jung's concept of the collective unconscious."

Sokal, even in captivity, remains unrepentant, and scornfully refers to the morphogenetic field as "the quantum counterpart of Einstein's gravitational field." Appearing before the opposing forces, Steven Weinberg labels the morphogenetic field a "crackpot fantasy."

The New Age reaction is swift: seizure of the administration building and the issuance of demands—the Academy should immediately be reconstituted as a personal growth center. Recalling the night of splitting universes, you experience a strong sense of déjà vu.

However, from a hundred and ten leagues a bird sees carrion far below; the very same bird near its end is unable to recognize a trap. It is true that the morphic (or morphogenetic) field is almost certainly a crackpot fantasy. But there are no natural laws that tell us where to draw the line. Biology has provided more than its share of analogies. The Earth as an organism, the Gaia hypothesis, championed by Vladimir Vernadsky, James Lovelock and more recently Lynn Margulis, has received both derision and praise. Al Gore transposes Gaia to the electronic realm, proposing that contemporary society might best be characterized as a "distributed intelligence." Lamenting that scientific concepts no longer inform public discourse, Gore sees "distributed intelligence" as an explanation of phenomena as diverse as the Internet and *America's Funniest Home Videos.* Biologist Richard Dawkins, refusing to cede place, has introduced the concept of "memes," ideas that behave like genes and propagate through the culture pool.

The evolutionary analogy has recently been taken to extremes by cosmologist Lee Smolin, who hypothesizes "natural selection of universes." Not mere species but entire universes evolve by natural selection. Smolin's mechanism requires baby universes sprouting from the interior of black holes, and fundamental "constants" that evolve from the parent universe to their progeny. Universes whose natural constants result in maximal production of black holes (and hence offspring) are those favored by natural selection.

Rupert Sheldrake's idea may be crackpot, but it has the virtue of being comparatively easy to falsify. Does Vernadsky's? Does Gore's? Does Dawkins's? Does Smolin's?

The dangerous face-off continues. Deepak Chopra appears, heading a column of supporters: "Through meditation," his soothing voice assures, "you will learn to experience the field of pure silence and pure awareness. In that field of pure silence is the field of infinite correlation, the field of infinite organizing power, the ultimate ground of creation where everything is connected with everything else."

The hedgehogs begin firing, but the bullets harmlessly pass through the prophet. Why are scientists so opposed to Chopra? His extension of scientific terminology and concepts beyond anything that could be called legitimate analogy is a genuine cause for concern. But apart from that, as Paul Davies has observed, "most scientists have a deep distrust of mysticism." They share this distrust with many nonscientists who also find implausible the mystics' claim "that they can grasp ultimate reality in a single experience, in contrast to the long and tortuous deductive sequence of the logical-scientific method of inquiry."

The subjective nature of the ascent to Reality is undeniably the primary reason skeptics remain hostile toward mysticism. But suppose we reworded Chopra's statement slightly: "Through observation you will learn to experience the harmony and order in the universe. In that state of appreciation of cosmic order, you will sense that this order is the sign of a divine plan, in which everything is adjusted to make life possible."

This, of course, is nothing more than the argument from design, the ultimate analogy, to which many contemporary scientists subscribe. Is belief in design more logical than belief in revelation?

Is Science Relative?

Peace negotiations between the foxes and the hedgehogs had broken down, the administration building remained occupied and no end to the crisis was in sight. As the stalemate dragged on, a gang of foxes unexpectedly sallied forth from behind the barricades and took the authors hostage, demanding that they fulfill their vow to consider the metaphorical transposition of concepts from the humanities to the sciences. As voiced earlier, the argument appeared to be a one-way street: science influences the humanities but not vice versa. There has been an undeniable and long-standing conceit among scientists that science's impact on the world has been far greater than any impact the world has had on science. Science, especially the physical sciences, advances by logic, experiment and mathematics, and its universality is impervious to cultural influences.

It is this conceit above others that has enraged the cultural critics and that lies at the heart of the current hostilities. One naturally wonders whether the Academy would at this moment be on the verge of disintegration had Einstein termed his theory "the theory of invariants," but history has settled on relativity and for the past decade the foxes have been keen to recast the products of science as social artifacts.

Social historian Andrew Pickering, a relativist, has claimed that the standard model of particle physics "should be seen as a culturally specific product . . . a communally congenial representation of reality." Furthermore, "the preponderance of mathematics in particle physicists' accounts of reality is no more difficult to understand than the fondness of ethnic groups for their native languages." The "communally congenial representation of reality" can be seen as no more than "opportunism in context."

Steven Weinberg, atop the opposing barricade, replies: "Whatever cultural influences went into the discovery of Maxwell's equations and other laws of nature have been refined away, like slag from ore. Maxwell's equations are now understood in the same way by everyone with a valid comprehension of electricity and magnetism. The cultural backgrounds of the scientists who discovered such theories thus become irrelevant to the lessons that we should draw from the theories."

Despite the fact that the authors are sitting at gunpoint, blindfolded among the foxes, with a few caveats they bravely agree with Weinberg. It is easy to imagine that had physics been created by third-world women instead of dead white European males, the subject might have evolved along very different lines. Rather than looking at the sky, perhaps physicists would have begun by looking at the hearth. Instead of trying to explain the solar system, physicists might have attempted to understand why heat is lost up a chimney. But ultimately someone would have asked why it is more difficult to boil an egg atop a mountain than at sea level. Regardless of what language they used to devise any theory to explain this phenomenon, if it correctly conformed to the observations, it would be translatable to our own language of gas laws and thermodynamics.

But here, climbing into the fox camp under a flag of truce, you object. The authors have spoken frequently of the shell model, which posited that within the nucleus resides another nucleus, just as the nucleus itself resides within the atom. Such a nucleus-within-the-nucleus does not exist. Is it not conceivable that a different culture could develop a model of the nucleus that bore no resemblance to the shell model and could not be thought of as equivalent to it? Is it not possible that a version of QED could have been invented that did not require renormalization?

You have a point. There is no obvious reason why another culture should come up with the nonphysical shell model; a reasonable culture probably never would conceive of it. And it is also possible that another culture might invent an "already renormalized" version of QED, after which our own version might—as Dirac claimed—be seen as a fluke.

But what all the models have in common, at least reasonable ones, is that they conform to the fundamental laws of nature. Regardless of what differences might exist in physics developed by third-world women, regardless of what language they expressed them in, they would arrive at the same physical laws. The laws are simply universal. It is often pointed out, for example, that the Hopi Indians lack verb tenses in their language and that the concept of causality is absent. Even if true, it does not exempt the Hopi Indians from any strictures of relativity, chaos theory or quantum mechanics.

A second caveat to agreement with Weinberg. Joseph Needham concluded that the birth of physics was intimately connected with the Western concept of God. Moreover, some areas of modern physics rely ever more on mathematical consistency and less and less on experimental evidence: cosmology and superstrings. To John Horgan these theories are "ironic"—divorced from Truth. If our theories are, or become, ironic, it is easy to imagine that cultural biases will help shape their content and form. Hartle and Hawking's no-boundary proposal, if it cannot make testable predictions, might be seen as some sort of cultural artifact, related to the Western concept of God, though we would be hard pressed to be more specific.

The foxes, baring their teeth, now threaten the authors with death; we have egregiously minimized the cultural effects on science, just as our predecessors have done. No. We have merely been talking about the results of physics, not the process. The fact that China did not invent modern science is probably the best example of the profound influence culture can exert on the course of science. Neither can one deny the influence of fashion and national styles on the course of research. The European school of physics is characterized by a strong blending of philosophy and science. It also supports more traditional areas of study that are looked on as somewhat backwater in America, for instance hydrodynamics and thermodynamics. Ilya Prigogine is probably the most famous exponent of the European school and his views of thermodynamics, presented in the Fifth Debates, trace their lineage directly to Max Planck. The American school, on the other hand, is characterized by preference for cutting-edge fields, of which particle physics has traditionally occupied pride of place, a reliance on sophisticated mathematics and a minimum of philosophy. Steven Weinberg is the most visible representative of the American school.

These debates have also been colored by the Indian "Vedantic" school of physics, not just "Thou Art That," but the belief that Reality cannot be perceived through science alone. What's more, the Upanishads are many splendored, but not a unified, integrated work. As a result, Indian philosophers tend not to offer overarching syntheses of the Vedanta; more characteristic is Shankara's Brahma Sutra, a thousand-page debate on the Upanishads, word by word. Yet a god of small things is still a god. To a large extent these debates have attempted to pinpoint gaps in our understanding that tend to get papered over and mistaken for solid ground. From these gaps, we hope, emerges a conviction that the universe is a richer and more ambiguous place than often acknowledged.

And now, having paused during the exposition, the foxes vote to dismember the authors. The previous exposition has been nothing more than an apology for the universality of science. Produce a genuine, specific cultural influence on science or die. Unfortunately, far less research has been done on the influence of culture on science than vice versa. One study by now famous in physics circles is Paul Forman's of 1971, *Weimar Culture, Causality and Quantum Theory.*

Forman argues that in the wake of World War I German scientists were utterly disillusioned with Germany's technological and military prowess and, with it, the exact physical sciences. Oswald Spengler's apocalyptic *Decline of the West* took Germany by storm, running through sixty editions in eight years. Everyone read it. Spengler declared that the decay of the West must bring with it a collapse of mathematics and the physical sciences. In particular, the "concepts of mass, space, absolute time, and causal natural laws" had reached the limits of their possibilities. According to Forman, Spengler exerted an enormous influence over Hermann Weyl and Erwin Schrödinger, who began to call for a radical restructuring of physics in which causality would be swept away. A few years later quantum mechanics was born.

None of this is to deny that crisis in physical theory itself was indispensable to the creation of quantum mechanics, but Forman makes a convincing case that social factors also played a major role. Fifteen years ago Freeman Dyson speculated that Capra's *Tao of Physics* might also lead to a revolution in physics. So far this appears not to have happened, but it may be that the New Age movement has helped turn scientists' attention away from the study of elementary systems and toward the study of complex systems, and to seek new basic laws in this domain. It is also quite clear, regardless of how the laws of quantum mechanics came to be, they exist independently of the West, declining or otherwise.

There are a few other examples of how extrascientific influences helped generate scientific discoveries. A student of Carl von Weizsacker reports that he got the idea for a theory of galaxy formation from watching cream swirling in a cup of coffee. The most famous story along these lines is how Friedrich Kekulé's dream of a serpent devouring its tail led to his discovery of the benzene ring. Hugh Everett is said to have received the inspiration for his many-worlds interpretation of quantum mechanics from Jorge Luis Borges's short story "The Garden of Forking Paths"; a passage from the story does preface the standard volume on the many-worlds interpretation. More substantially, J. V. Field argues that the Renaissance concern with perspective in art led directly to Girard Desargues's seventeenth-century invention of projective geometry. We mentioned in the Debates on Mathematics that observations of the stock prices on the Bourse led Louis Bachelier in 1900 to instigate the mathematical study of Brownian motion. It is also reasonable to suppose that the Japanese math tablets, the *sangaku,* hung as they were from Buddhist and Shinto temples, inspired mathematicians to tackle new problems and inspired young people to become mathematicians.

As Steven Weinberg said, one is entitled to draw inspiration from any well. What is more important to remember is that one *does* draw inspiration from *every* well. That is how civilization is created.

A Practical Exercise: Summarizing the Debates While Escaping the Academy

The authors' response had the effect of pacifying some of the more moderate foxes, but the radical cultural relativists remained intent on doing away with their hostages, and the foxes fell into argument. Taking advantage of the momentary confusion, you deftly slit the authors' bonds and the three of you made off down the street. The foxes were after you, but soon they encountered the antimob of hedgehogs and the collision produced many sparks.

As you fled the mêlée your only thought was to escape the Academy confines before the whole place burned down, but the authors suggested that this was the ideal moment to summarize the entire series of debates. You dimly recollected the moment when you passed through the Academy gates and found yourself embroiled in what was evidently an elementary dispute about scientific terminology. But counsel wrongly applied is like a vampire; it kills those who employ it, and the casual interchange of "law" and "principle" underscored the heavy reliance in science on the useful but untestable. The innocuous interchange of the words "theory" and "model" evidently feeds a confusion of models with the real world.

The distinction between models and Reality became a key theme throughout subsequent debates. The message that emerged after exhausting discourse is that scientific theory is much like art in progress. At each stage there are approximations, some disregard of features. The Earth is treated as a billiard ball until that idealization does not work, then more structure is brought in. Can this process result in the World Equation, the Equation of the Universe? No, it results in equations of the latest model.

Escape blocked by the scientific faction along one alleyway, you sped into another, only to face the humanist faction, and there you recollected another major theme that insidiously wound its way through the debates:

A working theory is not necessarily an understood theory.

Any number of working theories were examined, for example, quantum electrodynamics and general relativity, and it was found that basic questions remain about what the mathematical language describes.

This way! No, that way! Convenient or otherwise, physicists routinely claim that mathematics is the language of science. Bucking the general trend, the authors argued that mathematics is not the language of the World; mathematics is the language of models. Throughout your stay at the Academy, we were forced to point out instances where the mathematics did not describe anything that existed, or where the mathematics failed to distinguish among distinct physical systems. In almost all situations serious approximations were needed before one could get any sort of solution.

As you ran pell-mell through the marketplace, the authors took the opportunity to mention that the work of Desargues, whose name arose scant moments ago, is one of the best arguments against the equation of mathematics with Reality. With his invention of projective geometry Desargues showed that all theorems of Euclidean geometry remain valid if lines and points are interchanged. This proof was not only perhaps the first example of a duality transformation (a concept that later arose in discussion of the magical M-theory of everything), but it shows that the elements of a mathematical system have no absolute meaning.

Here, as the escapees rushed past the Knowledge Marker, on which was scrawled "Death Is Knowledge!," the idea that all knowledge can be boiled down to mathematics reminded them of another major artery of the debates: the conflict between reductionism and emergence.

Historically, reductionism has been the path of physics and chemistry, and it has had many successes, most obviously through Newtonian mechanics, relativity and quantum theory. Surely, if one knew all the ingredients of a system, it would be desirable to be able to reduce things to a few principles. But God has rarely provided lists and in some cases reductionism has failed. In those cases, scientists have retreated. For example, when attempts to reduce electromagnetism to mechanical phenomena failed, scientists declared dynamics to be more general than the dynamics of particles. Many of the debates were defined by a tension between reductionist forces and the nonreductionst opposition. The dispute over the direction of time was in part an argument between reductionists, who believe that the increase in entropy can be explained by an appeal to more basic laws, and antireductionists, who believe that time's arrow is in effect an emergent property of complex systems. The debate over whether chaos offers anything new or is merely a matter of not having sufficient time to do proper calculations also falls within this category. The use of symmetry principles, which surfaced time and time again, combined the notion that nature is mathematical with the notion that nature is simple enough to describe with symmetry groups. With equal frequency opponents found examples of asymmetry in nature or instances where nature was too complicated to be described by group symmetries.

Having gained the Library's spiral, it struck you that reductionism might best be viewed as a phase in theory building. It is useful as long as it is useful but must from time to time cede ground to other prescriptions. As Hermann Bondi wisely pointed out, the very success of science is due to its ability to say something without having to say everything.

For a moment you paused to catch your breath. On the field of battle below, a crowd of hedgehogs had collided with a phalanx of foxes, but the combined mob was now cornered by an influx of New Agers. You were reminded of the story of king Ajmal Hussein, who had a reputation of being wiser than all the scholars in his land and who thus incurred their envy and wrath. Convinced that true scholarship had long since disappeared from the face of the Earth and that only sham learning now existed, he invited questionnaires about his philosophy from fifty of his most famous professors and academicians. Ajmal Hussein answered them all differently, then convened a conference to allow the scholars to discuss his responses. At the conference, academicians presented fifty different versions of what he believed and each academician refused to give up his interpretation in favor of another. The result was the famous "brawling of scholars," which lasted five days. "And so," Hussein warned his acolytes, "you see that what matters to each one is his own opinion. They care nothing for truth. Do not become an authority. When you are alive, they torment you. When you are dead, they become experts on your work. Their highest motivation, though, is to vie with one another and oppose anyone outside their own ranks."

Nothing has changed, you perceived as the mob overturned the Knowledge Marker and you dashed up the Library ramp ahead of the authors. The basic intent of the New Age is to broaden the study of physics to include feelings, emotions and volition; hence the frequent attempts to connect quantum mechanics with phenomena like synchronicity. The desire to include such concepts, which have traditionally lain outside physicists' rather austere set of criteria, is not in itself bad. Eugene Wigner himself said on many occasions, "How can physics be complete when it doesn't concern itself with my happiness and my anxieties?"

Physical scientists have been a little like the person who, when remodeling the house, throws away a lot of old stuff in the attic, including valuable paintings; they have traditionally resisted the inclusion of mental phenomena in their theories. Indeed at this moment one sonic hedgehog is heard shouting above the mob, "Altered states do not exist. When your girlfriend goes to a hypnotist she is not hypnotized. She is merely falling prey to her own fantasies and delusions." Breathing hard up the ramp, the authors nevertheless find time to glance over their shoulders and return a parting shot: What nonsense! Even if she is merely fantasizing, that is an altered state and, as physicists, you should believe it is associated with some measurable change. If altered states do not exist, how did Wigner distinguish happiness from sadness?

The idea that mental states exist led to the authors' willingness to model the mind as a superfluid, for which the mob currently demands their heads, but the authors shout back that probably no scientist would object to a model of the mind as a collection of harmonic oscillators. In any case, the question is, What is gained by including feelings and so forth in a larger theory? Do feelings and consciousness have any role in structure of universe? We're not talking merely about the observer in quantum mechanics. Does the world have meaning? When you listen to a concert is there meaning in the music? Or is it just associations? Is there meaning in the pits on a compact disc? Does one's awareness alter the state of physical reality? If it does, the traditional program of physics has been unnecessarily constrained. If one could show that ESP or telekinesis existed, then we would certainly have to include them in a theory of physics. What part of us is not connected to the physical world?

The general program of the New Age, to come to grips with these questions, is admirable despite the incompetence of many New Age writings. In dismissing their claims out of hand, scientists run the risk of behaving like the physicists of the 1950s to whom parity violation was inconceivable. On the other hand, the New Age should not lightly dismiss accumulated scientific evidence in preference to something that glitters enticingly. In their attempts to put their beliefs on an acceptable scientific footing New Agers have oversimplified to the point of ridiculousness. Our attitude has been like that of the man from Missouri: show me. One of the main messages from the quantum debates is that quantum mechanics is more subtle than many things that have been said about it. Despite differences between quantum and classical physics, they share certain features, such as nonlocal phenomena, and even the beloved field is a model, not entirely free of ambiguities.

The fighting below had by now spilled onto the Library and you began to worry that in your precipitous flight you had unwittingly trapped yourself. The authors urged you to calm yourself and recontemplate the question of metaphors and analogies. Science does proceed by analogy. It is much like Miss Marple, who during an investigation frequently exclaims, "This reminds me of another case in my village." But as Freud says, analogies themselves prove nothing; they need to be confirmed by experiment. Miss Marple needs to prove her case. She always does; shouldn't we?

Scientists of course use analogies to model everything. Currently it is fashionable to view the brain as analogous to an electronic computer, but brain models have always been a reflection of their times; they have been mechanical, they have been digital, they have been analog, they have been neural nets, they have been holographic; the list is endless. Many feel that the modeling craze has been taken too far; when postmodern literary critics use mathematical equations in their texts, things have surely gone out of control. But one may at least ask, Are the New Agers worse?

Perhaps, it occurred to you, such transpositions should be viewed as metaphorical. Or maybe one should steer clear of metaphors altogether. That is the preference of scientists—anything nonscientists say about science, metaphorically or otherwise, is bound to be wrong. It is the precise issue that touched off the current fighting, which has got you fleeing to the upper reaches of the Library. All true. But if confining discourse of science to technical details would mean that art would be impoverished, we must accept the metaphorical transposition of science to other domains as natural and right, even while we speak out against the dangers that arise from such transpositions. The necessity of metaphor seems almost corollary to another of the authors' main contentions: that experience is meaningful, worthy of investigation, that Reality cannot be attained through science alone, that the universe is a rich place, infinitely richer than any theory or any work of art.

Weighing these words carefully, you pressed the authors, "Then you have nothing against applying the Heisenberg uncertainty principle to everyday life, or the second law of thermodynamics?"

As long as you don't use it to imprison political dissidents or misapply it to the design of aircraft. On the whole enlarging one's vocabulary and perception is desirable. It cannot be a bad thing to know that there are limits to precision, limits to knowledge. It cannot be a bad thing to know that miscommunication is not only common but inevitable. By the time an idea gets from brain to paper, some distortion has set in; perfection is impossible. The Indians have it that there are four stages from the perceived idea to the spoken word and something is bound to get screwed up along the way. The concept of entropy has been misapplied so often that it has become virtually meaningless. Yet, to be aware of the general idea that even when you recycle waste paper or aluminum some energy is lost could only be a positive influence on the environment.

"Hmm," you reflected, even as the contending forces followed to ever higher levels. "Tell me, we have discussed many things during these debates. Are there any other unanswered questions?"

"Oh, there are an infinite number of them," the authors assured you. "The academicians will never cease debating."

"Why don't you tell me some of them while we are waiting to be destroyed."

Very well. We have spent much time talking about particles and forces and fields. But in fact current theories are not able to describe ordinary radioactive particles. In quantum theory, decays are described as transitions from an initial state to a final state, both of which are assumed to be stable. But, as we mentioned in the context of universes tunneling from nothing, if the initial state is absolutely stable, then it must exist forever and so no decay is possible. So how radioactive atoms decay is actually not explained in the theory. We must enlarge the formalism.

"Hmm," you say, "this is very interesting. Why didn't someone tell me earlier? Anything else?"

Oh yes. The origin of mass is not understood. That is, in Newtonian physics, mass is a God-given property of an object. But modern theorists believe that this is a swindle and one should give a proper accounting of mass. Here is one of Steven Weinberg's jurisdictional matters; what was formerly a philosophical question has moved to the scientific domain. Nevertheless, the favored mechanism for doing this is a total kludge. The masses that come out depend on the parameters you've put in and it works only for certain particles; you need a different mechanism altogether to account for other particle masses.

"Anything else?"

We haven't spoken too much about consciousness, except to say that if you are going to bring consciousness into physics you must have a model, and in the Fifth Debates we gave a few general ideas. Much of our reticence stems from the fact that so little is known about consciousness that almost anything one says is bound eventually to prove naive. Most systematic expositions on consciousness relate to experience, not to explanation. The spiritual teacher Krishnamurti has said things like, "Consciousness is the totality of life," that "the intellect is part of thought," that he wants to know whether "there is a movement beyond consciousness." He asks, "Do you follow?"

No.

Roger Penrose has explicitly stated that any model of the brain must take into account quantum phenomena and he has suggested that even quantum gravity must be involved in an explanation of consciousness. The first proposal seems reasonable, the second is extremely difficult to credit, given the unbelievable weakness of the gravitational force. In the Quantum Debates Stephen Hawking made a cryptic remark evidently directed at Penrose's proposal: "I agree that we may introduce time asymmetry in the way we ask questions about observations. But I totally reject the idea that there is some physical process that corresponds to the reduction of the wave function or that this has anything to do with quantum gravity or consciousness." He concludes, "That sounds like magic to me, not science."

"I don't understand," you respond.

Neither do we, entirely, but Hawking's opinion of the Penrose hypothesis is clear.

"Anything else?"

Of course. We spent considerable time contemplating the origin of the universe. As a final exercise, why not contemplate something much simpler: the creation or destruction of a single particle of light—a photon.

Figure 10.1

"Don't I create photons when I turn on a flashlight?"

Yes, many photons. Or when you turn on a radio transmitter to speak to a friend over a two-way radio. These are merely longer-wavelength photons. But consider, How is this photon, which may have a wavelength meters long, absorbed by a tiny receiver? All at once.

"Like the photons being absorbed by the detector in the two-slit experiment."

"Precisely." But the entire photon, which is much larger than the detector, is absorbed instantaneously. We mentioned the phenomenon in the Quantum Debates. The mechanism by which such magic takes place is not explained by the theory. Much as is the case with the radioactive atoms, what we call the creation or absorption of a photon is just a transition from one state to another. We write down the initial state and the final state and see that the state of the system has changed. We associate this change of state with the creation or absorption of a photon. But it is just a transition. There is no mechanism in the theory for creating or destroying anything, let alone the universe.

"Do you mean the famous quantum jump is unexplained?"

The quantum jump is exactly the transition between one state and another. How that results in the creation or annihilation of a distinct particle or field is unexplained.

"Anything else?"

"Is the universe describable?"

Lost in the new round of questions and the old question, you were unaware that you had approached the summit of the Library and that the brawling scholars were now close at hand. "How are we going to escape our own destruction?" you asked.

The authors suggested that you still your mind and meditate on these new enigmas. Somewhat skeptical of the proposal, you asked if they would accompany you. No, they replied, it is time for our departure. Remember Confucius's definition of an annoyance: A young person who does not respect his elders, a person in full maturity who passes on nothing to the next generation and an old person who hangs around too long. All three apply in the current situation. Anyway, we must return to the fray for more debating . . .

With that you found a quiet spot near the summit of the Library and entered a profound meditation on the meaning of creation and annihilation. When you finally opened your eyes again, the Library, the authors and the Academy were nowhere to be found and you perceived that you had made a quantum leap of your own. Back in the Real World at last, you headed toward the nearest bed for a long sleep; along the way you noticed that a few things had changed . . .

CAN WE MAKE ANY MONEY OFF THIS?
Applied Quantum Mechanics

The Question

There was once a fellow of modest means who set up a number of bank accounts at various banks. Into the first account he deposited his entire savings, but a few days later he transferred it to the second account, and a few days after that he moved the money to the third account, and so on. He kept up this practice for some time, constantly transferring the funds from one account to the next. During the float—the few days after which he deposited a check in one bank but before it had cleared at the other bank—his savings would appear on the ledgers of both banks and accrue double the usual interest. By this trick he could convince his creditors that he was a wealthy man with active accounts in several banks.

One day, however, he fell ill and was unable to manage the accounts for some weeks. The transfer of funds naturally halted and it became clear to everyone that he had an account in one bank only, and a rather insignificant account at that. His creditors abandoned him and he was barred from the dinner club.

Similarly, after you had quit the Academy and had resided again in the Real World for many months, you began to read newspaper and magazine reports of miraculous advances in the use of quantum mechanics for electronic computation and even teleportation. The reports struck you as intriguing, if sometimes incomprehensible, and you resolved to see whether such matters could be clarified. There was only one thing for it.

And so it happened that you once more passed through the gates of the Academy to hear the academicians arguing:

"Yes."

"No."

"Maybe."

That is, was it possible to make money off quantum mechanics?

The Sphinx and the Cat

The key to all the newsworthy effects, you were informed immediately upon arrival, is a phenomenon with which you had become intimately acquainted in your earlier sojourn at the Academy: quantum superposition. The Sphinx could not be thought of as one-third bird, one-third woman and one-third lion; rather she was all of these together. An electron did not have a definite spin up or spin down state before measurement; rather it had both. Superposition was at the bottom of the two-slit experiment and resulted in the strange "nonlocal" correlations of the EPR experiment that occupied so much of the Seventh Debates.

Now, recall the ammonia molecule from the Ninth Debates, which the authors produced to illustrate the concept of tunneling. An ammonia molecule consists of three hydrogen atoms and a nitrogen atom in the shape of a pyramid, with the nitrogen atom at the apex. Each 30-billionth of a second, the nitrogen atom tunnels through the potential barrier created by the hydrogen atoms and inverts the pyramid. Back and forth, back and forth. The tunneling cannot be observed directly. Just as a measurement must find the Sphinx in one state or another, so must the nitrogen atom be observed either "up" or "down." No other location is allowed.

The quantum wave function itself, however, gives the probability of finding the system in a certain position. If you could observe the wave function of the ammonia, what would you see? At time equals zero, when the nitrogen atom is at the top of the pyramid, the wave function would also be concentrated in the up position. The probability of finding the nitrogen atom in that location would be nearly 100 percent. As time went on and tunneling progressed, the wave function would begin to slosh toward the other position, much like the money in the hapless banker's accounts. At half the tunneling time (about one 60-billionth of a second), the wave function would be equally distributed between nitrogen up and nitrogen down. If you performed measurements on the system, half the time you would find the nitrogen in the up location and half the time in the down location. (You have frozen the accounts.) Thereafter, the nitrogen wave function would be concentrated more in the down position, indicating that the odds of finding it there now exceed the probability of finding it up.

And so, if you had a very fast stroboscope and made repeated measurements at the proper time intervals you could find the pyramid in the down state more often than you found it in the up state; this would be evidence of an unequal superposition, more down wave function than up.

"But," you interrupt, impatiently, "you have often stressed that in quantum measurements—excluding those strange weak measurements you once talked about—you never observe the superposition itself. We have observed the Sphinx so often by now that I am convinced you can't see her as a superposition; you can view only one of her aspects at a time. The ammonia molecule seems no different."

"Thank you," says the Sphinx, joining the discussion.

You are correct, there is in principle no difference. There is a practical difference. The Sphinx is not a Real World animal. The ammonia molecule itself is too small and the tunneling too fast to observe such effects. But suppose you had a much larger molecule—one that you could almost see—then you would have an "almost macroscopic" demonstration of quantum superposition in real matter.

"Hmm," you reply. "Perhaps that would be impressive."

In the past seven years, investigators have produced such demonstrations in so-called Rydberg atoms, named after the nineteenth-century Swedish spectroscopist Johannes Rydberg. A normal atom is about a billionth of a meter in size. If one, however, hits it with controlled bursts of laser light, one can knock the electrons to much higher orbits and the atom swells by a factor of 1,000, to about a micron, roughly the size of a pit on a compact disc. To follow any given orbit an electron is constrained to have a certain energy. Now, from the discussion on the photoelectric effect we know that the energy of a particle corresponds to the frequency of the wave associated with it. Thus electrons may be thought of as waves of a given frequency traveling around the orbit. In the experiment one actually hit the atom with a laser pulse of several frequencies, effectively putting the electron in several closely spaced orbits (an orbital superposition). As the waves of various frequencies travel around the atom, in some places they will constructively interfere, and in other places destructively interfere. In the former places the electron becomes visible; in the latter places it disappears.

It turns out that the constructive interference can take place on opposite sides of the Rydberg atom. But since the waves correspond to a single electron, one is observing a single electron in the two different locations. This corresponds exactly to the case of the wave function of the nitrogen atom in the ammonia being equally distributed between up and down.

The first such experiment, using a potassium atom as the target, was carried out at the University of Rochester in 1991 by Carlos Stroud and John Yeazell. In 1996, Stroud and Michael Noel showed that one was in truth observing the same electron in two locations. Several other such experiments have been performed, with various techniques, but the basic objective is the same—to observe a quantum superposition at almost macroscopic separations. Such states have been dubbed "Schrödinger cat states," after Schrödinger's famous and tragic feline, who existed briefly as a superposition of live cat and dead cat.

"Every time I hear about Schrödinger's cat," says Stephen Hawking, "I want to reach for my gun."

The Sphinx agrees. Although the cat is a close relative of the Sphinx, both existing in superpositions, the Sphinx has also tired of the cat getting all the attention. She transforms herself into a lion and in one gulp swallows the cat whole. Now we can speak of Schrödinger's Sphinx, if desired.

Of Schrödinger's Sphinx experiments, physicist Serge Haroche declares, "This is the first time we can observe the progressive evolution of quantum to classical behavior."

A moment, you ask, perturbed, is that really what has been done? "I believe you misspoke a moment ago when you said we were observing a quantum superposition. You aren't, are you? With each snapshot, you're observing the electron in one place or another, yes?"

Correct. Neither has the collapse of the wave function been directly observed; here Haroche has exaggerated. What the experimenters observe—and can control—is how often, say, the nitrogen atom appears at top or bottom of the pyramid. This is fairly direct confirmation that a quantum superposition exists.

Somewhat pacified, you now put forth the question you returned to the Academy to ask: "What does this have to do with quantum computing?"

Nothing. Everything. Something. The principle of superposition is the basis for quantum computing, and experiments such as those involving the Rydberg atoms have displayed the ingredients necessary for elementary quantum computation. Whether it will ever be feasible to construct a practical quantum computer on the basis of such hardware is a question a few academicians are eager to take up. But you were interested in newsworthy developments and Schrödinger-lion experiments have recently captured much attention.

Before we go on to computing, there is another related phenomenon that has received a certain amount of attention in recent years. We refer to the "Zeno effect" introduced by Misra and Sudarshan in 1977. Consider again the ammonia molecule. At time equals zero the nitrogen atom is on top. The wave function is practically all concentrated in the upward configuration. As time goes on, the wave function oozes toward the bottom configuration. However, if one makes an observation at a very early time, the odds are high that one will observe the nitrogen atom at the top rather than the bottom. (Always bear in mind that the nitrogen is found *either* at the top or the bottom, nowhere in between.) The wave function "collapses" at the top configuration. But in that case, then by definition the wave function has ceased oozing toward the bottom; it has been "reset" to the zero position. And so, by repeated observations at short intervals (short compared to the tunneling time) one can prevent the nitrogen atom from ever leaving the top position. Hence "Zeno effect" after Zeno the Greek, who claimed that a runner could never cross a stadium because he would first have to cross half the stadium, then half of the remainder, then half of that and so on, and the sequence would never end. Zeno concluded that motion was impossible.

Contrast the quantum Zeno effect with a classical process, such as a horse running a race. In this case, you observe the horse at one second and it has gone a certain distance; after two seconds it has gone another distance. No matter how often you observe the horse its motion progresses. The Zeno effect was experimentally verified in 1990 by Wayne Itano and colleagues at the National Institute of Standards and Technology in an atomic system similar to the case of ammonia. One can really stop an atomic transition by repeatedly looking at it.

"Ageless bodies, timeless minds," confirms Deepak Chopra.

"Indeed," adds a fellow New Ager, tuning in, "doesn't this demonstrate that consciousness creates the material world?"

No, it does not demonstrate any such thing. Physical observations affect physical processes. The experiment could be carried out by computer and is in fact carried out by computer. Chopra's sales pitch notwithstanding, the Zeno effect seems to have nothing to do with either ageless bodies or timeless minds.

"Does it have anything to do with quantum computing?"

Maybe.

Quantum Computation and Related Mysteries

When the mathematician Ramanujan lay dying in the hospital, his colleague G. H. Hardy often went to visit him. It was during one of these visits that Hardy remarked, "I thought the number of my taxi-cab was 1729. It seemed a rather dull number." To which Ramanujan replied, "No Hardy! No Hardy! It is a very interesting number. It is the smallest number expressible as the sum of two cubes in two different ways."

Since the time of Ramanujan a favorite pastime of physicists when traveling through large cities is to take note of license-plate numbers and try to factor them. It is easy to multiply numbers together to get a product, but the reverse operation—to take a target number and find the prime numbers that multiplied together give back the target—is much more difficult. There are certain tricks that can aid you in the process. If a number ends in an even digit, then you know it is divisible by 2. If it ends on a 0 or 5, then it is divisible by 5. If the sum of the digits is divisible by 3, then the number is divisible by 3; if the sum of the digits is divisible by 9, then the number is divisible by 9.

Nevertheless, if you were confronted with a license-plate number, say, 172927, you would not have many obvious tricks at your disposal. Probably you would choose the brute-force approach: pick ever larger numbers and check whether they divide evenly into 172927. The one thing you are sure of is that you don't have to test any numbers larger than the square root of 172927 (just as 49 factors into 7×7).

If there are several people in the car, you can speed up the process somewhat by farming out duties: One person checks for division by 3, another checks for division by 9, and so on. This is an example of parallel processing. Nevertheless, it turns out that there is no good method—no good algorithm—for factoring large numbers, and the time it takes a computer to accomplish the task goes up exponentially with the number of digits in the target. Any number of more than about 150 digits becomes essentially impossible to factor with present-day computing power; a 250-digit number would take 800,000 years and a 1,000 digit number would take 10^{25} years, about 10^{15} ages of the universe.

All this has enormous practical implications: banks and security organizations rely on the effective impossibility of factoring 250-digit numbers for their cryptosystems. Were it possible to factor such numbers in a short time, many of the world's data encryption systems could easily be broken. With quantum computation that may happen.

The idea that one might use a quantum computer to simulate an elementary classical computer was first discussed by P. Benioff in 1980. Richard Feynman considered the reverse problem: how a classical machine might simulate a quantum computer. It was David Deutsch of Oxford, however, who in 1985 proved a general theorem that logical operations could be performed on quantum states and that quantum superposition would in principle allow a quantum machine to perform many calculations in parallel. Still, no one had a concrete demonstration of how that might be accomplished until 1994, when Peter Shor gave a quantum algorithm for factoring large numbers that was exponentially faster than any classical algorithm.

As most people know, for all the seeming miracles performed by classical computers, at bottom they do no more than manipulate bits of information in simple logical operations. The word "bit" originally stood for "binary integer," a 0 or a 1, which can stand for "yes," and "no," or "on" and "off." Because the electronic components of computers typically have two states, "on" and "off," 0 and 1, binary arithmetic has become synonymous with computation. But if you had another symbol, ?, for "maybe," the number of states available to the computer would increase and so would the speed of computation. After all, it is more efficient to tell a story with a large vocabulary at your disposal than a small one.

The kind of quantum superposition we have been talking about provides a computer with extra states. The spin of a particle is not necessarily up or "horizontal" but can be equal parts of both, effectively putting the spin at 45 degrees. Forty-five degrees is not the only option. There are an infinite number of angles between horizontal and up, and depending how you mix the spins, you can produce any one of them. Crudely speaking, a quantum computer's great advantage over a classical computer is that it would be put into a superposition of the two base states in an infinite number of ways, and the processing would then be carried out in parallel.

It is important to realize that a quantum computer, in any reasonable sense of the word, has not yet been built. Shor's algorithm is a mathematical prescription, not a functioning machine. A few of the atomic systems we have mentioned can be used to perform elementary logical operations on the superposed quantum bits ("qubits") but nothing that would seriously be called a computation. Furthermore, apart from factorization, the enormous quantum speed-up has been found for only one other similar application.

It is also important to realize that a quantum computer probably wouldn't resemble a conventional computer. From Shor's algorithm we can provide a hypothetical model. Shor showed that the problem of factoring a large number can be reduced to the problem of finding the period of a periodic mathematical function. (We give a simple example in the Commentaries.) The periodic function is just a function that repeats, but it may be much more complicated than an ordinary sine wave; perhaps it resembles more the wave that the Sphinx caused to ripple over the pond on page 165. Via Shor's algorithm, a classical computer would produce this complicated wave from the number to be factored.

At that point one could use the periodic function to modulate the intensity of a light source, for example a laser. (Modulate means to change the intensity in conformity with the periodic wave.) The light could then be passed through a slit. Recall from the Quantum Debates that light passed through a single slit makes a characteristic diffraction pattern on the detector or on a distance screen (see Figure 7.1b). If an unmodulated laser is used, then the bright spot is centered on the slit. However, if the beam is modulated by the function we are testing, the spot moves a distance that is simply related to its period.* And so, to calculate the period, we merely read off the distance the bright spot has shifted.

We see that this particular quantum computer is more similar to an analog computer (which calculates a process by duplicating it) than to a digital computer. More precisely, our "photonic computer" is almost identical to a classical optics experiment

---

* Strictly speaking, if the function is composed of many waves, each with its own period, a number of bright spots will be produced, one for each period.

of the type Academy undergraduates perform regularly. The light amplitude at each point in the optical train effectively represents the superposed qubits, and the wave as a whole is carrying out parallel processing in its propagation. That's all there is to it.

At perceiving the simplicity of quantum computation, a number of academicians bolt for the Academy gates to raise venture capital for their start-up companies.

"Not so fast!" cry a number of skeptical colleagues. "You have forgotten decoherence!"

So had you.

Crucial for the operation of the photonic computer is that light falling across the slit be coherent, so that waves emerging from the various parts of the slit can superpose to form the single-slit diffraction pattern. If the light is incoherent, no diffraction pattern will be produced and the result is lost. We have stressed that environmental factors can easily decohere a beam, destroying its ability to interfere. This is a serious problem for quantum computers, which must maintain coherence for the entire duration of the computation. Haroche and his colleague Jean-Michel Raimond claim that with current technology, coherence would have to be maintained for a full year.

"He that followeth me shall have the light of life, but not the light of quantum computers."

The academicians writing up business plans take pause.

"That's not the only problem," adds another naysayer. "The authors have lied about the optics experiment. To really perform a practical experiment with light would be virtually impossible."

The authors are blushing. In their pedagogical enthusiasm, they omitted a few details. To factor a 200-digit number in the manner we described would require modulating the laser with precision of one part in 10^{100} or maybe 10^{200}, which is not currently practical. This is not to say that factoring large numbers is ruled out. The optical experiment we described is a simple way of performing what mathematicians term a Fourier analysis, a method to break down a wave into its component parts. Many ways to perform Fourier analyses do not rely on optical techniques.

The skeptics remain unpacified. "There is the problem of chaos."

"Yes," reply those who believe chaos is the ultimate limitation to everything. It is true that any chaotic motion in the quantum computer would lead to decoherence and destroy the computation.

"No," ripostes David Deutsch: "[Chaotic] effects . . . occur only in classical physics—that is, not in reality, since reality is quantum mechanical." Deutsch believes that chaos in quantum systems does not exist and that one need not worry about it.

"Maybe," adjudicate the authors. Deutsch has a real point. It is not at all clear what quantum chaos means. Classical chaos refers to a situation in which the trajectories of two nearby objects diverge so rapidly that their behavior becomes unpredictable. But in quantum mechanics there are no trajectories of particles. Rather we observe, with a certain probability, that particles are "here" or "there." Particle trajectories don't figure.

But this does not imply that quantum chaos does not exist. What it does imply is that to talk about quantum chaos one must first define it. One definition currently under investigation is that the wave function oscillates with all possible frequencies—"white noise." This means that the wave function changes in a random fashion. Consequently, if two systems were started off in nearly the same state, after some time the shape of their wave functions might be very different, and when measurements were performed on the two systems, they might be found in very different states.

However, qubits are superpositions of two basic states, say, up or horizontal, each with its own frequency. Those are the only two frequencies available, and with two definite frequencies the wave function cannot oscillate in the random manner characteristic of white noise. The issue of quantum chaos is irrelevant. The real question is whether decoherence destroys the ability of the computer to retain a superposition long enough to perform a calculation.

"No," remarks one of the entrepreneurial academicians, "the real question is whether the computation takes place in one universe or many."

At this, the universe splits. The universe in which the authors find themselves seems dominated by many-worlders who, as you may recall, believe that at each quantum event, the universe branches. Extremists of this view interpret the existence of an interference pattern not as a single photon, which having passed through both slits is interfering with itself, they interpret it as the interference of two universes. The operation of the photonic computer, which requires interference, takes place in the "multiverse."

There can be no doubt about it. "The remarkable non-random interference phenomenon," declares David Deutsch, "is just as inescapable a piece of evidence for the existence of the multiverse as the phenomenon of shadows."

He goes so far as to challenge "cosmic Luddites" (those who cling to the obsolete notion of a single universe) to "explain how Shor's algorithm works. . . . When Shor's algorithm has factorized a number, using 10^{500} or so times the computational resources that can be seen to be present, where was the number factorized? There are only about 10^{80} atoms in the entire visible universe, an utterly minuscule number compared to 10^{500}. So if the visible universe were the extent of physical reality, it would not even remotely contain the resources required to factorize the number. Who did factorize it, then? How, and where, was the computation performed?"

We Luddites respond that David's question sounds a little like asking where the infinite number of integers in a mathematician's imagination take place. Infinity is much larger than 10^{80}, much larger than 10^{500} even.

"What I want to know," you say, curtailing the argument, "is when did the word 'universe' stop referring to everything."

It is true, the word "universe" now seems to refer to only that within the possible realm of our experience. Hence, "multiverse" to refer to Everything, including that which is beyond everything. Nevertheless, if you are willing to risk the Luddite classification, you may view the quantum computer as operating within a single universe. It will certainly be built within a single universe.

As the universe splits again, and half the participants invest in QMC Corporation (Quantum Multiverse Computers; "Every Result, Dead and Alive"), you remind the authors that they have yet to touch on the phenomenon of quantum teleportation, which has recently flashed through the media limelight. To the contrary, we reply, much of the Seventh Debates centered around the fundamentals of quantum teleportation: the EPR experiment.

The New Age has returned in force.

In the EPR experiment you and your partner measured spins of particles that were emitted from a decaying atom. The emitted particles were in no definite state, but rather in a superposition, consisting of up and down. Nonetheless, if you measured particle A to be up, then you know your partner must measure spin B to be down. This is true regardless of the distance separating A and B. Quantum teleportation, due to Charles Bennett and colleagues, is merely a variation on this theme; one might call it "tag-team EPR." Instead of using two particles we use three.

Suppose you want to transmit the state of a message particle, M, to your partner. M may be up or down or in a superposition, in which case it has no definite state and you do not even know the message you are transmitting. Now, by cleverly arranging your experimental apparatus, you let particle M interact with particle A. In doing so, M becomes "entangled" in a superposition with A, much as A and B are. It turns out there are four possible ways for this to occur; teleportation takes place only with one of them, call it the state MA.* In the particular superposition MA, whatever the state of M, A must be found in the opposite state. But B is prepared such that it must be found in the state opposite from A. This means a measurement must find B in the same state as M.

So, believe it or not, one can transmit the state of M instantaneously to a distant particle B. Quantum teleportation is not just a theoretical fantasy; in late 1997 Dik Bouwmeester and colleagues observed teleportation in a laboratory experiment involving photons. We hasten to point out that, as in the usual EPR experiment, no material object has been moved at faster-than-light velocities, no superluminal signaling has taken place. The states of the photons involved are correlated—

"But!" you remonstrate, silencing the authors, "in the basic EPR experiment, particles A and B were correlated since birth, right? Here you seem to have added another particle M to the mix, long after A and B were created. It seems quite extraordinary that you can instantaneously transmit the state of M to B and yet continue to claim that no superluminal signaling is involved."

Well, particles A and B in quantum teleportation are also correlated at birth, and remember they do not have definite states until a measurement is actually performed. In the plain vanilla EPR experiment a measurement determines the states of only A and B, which must turn out to be opposite. In the teleportation version, the same applies. A measurement on B must give the opposite state from A, except now the state of A includes information about a third particle M. So there is little difference between the old EPR experiment and teleportation and the same mysteries apply. It does, however, turn out that the state of M is destroyed during the teleportation process, and so it *is* teleportation rather than cloning—no copy remains.

Teleportation is intimately connected with quantum computation because it involves the "transmission" of qubits from one location to another, within the quantum

---

* The other three states can be transformed into the "teleportation state," MA.

computer. For example, it may be possible to teleport the contents of one register to another with no intervening wires—a perfect example of wireless transmission.

At this intelligence, the Knowledge Marker skyrockets and the stock in QMC instantaneously doubles.

But having now covered the latest developments in depth, you turn to the authors and announce, "I have only one thing to say about all this."

What's that?

"Beam me up, Scotty."

COMMENTARIES ON THE DEBATES

Prologue

Page

xv: The quotation from Pao Ching-Yen can be found in Colin A. Ronan's abridgment of Joseph Needham's *Science and Civilization in China* (Cambridge: Cambridge University Press, 1978), vol. 1, p. 294. (Henceforth *SCC.*)

Whether the Chinese actually laughed at Newton's laws, we don't know. However, a passage in Needham's *Grand Titration* (Toronto: University of Toronto Press, 1969), p. 308, indicates that the Chinese reacted to the concept with skepticism. In 1737 one d'Argens wrote that, according to the Chinese, if God established Laws, then it followed that all animals and plants must understand the Laws in order to execute them, "which is absurd."

The ancient Chinese concept of law was embodied in two words, *li* and *fa*, both of which referred only to human law. For the Confucians *li* corresponded roughly to Locke's natural law, the unwritten code of morals, customs and ethics. *Fa* corresponded to proactive, judicial or civil law. D'Argens's remarks show the difficulty which the Chinese had in extending their concept of law to the physical universe. That the modern Chinese translation of "laws of nature" is *tzu-jan fa*, or "spontaneous law," serves to emphasize how deeply paradoxical the Chinese regarded the very idea of physical law.

Steven Weinberg's remarks can be found in his *Dreams of a Final Theory* (New York: Vintage Books, 1993), p. 6. (Henceforth *Dreams.*)

xv: Roald Hoffmann's views on reductionism can be found in his *The Same and Not the Same* (New York: Columbia University Press, 1995), p. 20.

Unless otherwise noted, our quotations from the Upanishads and the Bhagavad Gita are in the translation by Eknath Easwaran (Tomales, CA: Nilgiri Press, 1987). This passage is from the Mundaka Upanishad, p. 109.

First Debates

Page

Epigraphs: "Thou inconceivable . . ." is from the Third Elegy of the *Lamentations of Gregor Narekatsi*. We have used the translation by Mischa Kudian, *Lamentations of Narek* (London: Mashtots Press, 1977), p. 20. "We shall not cease . . ." is from Eliot's *Four Quartets, Little Gidding*.

6: To propel a one-metric-ton UFO from Alpha Centauri to Earth at a 1-g acceleration, requires about 3×10^{21} joules (assuming perfect energy conversion) more than the annual world energy consumption. The figure goes up in direct proportion to ship mass and exponentially in terms of trip time or acceleration. For a slightly longer trip, or a larger UFO, the energy requirement easily exceeds human history's entire energy consumption.

8: Michio Kaku's declaration is in his book with Jennifer Thompson, *Beyond Einstein* (New York: Anchor, Doubleday, 1995), p. 100.

Weinberg's views on philosophy and physics are in *Dreams*, pp. 168–169.

For Cornford on Plato, see Plato's *Republic*, trans. with notes by Francis MacDonald Cornford (Oxford: Oxford University Press, 1970), p. 236.

11: Jean-Pierre Changeux's statement is in his book with Alain Connes, *Conversations on Mind, Matter and Mathematics* (Princeton: Princeton University Press, 1995), p. 46. (Henceforth *Conversations.*)

13–14: Dirac's views on renormalization can be found in *Scientific American* May 1963, p. 50.

15: Fritjof Capra, *The Tao of Physics*, 3rd ed. (Boston: Shambhala, 1991). (Henceforth *ToP.*)

15–16: Weinberg's remarks on dry wells are in "Sokal's Hoax," *New York Review of Books*, August 8, 1996. Easwaran's comments are in The Upanishads, p. 14.

For Mishra's views, see Rammurti S. Mishra, *Yoga Sutras, The Textbook of Yoga Psychology* (New York: Doubleday/Anchor Books, 1973), pp. x–xi.

Chopra's remarks are in Chopra, *The Seven Spiritual Laws of Success* (San Rafael: New World Library, 1994), p. 5. (Henceforth Chopra.)

Page 16–17: For Weinberg on the irrelevancy of measurement theory, see *Dreams*, p. 84.

Pages 18: The descriptions of meditative states are from Swami Lakshman Jee, *Kashmir Shaivism, The Secret Supreme* (Albany: SUNY Press under imprint of the Universal Shaiva Trust, 1988), p. 109.

Page 20: The passage from the Gita is verse 13 on page 38 of the translation by Eliot Deutsch (New York: Holt, Rinehart and Winston, 1968). The same passage is verse 13 on page 62 of Easwaran's translation.

For Hoffmann's remarks, see *The Same and Not the Same*, p. 78.

Second Debates

Page

Epigraphs: Hardy's statement is from his *A Mathematician's Apology* (Cambridge: Cambridge University Press, 1969), p. 123.

Leslie A. White, "The Locus of Mathematical Reality: An Anthropological Footnote," in James R. Newman, editor, *The World of Mathematics* (New York: Simon and Schuster, 1956), p. 2363.

22: The Fibonacci sequence, named after the thirteenth-century mathematician Leonardo Fibonacci, is a sequence in which each number is the sum of the previous two: 1, 1, 2, 3, 5, 8 . . . The ratio of the successive terms (1/1, 1/2, 1/3, 1/4 . . .) gives the angular position of the sunflower florets and chambers of the nautilus. The sequence in parentheses also converges to the famous "golden mean," $1/2(\sqrt{5}-1)$.

22: Wigner's famous essay, "The Unreasonable Effectiveness of Mathematics in the Natural Sciences," was published in *Communications in Pure and Applied Mathematics* **XII,** 1 (1960).

Dirac's views on beauty can be found in *Scientific American*, May 1963, p. 53.

24: The first book of *sangaku* problems in English was compiled by Hidetoshi Fukagawa and Dan Pedoe. It is entitled *Japanese Temple Geometry Problems* and is available from Prof. Charles Stanton, Charles Babbage Research Centre, P. O. Box 272, St. Norbert Postal Station, Winnipeg, Canada R3V 1L6 (e-mail: stanton@cc.umanitoba.ca). For more on the history of the *sangaku*, see T. Rothman's article "Japanese Temple Geometry" in *Scientific American*, May 1998, or on the *Scientific American* website at www.sciam.com. There are also a few web pages devoted to *sankagu*.

24–25: Hardy's remarks are on p. 123 of the *Apology*.

Connes's statement is in *Conversations* p. 36.

For Plato, see Cornford, pp. 241–243.

For more on the Pythagoreans, see G. S. Kirk and J. E. Raven, *The Presocratic Philosophers* (Cambridge: Cambridge University Press, 1971), p. 240.

25: The quotation of Einstein is a slight paraphrase of his views as quoted by White, op. cit., p. 2354.

26: White on etiquette is ibid, p. 2363.

27: Kepler's exultations are in Max Casper, *Kepler* (New York: Dover, 1993), p. 380.

28: James Jeans's remarks are in *The Mysterious Universe* (1930), as quoted by Stanley L. Jaki, *The Relevance of Physics* (Edinburgh: Scottish Academic Press, 1992), p. 114. (Henceforth, Jaki.)

Eddington's words (and number) open his famous 1938 Tarner lecture. See *The World of Mathematics*, vol. II, p. 1069. See also pp. 1071–1072.

30: Fermat's Last Theorem is so called because it was the last proposition of Fermat that was proved or disproved. The theorem states that there do not exist any integers a, b, c such that $a^n + b^n = c^n$ if n is any integer greater than 2. Fermat's Last Theorem was finally proved by Andrew Wiles in 1994.

33: Changeux's views on reductionism are in *Conversations*, p. 60.

38: Louis Bachelier's paper appeared in *Annales Scientifiques de L'Ecole Normale* **16** (1900), p. 21.

41: François Jacob's views are in his book, *The Logic of Life* (New York: Pantheon, 1973), p. 316. See also *Conversations*, p. 155.

Third Debates

Page

Epigraphs: Lewis Carroll, *The Hunting of the Snark*. Louis Pasteur before the French Academy. See, e.g., René Dubois, *Pasteur and Modern Science* (New York: Anchor/Doubleday, 1960).

44:   The story about the hotel Moscow seems to be widespread in Moscow, but we have not investigated the truth of it. The "deviations" of the building do exist.

Two lengths or any other quantity are said to be in a golden mean if their ratio is $1/2(\sqrt{5}-1)$. See commentary to Second Debates.

45:   The comments on the *Doctrine of the Mean* can be found in Li Fu Chen, *The Confucian Way* (New York: KPI, 1986), p. 214.

Weyl's remarks are at the opening of his book *Symmetry* (Princeton: Princeton University Press, 1960), p. 1.

46–47:   For Capra's views on symmetry, see *ToP*, p. 257.

Daisetz Suzuki's commentary on asymmetry in Japanese art is in his book *Zen and Japanese Culture* (Princeton: Princeton University Press, 10th printing, 1993), p. 27.

49:   Physicists routinely credit Dirac with the prediction of the positron, but Dirac himself makes quite clear that he thought the new particle was the proton. See Dirac's Oppenheimer Memorial Prize acceptance speech, published as *The Development of Quantum Mechanics* (New York: Gordon and Breach, 1971).

Anthony Zee's remarks can be found in his *Fearful Symmetry* (New York: Macmillan, 1986), p. 3.

Also, see Kaku, op. cit.

50:   Much nonsense has been and continues to be written about the life of Evariste Galois (1811–1831). Although he did die in a duel at the age of twenty, it was not for political reasons, or on account of an agent provocateur or a prostitute; nor did he create the theory the night before he was shot. See "Genius and Biographers: The Fictionalization of Evariste Galois" in T. Rothman's *Science à la Mode* (Princeton: Princeton University Press, 1989) or on his website (current address) http://www.iwu.edu.

52:   The group of actions that governs isospin invariance between neutron and proton goes by the name SU(2), which is read much as it is written: "S-U-two." The SU stands for "special unitary group." It is best not to inquire what this name means.

53:   The term "hypercharge," encountered equally often, is equivalent to strangeness. The group that governs the behavior of strange particles is termed SU(3), which shows that the number in the parentheses does *not* refer to the number of particles under consideration.

54:   The group that governs quark symmetries is once again SU(3).

55:   Physicists term the spliced-together symmetry SU(2) X U(1) (read "SU-two cross U-one"). U(1) is the symmetry that governs the single boson of electromagnetism: the photon.

The theory of the strong force is known, after the quark colors, as quantum chromodynamics, another example of physicist humor. Also, like quarks, gluons are governed by the symmetry called SU(3). The big symmetry of the standard model is obtained by splicing together the electroweak and quark symmetries to get SU(3) X SU(2) X U(1).

59:   The baryon symmetries are described by the group SU(3).

63:   Chopra on least effort can be found in Chopra, p. 53.

In contemporary elementary physics the action is merely the integral over time of the kinetic minus the potential energy, in other words the time integral of the Lagrangian. Maupertuis was incorrect in thinking the action is always minimized. In some instances it may be a maximum. See Fourth Debates.

65:   Roald Hoffmann on the schizophrenia of physics is from "Qualitative Thinking in the Age of Modern Computational Chemistry," preprint courtesy the author.

66:   Maxwell's remark is from Jaki, p. 330.

Fourth Debates

Page

Epigraphs: Russell's first remark is from *Religion and Science* (New York: Holt, 1935), p. 153, as quoted by Jaki, p. 360. Russell's second remark is from "On the Notion of Cause," which can be found in his *On the Philosophy of Science* (New York: Bobbs-Merrill, 1965), p. 163. The full quotation reads: "The law of causality, I believe, like much that passes muster among philosophers, is a relic of a bygone age, surviving, like the monarchy, only because it is erroneously supposed to do no harm."

70–71: Chopra on Karma is Chopra, p. 39.

79: Euler's views on the action are as found in John Barrow and Frank Tipler, *The Cosmological Anthropic Principle* (Oxford: Oxford University Press, 1986), p. 150. (Henceforth Barrow and Tipler.)

82: For the theory of Tetrode and Fokker, see H. Tetrode, *Zeitsch. f. Physik* **10,** 317 (1922); A. D. Fokker, *Zeitsch. f. Physik* **58,** 386 (1929); A. D. Fokker, *Physica* **9,** 33 (1929) and **12,** 148 (1932).
The theory by Wheeler and Feynman was published in *Reviews of Modern Physics* **17,** 157 (1945).

84: Koestler's "library angel is discussed in Allan Combs and Mark Holland, *Synchronicity* (New York: Marlowe and Company, 1996), p. 21. (Henceforth Combs and Holland.)

85: Jung's remarks were made in "Richard Wilhelm: in Memoriam." We have quoted them from Harold Coward, *Jung and Eastern Thought* (Albany: SUNY Press, 1985), p. 43.

85: The famous scarab story can be found in Carl Jung, "Synchronicity: An Acausal Connecting Principle," in *The Interpretation of Nature and the Psyche* (New York: Pantheon Books, 1955), p. 31.

86: Koestler's story of the paper mill is found in A. Hardy, R. Harvie and A. Koestler, *The Challenge of Chance* (New York: Random House, 1973), pp. 194–195 as well as in Combs and Holland, p. 156.

87–88: One finds conflicting dates for the *I Ching.* Our account is taken from SSC, vol. I, pp. 182–185.

88: Carl Jung's interpretation of the *I Ching* is from his Foreword to the *I Ching* (Princeton: Princeton University Press, 1983).

Fifth Debates

Page

Epigraphs: Eddington's famous remark on entropy is from *The Nature of the Physical World* (New York: Macmillan, 1929), p. 74. Einstein's famous remark on time is from a letter written on the death of his best friend, Michele Besso, to Besso's sister.

97: For Chopra on waste, see Chopra, p. 55. For Rifkin, see his *Entropy* (New York: Viking, 1980), p. 260.

97: More on the history of entropy can be found in "The Evolution of Entropy" by T. Rothman in *Science à la Mode* (Princeton: Princeton University Press, 1989).

98–102: The history of Boltzmann's H-theorem is extremely tangled even by the usual standards of the history of science, and most texts seem to lump all versions of the theorem together. We have consulted a number of sources to put together this debate, including Ludwig Boltzmann, *Lectures on Gas Theory* (Berkeley: University of California Press, 1964); P. C. W. Davies, *The Physics of Time Asymmetry* (Berkeley: University of California Press, 1977); Richard Tolman, *Statistical Mechanics* (Oxford: Oxford University Press, 1938); Stephen G. Brush, editor, *Kinetic Theory* vol. 2 (Oxford: Pergamon Press, 1966). This last volume contains some original papers.
The "quotations" of Loschmidt, Zermelo and Boltzmann are fabricated but, we hope, reflect their positions reasonably well. However, Boltzmann's anthropic argument is a paraphrase from "On Zermelo's Paper 'On the Mechanical Explanation of Irreversible Processes,'" to be found in Brush p. 242.
Poincaré's objection is genuine and is taken from "Mechanics and Experience," in Brush, p. 206.

104: Murray Gell-Mann's remark on peanut butter is in *The Quark and the Jaguar* (New York: W. H. Freeman, 1994), p. 218.

104–105: Planck's defense of thermodynamics is in his *Treatise on Thermodynamics* (New York: Dover, 1945), p. 106.

105–106: Weinberg's position is spelled out in *Dreams*, p. 41 and p. 9.

106–107: For Penrose's cosmological solution to the arrow of time and Weyl-tensor hypothesis, see his *Emperor's New Mind* (Oxford: Oxford University Press, 1989), chapter 7. For a technical critique of the hypothesis, see T. Rothman and P. Anninos, *Physical Review* **D55,** 1948 (1997), and references therein.

107–108: Hawking's objections to Penrose are found in Hawking and Penrose, *The Nature of Space and Time* (Princeton: Princeton University Press, 1996), chapters 5 and 7. (Henceforth, *NST.*)

109: Penrose's "reasonable differences in viewpoint" remark is in *The Emperor's New Mind*, p. 310.

110: Rudolf Peierls's statement can be found in *Physics Today*, November 1994, p. 115.

112: Henri Poincaré's prediction about kinetic theory is from his book *The Value of Science* (New York: Dover, 1958), as cited by Ilya Prigogine and Isabelle Stengers in *The End of Certitudes* (English language preprint, courtesy of the authors).

A slightly more biographical article about Prigogine's work is "Irreversible Differences," by T. Rothman in *The Sciences,* July/August 1997. Prigogine's remarks in this chapter were taken from interviews for that article.

113: Berry's model is presented by David Ruelle in his book *Chance and Chaos* (Princeton: Princeton University Press, 1991), p. 76.

115–116: John Horgan's Op-Ed piece was in the *New York Times,* July 16, 1996.

116: Penrose's original list of seven arrows was published in Roger Penrose, "Singularities and Time Asymmetry," in S. W. Hawking and W. Israel, editors, *General Relativity, An Einstein Centenary Survey* (Cambridge: Cambridge University Press, 1979).

118: Lighthill's *mea culpa* is in "The recently recognized failure of predictability in Newtonian dynamics" by Sir James Lighthill, *Proceedings of the Royal Society of London* A407 (1986), pp. 35–50.

120–121: Capra quotes Suzuki on p. 179 of *ToP.*

122: See "Speeding up an Internal Clock in Humans?" by John H. Wearden at http://www.psy.man.ac.uk/ResearchFolder/PostFold/WeardenPoster.

123: The verse is a hymn to Dakshinamurti in *Hymns by Shankara,* edited by T. M. P. Mahadavan (Madras: Ganesh and Company, 1970), p. 5.

122–124: For more on time in the Indian tradition, see Mircea Eliade, "Time and Eternity in Indian Thought," in *Man and Time,* papers from the Eranos Yearbooks (Princeton: Princeton University Press, 1973).

The first of the Chinese astronomical clocks seems to have been constructed by Chang Su-hsun in A.D. 979, although the more famous one is Su Sung's of 1094. See David S. Landes, *Revolution in Time* (Cambridge: Harvard University Press, 1983), chapter 1.

For a detailed discussion of time in Chinese tradition, see Joseph Needham's "Time and Eastern Man" in *The Grand Titration.* His view is quite categorical: "Western man had no monopoly of the sense of linear time, and [the idea of] the 'timeless Orient' is nonsense." He also considers, but rejects, the idea that Chinese notions of time were responsible for the fact that modern science did not develop in that culture.

124: The verse is Sudarshan's paraphrase of the Ramayana of Valmiki, vol. II, chapter 35, verse 8.

Sixth Debates

Page

Epigraphs: We have taken the remark of Descartes from Abdus Salam, *Journal of Molecular Evolution* **33,** 105 (1991). The right- and left-handed tantras are traditional.

126: We do not know where the rumor started that medieval castles have left-handed spiral staircases for the advantage of the defenders, but John Galloway of the Cancer Research Campaign, London, reports that he and the Scottish Office of National Monuments conducted a large survey of castles and showed that there is no preference for one handedness over the other. See *Biological Asymmetry and Handedness,* CIBA Foundation Symposium 162 (Chichester: Wiley Interscience, 1991), p. 34. (Henceforth *BAH.*)

On the other hand, the Kerr family stronghold in Ferniehirst on the border between England and Scotland evidently shows a prevalence of right-handed staircases that has been attributed to the predominant left-handedness of the family. This attribute of the Kerrs was so marked that throughout Scotland "Kerr-handed" is a synonym for "left-handed." See http://gps.leeds.ac.uk/genetics/internal/Gene1015/Usdin/Q&As.html.

126–127: For information on the painting of Adriaen van der Werff (1659–1722) we are indebted to Roald Hoffmann. See also chapter three of his book with Shira Leibowitz Schmidt, *Old Wine, New Flasks* (New York: W. H. Freeman, 1997).

The relevant biblical passage is Genesis 48: 9–20.

128—129: Much of the information about chirality for the Sixth Debates was provided by Dilip Kondepudi of Wake Forest University. See in particular his article with Roger Hegstrom, "The Handedness of the Universe," in *Scientific American,* January 1990. The experiment with bacteria was carried out by Neil H. Mendelson and coworkers in the 1970s.

130: For those unfamiliar with the concept of polarized light, light consists of an oscillating electric and magnetic field. The direction along which the electric field oscillates is termed the polarization axis. Usually, the electric field's direction of oscillation is random and we say the light is unpolarized. However, for various reasons the field may oscillate along a fixed axis (say, up and down). In this case the light is said to be polarized in the vertical direction. A polarizing filter (such as Polaroid sunglasses) has molecules aligned in a certain direction. If the filter's axis is aligned with the polarization axis of the light, all the light gets through. If the filter axis is crossed at 90 degrees to the light's polarization axis, then no light gets through. Polaroid sunglasses work on the assumption that only some of the randomly polarized sunlight will pass through the sunglass axis. When we say that optically active crystals rotate polarized light, we mean they rotate the axis of polarization.

Pasteur's discovery is described in detail by René Dubos in *Pasteur and Modern Science* (New York: Doubleday/Anchor, 1960), chapter 2. The words we have given Pasteur on his discovery of the optical activity of paratartaric acid are paraphrases from that chapter. His remarks on the prepared mind, the most distinctive chemical characteristic of living organisms, and his statement before the Academy are all genuine.

For more on the handedness of amino acids, see Gregory Petsko, "On the Other Hand . . .," *Science* **256,** 1403 (1992).

132: The famous paper by Lee and Yang is in *Physical Review* **104,** 254 (1956).

The famous paper by Madame Wu and her colleagues is C. S. Wu et al., *Physical Review* **105,** 1413 (1957).

E. C. G. Sudashan and Robert Marshak's theory of the weak force (V-A theory) can be found in *Proceedings of the Padua–Venice Conference on Mesons and Newly Discovered Particles* (Bologna: Zanichelli, 1958); reprinted in P. K. Kabir, editor, *Development of Weak Interactions* (New York: Gordon and Breach, 1964).

133: For more on the distinction of left and right in art, see *The Same and Not the Same,* p. 41, and *Old Wine, New Flasks.*

For the competing view of perceptual psychologists, see *BAH,* p. 35, and Blount et al., *Perception* **4,** 385 (1975).

The predominance of left cheeks was reported by Nick Humphrey and Chris McManus in *New Scientist* **59,** 437 (1973). Of 1,471 portraits from various English galleries, they found that 891 showed more of the left cheek and 583 more of the right. Assuming this was a random occurrence, it would be expected to take place less than once in 10,000 times. The authors discounted the hypothesis that right-handed artists might find it easier to paint left-cheeked portraits, largely because when portraits are divided by sex of subject, 68 percent of the women showed more of the left cheek, but only 56 percent of the men, a statistically significant result at the .001 level. That the distinction seems to be based on sex, along with the fact that Rembrandt, like other artists, tended to paint himself and family members with the right cheek and strangers and women with the left, led Humphrey and McManus to suggest that (predominantly male) artists identified the right cheek with the self and close relations, and the left cheek with non-self and more distant relations, including women. One wonders whether the results would hold up in light of the discovery that many Rembrandts are now known to be fakes.

135: For more on the chirality of atoms, see Hegstrom and Kondepudi, op. cit.

The synthesis of the HIV molecule is described by Petsko, op. cit., and Milton, Milton and Kent in *Science* **256,** 1445 (1992).

135–136: For more on the *Sonic hedgehog* gene etc., see *Nature* **377,** 103 (1995), *Science* **269** (1995) and Levin et al., *Cell* **82,** 803 (1995).

For more on the theoretical mechanisms to connect handedness at various levels, see N. Brown and L. Wolpert, *Development* **109,** 1 (1990).

137: The excess of L-amino acids in the Murchison meteorite was reported by John Cronin and Sandra Pizzarello in *Science* **275,** 951 (1997).

For more about the symmetry breaking in sodium chlorate crystals, see Dilip Kondepudi et al., *Science* **250,** 975 (1990).

139: Right (left) circularly polarized light is light in which the electric-field axis spirals like a right-hand screw (left-hand screw) as the light travels. (See Commentaries to p. 130.)

Early proposals based on circularly polarized light are reported by Stephen F. Mason in "Origins of the Handedness of Biological Molecules" in *BAH.* For Kuhn and associates and the Vester-Ulbricht hypothesis, see the review article on the origin of chirality by William A. Bonner, *Origins of Life and Evolution of the Biosphere* **21,** 59 (1991). Apparently, in 1968 Garay reported a positive observation of the Vester-Ulbricht effect, but the results were never confirmed.

141: For the neutron-star model, see W. A. Bonner and E. Rubenstein, *Biosystems* **20,** 99 (1987).

141–143: The banquet discussion is indeed (with obvious inclusions excepted) an abridged version of that found in *BAH,* pp. 312–315.

144: A little more on algorithmic complexity can be found in Ruelle's *Chance and Chaos,* chapters 21–23.

145: Schrödinger's passage on emergence is from *What Is Life?* (Cambridge: Cambridge University Press, reprinted 1980), chapter 7.

145: Roald Hoffmann's remarks are from a private communication.

146–147: Penrose's questions about the Möbius strip, etc., are from "Must Mathematical Physics Be Reductionist?" in *Nature's Imagination* (Oxford: Oxford University Press, 1995).

We have, of course, endeavored to accurately reflect Weinberg's position on reductionism, but his views are not easily summarized and we refer the reader to *Dreams*, pp. 40–41 and chapter 3, from which we have excerpted all his remarks.

148: Philip Anderson's statement is in *Science* 177, 393 (1972).

Both of Wesson's remarks, as well as those of Mayr and Mario Bunge, are from Robert Wesson, *Beyond Natural Selection* (Cambridge: MIT Press, 1994), pp. 26–27.

149–150: For Weinberg, see note to pp. 147–148; for Schiller, see Bernard M. Baruch's foreword to Charles Mackay, *Extraordinary Popular Delusions and the Madness of Crowds* (New York: Noonday Press, 1932).

150: Gore's words on Prigogine are from a speech entitled "What Is the Role of Science in American Society?" delivered 12 February 1996. Reprinted with abridgements in *Science* 272, 177, April 12, 1996.

Freeman Dyson's remarks are from "The Scientist as Rebel," in *Nature's Imagination*.

Seventh Debates

Page

Epigraphs: R. Blackmore, *Creation, A Philosophical Poem*, as quoted by G. N. Cantor, *Optics After Newton* (Manchester: Manchester University Press, 1983). Francis Bacon, *On the Dignity and Advancement of Learning*, as quoted by Jaki, p. 420.

153: Francis Bacon's story can be found in "The Sphinx," in *Great Essays in Science*, edited by Martin Gardner (New York, New American Library, 1984).

The story behind Young's celebrated experiment is confused. Generally, it is said that he performed the two-slit experiment in 1800 or 1801, but no mention of it appears in his papers delivered before the Royal Society during the years 1800–1803. Young's first account of the experiment, including the famous diagram, appears in his *Lectures* of 1807 with no date given for the actual performance. Moreover, the description is sufficiently abstract that at least one historian has conjectured that Young never carried it out at all. Nonetheless, Young's description does lead one to believe he must have made some estimate of the wavelength of light. For more details, see Cantor, op. cit., chapter 5.

157: The remarks by Einstein's colleagues have been taken from Abraham Pais, *Subtle Is the Lord, The Science and Life of Albert Einstein* (Oxford: Oxford University Press, 1982), p. 383.

160: More on the delayed-choice experiment can be found in John Wheeler's *Frontiers of Time*, his 1978 Varenna lecture series, which were also published by the Center for Theoretical Physics at the University of Texas, Austin, 1978. See as well, "A 'Delayed Choice' Quantum Mechanics Experiment" by William C. Wickes et al. in *Quantum Theory and Measurement*, edited by John A. Wheeler and Wojiech H. Zurek (Princeton: Princeton University Press, 1983).

162: Amit Goswami is as quoted in the magazine *What Is Enlightenment?* 6, 1 (Spring/Summer 1997).

163–165: Heisenberg's statement is from one of his essays on the uncertainty principle, as quoted by David C. Cassidy, *Uncertainty, the Life and Science of Werner Heisenberg* (New York: W. H. Freeman, 1992), p. 228. The statement of Niels Bohr reflects his opinion but is invented.

165: On the Schrödinger uncertainty relationship: Heisenberg's principle tells you only that the product of the uncertainties in momentum and position must always exceed a certain minimum value ($h/4\pi$). It does not tell you what the *actual* minimum is for a particular situation, and for even the simplest quantum systems—the perenially favorite harmonic oscillator—the actual minimum oscillates with time. In 1930 Erwin Schrödinger and H. P. Robertson independently devised stronger uncertainty principles that are constant in time and pinpoint the actual minimum value more closely. But "Schrödinger and Robertson May Have Slept Here" bumper stickers never caught on with the public. See E. Schrödinger, *Ber. Kgl. Akad. Wiss* 296 (1930); H. P. Robertson, *Physical Review* A35, 667 (1930).

166–167: Some remarks of Bohr, Wheeler, Eddington and Hinshelwood were conveniently collected at website http://marlowe.wimsey.com. However, many of Wheeler's and Bohr's remarks can be found in Wheeler's *Frontiers of Time* (see reference to page 160).

168–169: For Goswami, see Commentary to page 162.

169: The verse is Kena Upanishad, I, 7.

169: Deepak Chopra's remarks are in Chopra, p. 69.

169–170: The standard work on the many-worlds interpretation is *The Many Worlds Interpretation of Quantum Mechanics*, edited by DeWitt and Graham (Princeton: Princeton University Press, 1973). DeWitt's remarks can be found there in "Quantum Mechanics and Reality," p. 161.

Gell-Mann talks about many-worlds in *The Quark and the Jaguar,* p. 138.

170:   On whether the many-worlds interpretation is dealing with a quantum problem: Some of our colleagues would have our heads for this remark. They point out that a quantum coin is not the same as a classical coin; when we flip a classical coin we know that it comes up heads or tails. A quantum coin, on the other hand, is in a superposition of states and the whole measurement problem asks how it gets from a superposition to a definite state. We reply that the many-worlds interpretation doesn't tell you. Consider a photon passing through the two slits, as discussed on pp. 157–159 of the main text. Incoherent light consists of photons behaving like classical coins—they go definitely through the head slit or the tail slit. Coherent light consists of photons that behave like quantum coins; they can pass through both slits and interfere with themselves. Our colleagues' point is that it is precisely this interference that distinguishes the quantum from the classical case. But you are interested in deciding whether the photon came out of the head or the tail slit (whether it landed in the head or the tail universe). As detailed in the main text, if you make such a determination you can have no knowledge of the interference pattern. Thus, as far as asking the heads or tails question goes, you are dealing with classical coins.

171:   Dirac's remark that a photon can interfere only with itself raises the interesting question about what constitutes a single photon. One can, for instance, put two phase-locked lasers (a "phaser") behind the slits and obtain an interference pattern. In this case one must regard the two lasers as producing a single photon. To this day some academicians reject the idea that a single photon can be produced from two different sources, but it is necessary only to listen to music from one radio station interfering with music from another to hear the process at work.

"Ah," you say.

171:   More on decoherence can be found in "Bringing Schrödinger's Cat to Life" by Philip Yam in *Scientific American,* June 1997, as well as in "Decoherence and the Transition from Quantum to Classical" by Wojciech Zurek in *Physics Today,* October 1991.

The original paper on weak measurements was Y. Aharonov, D. Z. Albert and L. Vaidman, *Physical Review Letters* **60,** 1351 (1988). The experiment was proposed by I. M. Duck, P. M. Stevenson and E. C. G. Sudarshan in *Physical Review* **D40,** 2112 (1989), and the successful test was reported by J. G. Story, N. W. M. Ritchie and R. G. Hulet in *Physical Review Letters* **66,** 1107 (1991). A short review of weak measurements by the same authors can be found in *Modern Physics Letters* **B5,** 1712 (1991).

173:   The famous EPR paper is A. Einstein, B. Podolsky, and N. Rosen, *Physical Review* **47,** 777 (1935).

178:   John Bell's famous paper appeared in *Physics* **1,** 195 (1964).

178:   For Aspect's famous experiment, see A. Aspect, P. Grangier, and G. Roger in *Physical Review Letters* **47,** 460 (1981). In this experiment the polarizers measuring the spin of the photons were held in a fixed position for the entire run. Thus one might object that the measurements were not really independent. However, in a second experiment, Aspect's group varied the position of the polarizers while the photons were in flight; therefore there was no way that one polarizer could "know" ahead of time the position of the other. This is effectively a delayed-choice experiment, except that the polarizer settings were not chosen at random. For details, see A. Aspect, J. Dalibard and G. Roger, *Physical Review Letters* **49,** 1804 (1982). More recently the predictions of quantum mechanics have been verified to 22 standard deviations. See T. E. Kiess et al., *Physical Review Letters* **71,** 3893 (1993).

178:   "The conclusions . . . are philosophically startling. . . ." is from J. F. Clauser and A. Shimony, *Reports on Progress in Physics* **41,** 1881 (1978).

178–179:   Capra's remark is in *ToP,* p. 313.

For Victor Mansfield's views, see *Synchronicity, Science and Soul-Making* (Chicago: Open Court, 1995), p. 106.

We don't know of anyone in quantum mechanics who refers to anything as synchronicity, but Combs and Holland's remark may be found on p. 12 of their book.

Goswami's views are found in his interview in *What Is Enlightenment?*

180:   The weakness of the field picture in certain circumstances can be illustrated by an example that has actually been calculated. Consider two identical atoms, Alice and Bob. Alice is in an excited state, but Bob is cool. Alice deexcites, emitting a photon that is absorbed by Bob. If a photon is viewed as a field propagating at the speed of light, one would conclude that the earliest time Bob can be excited is at a time equal to their separation divided by the speed of light. However, one can show that at *any* time greater than zero there is a small probability that Bob will excite. One must either conclude that a field is not being propagated causally or not regard the two atoms and the photon as separate systems. See G. Hagerfeldt, *Physical Review Letters* **72,** 596 (1994).

181:   The fish story. First consider positive fish. It is easy to show that $y = 2/3(x-1)$, where x is the number of fish each fisherman finds upon waking, and y is the number he leaves behind. The lowest integral value of y that works for all three fishermen is 25.

Note that -2 is a good solution to the equation $y = 2/3(x - 1)$. Each fisherman wakes to find -2 fish, throws one overboard, leaving -3 fish, of which he takes -1, leaving -2 again, and the process repeats indefinitely. This was, perhaps, Dirac's reasoning.

181–182:   For a fuller discussion of the negative probabilities, see E. C. G. Sudarshan and T. Rothman in the *International Journal of Theoretical Physics*, **32**, 1077 (1993). Others have had the same idea. W. Mückenheim et al. have written a review article on "extended probabilities" in *Physics Reports* **133**, 337 (1986).

We can give here the basic idea, although the material is more difficult than that of the main text. Recall that Bell's inequalities deal *not* with the probability that A and B were emitted with spin up or spin down, but with what we referred to as "joint" probabilities such as $P(x+; y+)$, which is the probability that particle A was emitted with spin up along the x-direction, while particle B was emitted with its spin up along the y-direction. This led to the Bell inequality $P(x+; y+) < P(x+; z+) + P(z+; y+)$.

How are the P's computed?

In exactly the way you usually compute odds. To find $P(x+; y+)$, for example, we first count all possible ways particle A can be emitted with spin up along the x-direction while particle B is emitted with spin up along the y-direction, then divide this number by the total number of possibilities for all directions. The situation is identical to one in which each of you have three coins labeled x, y, z and flip them. By conservation of angular momentum, however, these coins are loaded so that each time you get a head $(+)$ on the x coin, your partner's x coin must show a tail $(-)$. One possible outcome is that you get $(+ + +)$ while your partner gets $(- - -)$. [Here $(+++)$ means $(x = +; y= +; z= +)$, and similarly for $(- - -)$.]

There are eight such combinations in all:

1. $(+ + +)$ $(- - -)$
2. $(+ + -)$ $(- - +)$
3. $(+ - +)$ $(- + -)$
4. $(- + +)$ $(+ - -)$
5. $(+ - -)$ $(- + +)$
6. $(- + -)$ $(+ - +)$
7. $(- - +)$ $(+ + -)$
8. $(- - -)$ $(+ + +)$

Notice that only combinations **3** and **5** give you a spin up along the x-direction and your partner a spin up along the y-direction. Thus the probability $P(x+; y+)$ is just $P_3 + P_5$, where by definition $P_3 + P_5$ is the number of particles given off with combinations 3 and 5 divided by the total number of particles given off in all eight combinations. (Symbolically, $P(x+; y+) = P_3 + P_5 = (N_3 + N_5)/N$, where N_3 and N_5 are the number of particles given off with combination 3 and 5, respectively, and N is the total.) If equal numbers were emitted in each combination, the probability would be just two divided by eight, or 1/4.

The exact numbers are unimportant. What is important is to notice that to compute $P(x+; y+)$, which concerns measurements in *two* directions, we first considered all possible measurements in *three* directions. But we didn't care about one direction (z) in combinations **3** and **5**, and so it disappeared when computing $P(x+; y+)$, the probability that actually figures in the Bell inequalities. That's the single reason we have gone through all this.

(In fact it is now very easy to derive a Bell inequality. Clearly, the number of particles should be positive. Then, obviously,

$$N_3 + N_5 < (N_2 + N_5) + (N_3 + N_7)$$

since we have just added the positive numbers N_2 and N_7 to $(N_3 + N_5)$. **But if we divide each of these numbers by N** and refer to the chart, you see that this is no more or less than

$$P(x+; y+) < P(x+; z+) + P(z+; y+),$$

the above Bell inequality.)

What we have done is pure classical probability—just counting—and is thus applicable to the hidden-variable theory. In quantum mechanics probabilities must be computed differently. If you have seen the classroom demonstration in which a light beam is shone through a Polaroid filter or sunglasses, you can get an idea of how this is done. (See Commentary to p. 130 of the Sixth Debates.) Suppose the photons in the beam had definite spin up along the filter axis. Then by rotating the filter by 45 degrees you might expect that the beam would be blocked and no light would come through. However, this is not what happens. As you rotate the filter, the beam's brightness diminishes, but does not become zero until the filter has been rotated 90 degrees. That you still get some light at 45 degrees is a reflection of the quantum-mechanical fact that photons are in a superposition; they do not have exclusively spin up, which would cause them to be blocked. Rather, there is a certain probability that the photon will go through the filter at 45 degrees.

Recognizing this, one can use standard quantum mechanics to compute combinations analogous to those above, that is, probabilities that you will measure $(+ + +)$ while your partner measures $(- - -)$. It is these probabilities that sometimes come out negative. However, just as in the classical case, one of the directions always drops out when we compute the "two-directional" quantum P's, so the sometimes-negative "three-directional" probabilities, are not what you actually measure. The quantum-mechanical P's are always positive, as conventional probabilities are.

183:   The various quotations for the fractured debate are taken from Weinberg's "Sokal's Hoax," *New York Review of Books,* August 8, 1996, p. 12; Amit Goswami, *The Self-Aware Universe* (New York: Tarcher/Putnam, 1993), p. 127; Combs and Holland, pp. 11–13; Erwin Lazlo in his Foreword to Combs and Holland, XIV; Chopra, p. 67; Wilber in *What Is Enlightenment?*, p. 49; Weinberg's *Dreams,* pp. 83–84.

185:   The passage is from Schrödinger's *My View of the World,* which is available in translation from Cambridge University Press, 1964; however, we have used the better translation from Moore, pp. 252, 477.

186:   David Bohm's views on Eastern philosophy can be found in his *Wholeness and the Implicate Order* (London: Ark paperbacks, 1983), p. 23. His discussion of the holomovement and implicate order are mostly in chapter 6. The Chinese achievements in time measurement are detailed in *SCC,* vol. 2. See also Commentary to p. 124.

187:   Capra discusses Bohm on pp. 319–321 of *ToP.*

188:   Lewis Thomas's description of the ant colony is in *The Lives of a Cell* (New York: Bantam Books, 1975), pp. 12–13.

Eighth Debates

Page

Epigraphs: John Webster, *The Duchess of Malfi,* c. 1613. The stanzas by Roald Hoffmann are from "Grand Unification" in his collection of poems *Gaps and Verges* (Orlando: University of Florida Press, 1990).

192:   The passage from the Mundaka Upanishad is in the translation by Patrick Olivelle (Oxford: Oxford University Press, 1996), p. 268.

194:   The original GUT paper is H. Georgi and S. Glashow, *Physical Review Letters* 32, 438 (1974).

195:   Sakharov's prediction of proton decay was made in A. D. Sakharov, *JETP Letters* 5, 24 (1967), reprinted in A. D. Sakharov, *Collected Scientific Works* (New York: Marcel Dekker, 1982).

196:   For Hertz's somewhat obscure theory, see Heinrich Hertz, *The Principles of Mechanics* (New York: Dover, 1956), especially p. 22 of the Introduction and section 594 et seq. In Hertz's theory one regards a body's change of motion as due to a series of collisions with "concealed masses." Then one can dispense with forces in favor of mass and the conservation of energy and momentum.

In Compton scattering, a photon collides with an electron, the electron becomes a virtual electron which then decays back into a real electron emitting a photon.

200–201:   Hawking's remarks on mathematical models are in *NST,* p. 4. His famous paper on black hole radiation is S. W. Hawking, *Communications in Mathematical Physics* 43, 199 (1975). The original suggestion that black holes had an entropy associated with them was made by Jacob Beckenstein, *Physical Review* D7, 2333 (1973).

The fact that black holes radiate energy is, at least with hindsight, not so strange. Investigators before Hawking had recognized that an observer accelerated through the vacuum would see particles with a blackbody spectrum. According to the basis of general relativity, the principle of equivalence, acceleration is indistiguishable from a gravitational field. Since a gravitational field represents curvature, one would therefore expect curved space to produce particles. A black hole can be the most extreme example of spatial curvature.

202:   Kaluza's paper is T. Kaluza, *Preussischen Akad. d. Wissenschaften,* 966 (1921). Klein's is O. Klein, *Zhurnal Physik* 37, 895 (1926).

203:   Paul Ginsparg and Sheldon Glashow's article "Desperately Seeking Superstrings" appeared in *Physics Today,* May 1986, p. 7.

203–204:   Hawking's views are in *NST,* pp. 4–5.

204:   The calculation of black-hole entropy from string theory is A. Strominger and C. Vafa, *Physics Letters* B379, 99 (1996). It can also be found on the Internet at the Los Alamos National Lab (LANL) preprint archive [http://www.lanl.gov], under hep-th/9601029. See also Gary Horowitz's review on the LANL archive, gr-qc/9704072. These articles are for specialists only.

205–206:   A good, recent article on strings is Michael Duff's, "A Theory Formerly Known as Strings," in *Scientific American*, February 1998. More technical discussions on strings and duality, suitable for physicists, can be found in Edward Witten, "Duality, Spacetime and Quantum Mechanics," *Physics Today* May 1997, p. 28, and "Reflections on the Fate of Spacetime," *Physics Today*, April 1996, p. 24.

208:   For Weinberg's remarks on the final theory, see *Dreams*, p. 236. For his remarks on symmetry, see p. 212.

208:   For Bondi's views, see "The Lure of Completeness," in *The Encyclopedia of Ignorance*, Ronald Duncan and Miranda Weston-Smith, editors (Elmsford, NY: Wallaby Pocket Books, 1977), pp. 7–8.

209:   The rare earths are unusual in that, unlike most elements, their outer electron shells fill up before the inner shells. It is the outer shells that determine an element's chemistry and so all the rare earths have similar chemical properties. Consequently they are grouped together on the periodic table. From the periodic table, or symmetry considerations, you would probably conclude that all their structures are the same, which is not the case.

210–211:   For the two views of waves, see Eddington, op. cit., p. 316.

211–212:   The quotations of Shankara can be found in the Brahma Sutra of Shankaracharya, translated by Swami Gambhirananda (Calcutta: Advaita Ashrama, 1993), section 1.

Ninth Debates: Part I

Page

Epigraphs: Homer's is from the opening verse of the *Iliad*. Popper is as quoted by Barrow and Tipler, p. 367.

217:   The passage from Tolman is the concluding paragraph of R. C. Tolman, *Relativity, Thermodynamics and Cosmology* (Oxford: Oxford University Press, 1934), p. 488.

220:   For Capra's remarks, see *ToP*, p. 209.

221–222:   Poincaré's statement that no geometry is truer than another is from his *Science and Hypothesis* (New York: Dover, 1952), pp. 71–72. His two interpretations of the shipping routes are our invention but, we think, reflect his position.

222:   Weinberg's views on the geometric "analogy" are from his *Gravitation and Cosmology, Principles and Applications of the General Theory of Relativity* (New York: John Wiley and Sons, 1972), p. 147.

223:   Poincaré's challenge is genuine. See Poincaré, op. cit., p. 75.

223:   Ian Roxburgh's remarks can be found in his article, "Is Space Curved?," in *The Encyclopedia of Ignorance*, p. 85.

224:   Work on Ellis's averaging problem is ongoing. One technical paper is T. Buchert and J. Ehlers, *Astronomy and Astrophys* **320**, 1 (1997). It is also posted on the LANL site at astro-ph/9510056.

225–226:   Newton's and Mach's remarks are paraphrases. See Ignazio Ciufolini and John A. Wheeler, *Gravitation and Inertia* (Princeton: Princeton University Press, 1995), pp. 384 ff. Pais's remarks are in Pais, op. cit., p. 288.

The empty models that display inertial properties, include, for instance, models that rotate relative to a gyroscope. What is rotating if the model is empty? One can say that spacetime itself is rotating or that a gravitational wave is circulating around the universe. A prediction of GR that seems to support Mach's principle is in the "dragging of inertial frames," or Lense-Thirring effect. The rotation of a massive object, for instance a black hole, will cause spacetime itself to rotate, in turn causing a gyroscope to precess. This effect has been observed by the LAGEOS satellites and more recently in X-ray binary systems.

226:   Martin Rees's offer is to be found in *Before the Beginning* (Reading, MA: Addison-Wesley, 1997), p. 1. (Henceforth, Rees.) This is also a good, judicious view of recent developments in cosmology.

227:   Hubble's law says that the velocity with which a galaxy is receding from the Milky Way is directly proportional to its distance. Assume galaxy A is receding from us with velocity v. If galaxy B is twice as far from us as galaxy A, then it is receding with velocity $2v$; and if galaxy C is three times as far as A, it is receding with velocity $3v$. In symbols, $v = Hd$, where H is a constant, the famous Hubble constant, which must be determined by astronomical observation. (H is merely the slope of the graph of recessional velocity versus distance.)

Note that the form of Hubble's law means that all galaxies are receding from each other. Suppose that instead of being located in the Milky Way, we are located on galaxy A. Then galaxy B will be seen to be receding at velocity v, the velocity at which A appeared to be receding from the Milky Way. Similarly, galaxy C will now be receding from A at velocity $2v$, the velocity at which B had appeared to be receding from the Milky Way. Hubble's law implies that no galaxy is at the center of the universe.

228:   The CMBR had actually been predicted by Ralph Alpher and Robert Herman in 1948, but the prediction was forgotten. More details on the tangled history can be found in Steven Weinberg's classic *The First Three Minutes* (New York: Basic Books, 1977).

229:   For more on nucleosynthesis, see Weinberg, ibid.

229:   A discussion of structure formation is given by Rees, chapter 7, and by Michael Rowan-Robinson, *Ripples in the Cosmos* (New York: W. H. Freeman, 1993), chapter 10.

All structure-formation models require that the primordial fluctuations be of a certain size at about 300,000 years after the big bang. Earlier than this time (which depends on the actual model) the universe resembled a dense fog. A light beam could not travel far but was scattered off the loose electrons in the primordial soup. During this time structures could not grow because the pressure due to the light itself acted against the force of gravity and prevented matter from clumping. However, at about 300,000 years the universe had cooled enough for electrons to finally attach to atoms (a process rather strangely called "recombination") and the fog lifted. From then on photons no longer appreciably interacted with matter, the universe became transparent, and light was free to stream across the universe. Since the universe was now transparent to light, light pressure on matter became negligible and gravitational clumping could proceed, resulting in the structures we observe today. Because the universe was opaque to electromagnetic radiation before 300,000 years, the CMBR as we observe it actually dates from that epoch, not from the moment of the big bang itself.

231:   Horizons. One might think that if the big bang took place, say, fifteen billion years ago, then the horizon distance would be fifteen billion light-years, the distance light travels in fifteen billion years. Due to the expansion of the universe, however, the horizon distance is actually three times this value, or forty-five billion light-years. For a discussion of horizons accessible to those familiar with special relativity, see G. F. R. Ellis and T. Rothman, *American Journal of Physics* **61,** 883 (1993), also *American Journal of Physics* **63,** 87 (1995).

231–232:   The closed, negatively curved universes are technically referred to as compact hyperbolic manifolds. More about them on the graduate mathematics level can be found in John G. Ratcliffe, *Foundations of Hyperbolic Manifolds* (New York: Springer Verlag, 1994). The Seifert-Weber dodecahedron first appeared in C. Weber and H. Seifert, *Math Zeit* **37,** 237 (1933).

233:   For the "Dicke coincidences," as the horizon and flatness problem were originally called, see R. H. Dicke, *Gravitation and the Universe* (Philadelphia: American Philosophical Society, 1970); R. H. Dicke and P. J. E. Peebles in Hawking and Israel, op. cit., chapter 15; and P. J. E. Peebles, *Principles of Physical Cosmology* (Princeton: Princeton University Press, 1993), pp. 101, 361.

In terms of the universe growing away from flatness, we should point out that as the standard model expands it always gets flatter in absolute terms, just as an expanding balloon becomes flatter. However, in the Einstein equation governing the expansion, the term representing curvature gets larger than the term representing matter. Hence the universe becomes more curved in a relative sense. This is sufficient to drive the universe away from the critical density, because a universe at the critical density is defined to be one with matter alone, regardless of the numerical value of its density.

234: The anthropic principle is by now the subject of an extensive literature. The Bible of anthropic lore is Barrow and Tipler, although much of it is fairly technical and the authors do hold extreme positions. An easier introduction to the subject is T. Rothman, "A What You See Is What You Beget Theory," *Discover* magazine, May 1987; reprinted with revisions as "The Measure of All Things," in *A Physicist on Madison Avenue* (Princeton: Princeton University Press, 1991). For a slightly more technical introduction, see Bernard Carr and T. Rothman, "Coincidences in Nature and the Hunt for the Anthropic Principle," in *Isaac Asimov's Science Fiction Magazine,* October 26, 1981, or in T. Rothman et al., *Frontiers of Modern Physics* (New York: Dover, 1985). Rees also has a brief survey.

Contrary to widespread belief, anthropic reasoning has resulted in a number of predictions. We mention two. In the nineteenth century, Lord Kelvin, an opponent of Darwin, argued that the Sun could never have burned long enough for life to evolve on Earth. Geologist Thomas Chamberlain replied in 1899 that Kelvin's argument merely implied that the Sun must burn by a hitherto undiscovered source of energy. This was over half a century before the explosion of the first hydrogen bomb.

In 1954 astrophysicist Fred Hoyle argued that a certain reaction producing carbon must take place "resonantly," or with extreme efficiency; otherwise not enough carbon would be produced in stars to support life. At the same time, a second reaction, transmuting carbon into oxygen, must *not* take place resonantly, or else all the carbon formed by the first reaction would be depleted. The resonance in the first reaction was shortly found exactly where Hoyle predicted; the second reaction was found to be nonresonant. (The reactions in question were $3He^4 \rightarrow C^{12} + 2\gamma$ and $C^{12} + He^4 \rightarrow O^{16}$.)

Opponents of the anthropic principle argue that these predictions could have been made by other means. Had Chamberlain in 1899 known the true age of the Sun, he would not have needed to invoke life to reach his conclusion. Hoyle might have merely said that to produce the observed abundance of carbon in the universe, the relevant reactions must proceed as described. But the fact is Chamberlain and Hoyle did use anthropic reasoning to arrive at their conclusions. One might compare their use of the anthropic principle to the use of symmetry arguments to predict the Omega-minus particle, which might also have been predicted by other means.

As a final remark, the anthropic principle is introduced basically to deal with the problem of boundary conditions: Why did the universe begin in such a way to produce the cosmos we observe? The necessity to introduce boundary conditions is, as we have said, a result of time-symmetric laws of physics. If the laws of the universe were time-irreversible, and always led to the same outcome, then the anthropic principle would be unnecessary.

235:   The horizon distance is actually not $h = ct$ but $h = 3ct$, due to the expansion of the universe. See Commentary to page 231.

Ninth Debates: Part II

Page

Epigraphs: Leibnitz is as quoted by John Wheeler, *A Journey into Gravity and Spacetime* (New York: W. H. Freeman, 1990), p. 3. St. Augustine's reflection on eternity is from his *Confessions*, translated by R. S. Pine-Coffin (New York: Barnes and Noble, 1993), p. 261.

237:   Alan Guth's famous paper on inflation is A. Guth, *Physical Review* **D23**, 347 (1981). For an excellent, thorough, if naturally somewhat partisan, account of inflation and other recent developments in cosmology, see Guth, *The Inflationary Universe* (Reading, MA: Addison-Wesley, 1997). (Henceforth, Guth.)

244:   The age problem. The time for a galaxy receding at velocity v from the Milky Way to reach a distance d is merely $t = d/v$. From the Commentary to page 227, Hubble's law states that $v = Hd$. Substituting Hd for v gives $t = 1/H$. In other words, at a time $t = 1/H$ in the past, regardless of how far apart they are today, all galaxies were on top of each other. The "Hubble time" $t = 1/H$, then, is the time since the big bang, the age of the universe. Hence the rush to measure the Hubble constant.

But the above expression assumes that v is constant, whereas due to the mutual gravitational attraction of its contents, the recessional velocity of galaxies and hence the expansion rate of the universe is decreasing. Thus today's value of H (which is the expansion rate) must be a *minimum* and $t = 1/H$ a *maximum*. The true age of the universe must be less than the Hubble age, the exact value depending not only on H but also on the density (Omega) and whether the contents behave like matter, radiation, or cosmological constant. The figure of fifteen billion years in the main text is the approximate age for an Omega = .2 with $H = 55$ in the traditional units of astronomers (km per second per megaparsec). However, newer measurements for H have suggested that it is somewhat higher, say between 65 and 80. These figures give, respectively, about 15 and 12.5 billion years for $1/H$. Note that the latter is already in condradiction to the usual age estimate for the oldest stars of about fifteen billion years.

For an inflationary universe in which the content is ordinary matter, the age is $t = 2/(3H)$, two-thirds the Hubble time. Thus inflation tends to exacerbate any contradiction between the age of the universe and stellar ages. $H = 80$ gives $t = 8.3$ billion years. As mentioned in the main text, however, the age estimates for the oldest stars (in globular clusters) may also be off because the technique used to calculate the stellar ages requires knowing the distance to these stars, which is also uncertain, perhaps by as much as 30 percent. Feast and Catchpole (below) argue that the *Hipparcos* satellite data indeed indicates that the true age of globular clusters may be only eleven billion years (down from fifteen).

Independently of the *Hipparcos* analysis, in early 1998, five research groups announced that the Hubble constant had gone down again (to about 57) and hence the age of the universe was back at about fifteen billion years. These groups also claimed that Omega was significantly below 1 and that therefore the universe would expand forever. The results are quite recent and few publications have appeared. A website devoted to the findings is http://www-supernova.lbl.gov (note hyphen).

For a popular account of last year's news on the Hubble constant and ages, see http://da.saao.ac.za/news/index.html. For technical articles, see M. Feast and R. Catchpole, *Monthly Notices of the Royal Astronomical Society* **286,** L1 (1997); M. Feast, *Monthly Notices of the Royal Astronomical Society* **284,** 761 (1997); M. Pierce et al., *Nature* **371,** 385 (1994); Wendy Freedman, "Determination of the Hubble Constant," in Neil Turok, editor, *Critical Dialogues in Cosmology* (Singapore: World Scientific, in press), also available from the LANL website under astro-ph/9612024.

245:   The chaotic inflationary scenario was introduced by A. D. Linde, *Physics Letters* **B129,** 177 (1983).

The first suggestion that particle creation, by its reaction on the gravitational field, might isotropize the universe was made by Ya. B. Zel'dovich, *JETP Letters* **12,** 307 (1970).

246:   For Rees's remark, see Rees, p. 171.

247:   Vilenkin's paper on creation of the universe from nothing is A. Vilenkin, *Physics Letters* **B117,** 25 (1982). An earlier suggestion that the universe might have been created out of the quantum vacuum (more than nothing) was made by Edward Tryon, *Nature* **246,** 396 (1973).

St. Augustine's famous questions are op. cit., p. 261.

248:   Gamow's original paper on tunneling is G. Gamow, *Z. Physik* **51,** 204 (1928).

Schwinger's paper is J. Schwinger, *Physical Review* **82,** 664 (1951).

249:   Why is imaginary time like distance? From the Pythagorean theorem, the distance d between any two objects in three dimensions is given by $d^2 = x^2 + y^2 + z^2$, where x, y and z are the distances in the x, y and z directions. In relativity theory, we deal with the spacetime separation between two events, which is given by $d^2 = -t^2 + x^2 + y^2 + z^2$. Notice that if you let $\tau = i\,t$, where $i = \sqrt{-1}$, then you get $d^2 = \tau^2 + x^2 + y^2 + z^2$, which gets rid of the minus sign and makes the formula look like the ordinary Pythagorean one. Such a version of the spacetime interval is actually termed "Euclidean," since it is identical to the distance formula in plane, or Euclidean, geometry. It is in this sense that imaginary time resembles a distance. Providing a physical interpretation for imaginary time, however, remains problematic.

The no-boundary proposal is found in J. B. Hartle and S. W. Hawking, *Physical Review* **D28,** 2960 (1983).

252:   Hawking's claims for the no-boundary proposal are in *NST,* pp. 79, 86, 103.

256–257:   Freeman Dyson's argument from design is in his *Disturbing the Universe* (New York: Harper and Row, 1979), p. 251.

The quotations from the various scientists can be found on the Witness Box Ministry website http://witnessbox.com/page8.html. Those we have been able to confirm are completely accurate but because we could not trace all of them we cited them as anonymous.

257:   For Martin Rees's feelings on physics and theology, see Rees, p. 6.

257–258:   Ellicott and Paley are quoted by Jaki, pp. 434 and 453.

Yakov Zel'dovich's remark was made to T. Rothman.

258:   For Alan Guth's remarks, see Guth, pp. 249, 252, 26.

258–259:   The dialogue is from P .C. W. Davies, *God and the New Physics* (New York: Simon and Schuster, 1983), p. 128. In his newer book, *The Mind of God* (New York: Simon and Schuster, 1992), he seems to have modified his position on some issues.

259:   For St. Augustine on time, see Augustine, op. cit., pp. 263–264.

259:   For Atkins's version of genesis, see P. W. Atkins, *Creation Revisited* (New York: W. H. Freeman, 1992), pp. 129, 149.

260:   Davies on God and time is op. cit., p. 133.

260:   Epicurus is from Jaki, p. 415.

261:   The hymns are from the *Rig Veda,* translated by Wendy Doniger O'Flaherty (London: Penguin Books, 1981), pp. 25, 28.

261:   Needham is as quoted by Jaki, p. 419. See also *SCC,* pp. 305–306.

262:   Tolman's warning is Tolman, op. cit., p. 488. However, Tolman is incorrect when he makes the common assertion that the (standard) universe models emerge from a point of zero volume. This is true in the closed case, but the flat and "open" FLRW models have infinite volume, and had infinite volume even at the big bang.

Silk is as cited by Rees, p. 6.

Tenth Debates

Page

Epigraphs: Paul Valéry, *Introduction to the Methods of Leonardo da Vinci*, 1895. Sigmund Freud, *New Introductory Lectures on Psychoanalysis*, 1932.

266–268: All the examples of the Newtonian impact on philosophy and politics are from chapter 1 of I. Bernard Cohen, *Science and the Founding Fathers* (New York: W. W. Norton, 1995). For the exchange between Franklin and Adams, see p. 229. The version here is slightly edited.

Adams's description of Franklin is from Carl van Doren, *Benjamin Franklin* (New York: Penguin, 1991), p. 600.

Information about the iatrophysical school is from John Duffy, *The Healers: A History of American Medicine* (Urbana: University of Illinois Press, 1979), p. 27.

David Bohm's analogy between quantum processes and thoughts can be found in his *Quantum Theory* (New York: Dover, 1989), pp. 169–172.

Combs and Holland's views on cultural denial of symbolic connections are p. 101 of their book.

268: Chopra's remarks are from Chopra, p. 21.

269: Sokal's parody was entitled "Transgressing the Boundaries: Towards a Transformative Hermeneutics of Quantum Gravity." It originally appeared in *Social Text*, Spring/Summer 1996. The quotations here were taken from the article as posted on Sokal's website http://www.physics.nyu.edu/faculty/sokal. At this site are also posted numerous other articles relating to the hoax, including Steven Weinberg's.

269–270: Weinberg's views were originally stated in "Sokal's Hoax," *New York Review of Books*, August 8 and October 3, 1996. His remarks about the product of the scientific worldview are as quoted by Silvan Schweber, *Physics Today*, March 1997, p. 73.

271: President Jiang's coopting of relativity appeared in the *Washington Post*, October 29, 1997. We cite Jiang's remarks as quoted by Ke-Chiang Hsieh in *American Physical Society (APS) News*, December 1997, p. 4.

271: The excerpts from the Bhagavad Gita are chapter 4, verse 18 and chapter 5, verse 14.

272: The impact of the fourth dimension on art is the subject of Linda D. Henderson's *The Fourth Dimension and Non-Euclidean Geometry in Modern Art* (Princeton: Princeton University Press, 1983). For the general history, see the Introduction and chapter 1. For Hinton, see pp. 27–29. For Ouspensky, see pp. 249–251.

275: Sheldrake is quoted by Combs and Holland, p. 24.

275–276: Gore's metaphor of distributed intelligence appears in Gore, op. cit.

Smolin's speculation first appeared in *Classical and Quantum Gravity* **9**, 173 (1992). A scientific critique of the hypothesis can be found in T. Rothman and G. F. R. Ellis, *Quarterly Journal of the Royal Astronomical Society* **34**, 201 (1993). Smolin replies in his *The Life of the Cosmos* (New York: Oxford, 1997). See also Rees for a discussion of the idea.

277: Andrew Pickering's views are presented in *Constructing Quarks: A Sociological History of Particle Physics* (Chicago: University of Chicago Press, 1984), p. 413. According to Kurt Gottfried in *Physics Today*, January 1997, p. 61, Pickering has somewhat modified these views though "he appears to adhere to the same bottom line."

279: The full title of Forman's famous study is *Weimar Culture, Causality and Quantum Theory, 1918–1927: Adaptation by German Physicists and Mathematicians to a Hostile Intellectual Environment*, vol. 3 of *Historical Studies in the Physical Sciences* (Philadelphia: University of Pennsylvania Press, 1971). See also Freeman Dyson's review of Forman in *From Eros to Gaia* (New York: Pantheon, 1992), chapter 24. Gottfried, op. cit., also has thoughts on the thesis.

281–282: The Ajmal Hussein story is a traditional Sufi tale. We have adapted it from Idries Shah, *Wisdom of the Idiots* (London: Octagon Press, 1970), pp. 51–52.

284: Krishnamurti's remarks are from "What Is Consciousness?" in Pupul Jayakar, *Fire in the Mind; Dialogues with Krishnamurti* (New Delhi: Penguin, 1995).

Hawking's remarks on Penrose's idea are, once again, *NST*, p. 124.

Supplemental Debate

Page

289: For more on the recent Schrödinger cat experiments, see Yam, op. cit. Haroche is as quoted by Yam.

290: The paper on the Zeno effect was B. Misra and E.C.G. Sudarshan, *Journal of Mathematical Physics* **18,** 756 (1977). The experimental verification was reported by Wayne Itano et al., *Physical Review* **A41,** 2295 (1990).

290–291: The famous Ramanujan story is from C. P. Snow's foreword to Hardy, p. 37.

291: For more on quantum computing, see Yam. Somewhat more technical articles are Charles Bennett, "Quantum Information and Computing," *Physics Today,* October 1995, p. 24; Serge Haroche and Jean-Michel Raimond, "Quantum Computing: Dream or Nightmare?" *Physics Today,* August 1996, p. 51; Samuel L. Braustein, "Quantum Computation," preprint, SERCS, University of Wales, Bangor, Gwynedd LL57 1UT, UK.

293: David Deutsch's remarks can be found in his book, *The Fabric of Reality* (New York: Allen Lane, 1997), chapter 9.

How can factoring a large number be related to finding the period of a function? To illustrate, take the number 15. Its prime factors are 3 and 5, but we can find them by a more complicated method. In the chart below, we first write out the first 7 integers, starting with 0. In the second line we write out the first seven powers of 2: 2^0, 2^1 and so on. In the third line, we write out what is called 2^n (mod 15). For those who have not seen modular arithmetic, x (mod y) is shorthand for (x − nearest smaller multiple of y). For example, 2^0(mod 15) = 1 because 2^0 = 1, the nearest smaller multiple of 15 is 0, and 1 − 0 = 1. Similarly, 32 (mod 15) is 32 − 30 = 2. Therefore, we have:

n:	0	1	2	3	4	5	6	etc.
2^n:	1	2	4	8	16	32	64	etc.
2^n(mod 15):	1	2	4	8	1	2	4	etc.

Notice that the pattern in the third line begins repeating after 2^4 or 16. Thus 2^n (mod 15) is a periodic function. To find the factors we note that 16−1 = 15, or 2^4 − 1 = 15. The left-hand side can be rewritten as $(2^2 − 1)(2^2 +1)$, or 3 × 5. Voilà.

Thus the trick in factoring a number N by Shor's method is to find a number a (in our case 2), such that a^r (mod N) = 1, where r, the period, is an even number. Then $(a^{r/2} − 1)(a^{r/2} + 1)$ = 1, as we have just written for the example. We see that knowing r is equivalent to allowing N to be factored.

294–295: Quantum teleportation is described by Bennett. The first experimental confirmation is reported by Dik Bouwmeester et al., "Experimental Quantum Teleportation," preprint Institut für Experimentalphysik, Universität, Innsbruck, Technikst. 25, A-6020, Innsbruck, Austria (1997).

INDEX